World Geomorphological Landscapes

Series editor

Piotr Migoń, Wroclaw, Poland

For further volumes:
http://www.springer.com/series/10852

Vishwas S. Kale
Editor

Landscapes and Landforms of India

Springer

Editor
Vishwas S. Kale
Department of Geography
University of Pune
Pune
India

Every effort has been made to contact the copyright holders of the figures and tables which have been reproduced from other sources. Anyone who has not been properly credited is requested to contact the publishers, so that due acknowledgment may be made in subsequent editions.

ISSN 2213-2090 ISSN 2213-2104 (electronic)
ISBN 978-94-017-8028-5 ISBN 978-94-017-8029-2 (eBook)
DOI 10.1007/978-94-017-8029-2
Springer Dordrecht Heidelberg New York London

Library of Congress Control Number: 2014939066

Printed on acid-free paper

Springer is part of Springer Science+Business Media (www.springer.com)

Series Editor Preface

Landforms and landscapes vary enormously across the Earth, from high mountains to endless plains. At a smaller scale, nature often surprises us by creating shapes which look improbable. Many physical landscapes are so immensely beautiful that they received the highest possible recognition—they hold the status of World Heritage properties. Apart from often being immensely scenic, landscapes tell stories which not uncommonly can be traced back in time for tens of million years and include unique events. In addition, many landscapes owe their appearance and harmony not solely to the natural forces. For centuries, or even millennia, they have been shaped by humans who modified hillslopes, river courses, and coastlines, and erected structures which often blend with the natural landforms to form inseparable entities.

These landscapes are studied by Geomorphology—"the Science of Scenery"—a part of Earth Sciences that focuses on landforms, their assemblages, surface, and subsurface processes that moulded them in the past and that change them today. Shapes of landforms and regularities of their spatial distribution, their origin, evolution, and ages are the subject of research. Geomorphology is also a science of considerable practical importance since many geomorphic processes occur so suddenly and unexpectedly, and with such a force, that they pose significant hazards to human populations and not uncommonly result in considerable damage or even casualties.

To show the importance of geomorphology in understanding the landscape, and to present the beauty and diversity of the geomorphological sceneries across the world, we have launched a new book series *World Geomorphological Landscapes*. It aims to be a scientific library of monographs that present and explain physical landscapes, focusing on both representative and uniquely spectacular examples. Each book will contain details on geomorphology of a particular country or a geographically coherent region. This volume, the third in the series, introduces the geomorphology of India—a vast country with highly diverse landscapes, from the lofty Himalaya to the world's largest delta, and from subtropical mountains in the south to the deserts in the north-west. To do justice to the enormous geomorphological diversity of India in one book is nearly impossible. However, the reader will be helped by an extended Part II of the book in which, after general presentations of tectonic background, timescales involved in landforms evolution, and the climate past and present of the subcontinent, five main geomorphic provinces of India are presented. These are followed in Part III by 18 specific examples of great landscapes stretching from the frigid Ladakh to the subtropical Andaman Archipelago. Nearly four dozen geomorphosites are listed in Part IV of the monograph. Each is worth a special visit and the incoming IX International Conference on Geomorphology in India in 2017 will be a perfect opportunity to do so.

The World Geomorphological Landscapes series is produced under the scientific patronage of the International Association of Geomorphologists (IAG)—a society that brings together geomorphologists from all around the world. The IAG was established in 1989 and is an independent scientific association affiliated with the International Geographical Union and the International Union of Geological Sciences. Among its main aims are to promote geomorphology and to foster dissemination of geomorphological knowledge. I believe that this lavishly illustrated series, which sticks to the scientific rigour, is the most appropriate means to fulfill these aims and to serve the geoscientific community. To this end, my great thanks go to the Editor, Professor Vishwas S. Kale, who coordinated the work of many authors from different countries with extreme dedication, enthusiasm, and patience. I am also grateful to all individual contributors who agreed to add the task of writing chapters to their busy agendas and delivered high quality final products.

Piotr Migoń

Foreword

The Indian subcontinent covers approximately 4.5 million km^2 area. In spite of the fact that the Indian landmass is geographically a part of the Eurasian continent, the Indian Peninsula and the Himalayan sector together represent and make up a distinct more than 3 billion years of geologic history. This monograph provides a synoptic view of this diverse, scenic, and ancient landmass.

The Indian subcontinent is composed of incredible variety of landscapes and landforms. It includes the Himalaya—the loftiest mountain range in the world, with some of the longest glaciers, the vast interminable alluvial plains of the Indus, Ganga, and Brahmaputra, the Western Ghat (*aka Sahyadri*)—one of the most spectacular great escarpments of the world, the Deccan Traps—one of the largest igneous provinces in the world, and the Ganga–Brahmaputra Delta—the world's largest delta. The Lonar Crater, one of the youngest and best-preserved impact craters in the world, is located within the Deccan Traps country in the Indian Shield.

Some of the other outstanding landforms of the Indian region are the intermontane basins and valleys of Himalaya (Kashmir, Pinjaur Dun, Dehradun, etc.), the megafans at the foot of Himalaya (such as those of Kosi, Tista, and Gandak), Ladakh—the cold desert north of Himalaya, and the Great Indian Desert known as the Thar Desert. Like other shield areas in the tropics, the Indian Shield is also characterized by vast undulating plains with solitary, dome-shaped (inselbergs) or boulder-strewn (nubbins) hills. Another distinctive landscape located in western India is the Great Rann of Kachchh, an unusually large, saline playa spanning an area of more than 16,000 km^2. The $\sim$7,000 km long coastline of India also displays numerous distinct features. Whereas rocky features dominate the west coast, the eastern seaboard is delta-studded.

The landforms of the Indian subcontinent have been shaped primarily by three factors through time—lithology, tectonic history and climate. The subcontinent is dominated by a large variety of rocks ranging from Archean to Holocene in age. The spectacular granite landforms of the Indian Shield, the stepped landscape of the Deccan Traps region, and the duricrusted landforms display the strong control of lithology. On the subcontinental scale, however, the mega-landforms and drainage owe their present form and disposition largely to the geological and tectonic processes that began with the northward drift of India more than 150 million years (Ma) ago. Once a part of the Gondwanaland, India broke away from it, made a solitary northward journey, and finally collided with Eurasia resulting in the formation of the mighty Himalaya and the vast foreland basin. At least during the last 8–10 Ma the landforms and landscapes of the Indian subcontinent have been fashioned by the Indian monsoon. The unique behavioral characteristics of the Indian rivers could be attributed to the monsoon climate of the region.

The Indian subcontinent also has significant and rich cultural heritage resources. Archaeological and architectural remains indicate that the region has been occupied since early Acheulian times ($\sim$1 Ma). The vast landscape is dotted with countless Palaeolithic, Mesolithic, Neolithic, and Chalcolithic sites and historical settlements as well as numerous ancient temples, monuments, caves, rock sculptures, inscriptions, or monoliths, and Medieval forts. The Harappan civilization, one of the three oldest civilizations of the world, flourished in the Indus Valley ~3.5–1.5 ka BCE (ka = thousand years; BCE = Before the Common Era). Presently (January 2014), there are 30 World Heritage Properties in India, out of which 6 are Natural Properties and 24 are Cultural Properties. In addition, there are over 3,500 protected monuments under the protection of Archaeological Survey of India.

This book with a focus on the landforms and landscapes of India is a part of the recently launched book series *World Geomorphological Landscapes* and is produced under the scientific patronage of the International Association of Geomorphologists (IAG).

The contents of this book are divided into four parts. Part I contains the background about geology and tectonic framework (S. K. Tandon et al.), present and past climate (A. K. Singhvi and R. Krishnan), and the geomorphic provinces and geomorphological history (V. S. Kale). The landscapes of the four major geomorphic provinces are described in slightly greater detail in Part II of the book. These include—Himalaya (L. A. Owen), the Indus–Ganga–Brahmaputra Plains (R. Sinha and S. K. Tandon), the Indian Peninsula (V. S. Kale and R. Vaidyanadhan), and the Thar Desert (A. Kar). The focus of the next chapter (R. Mukhopadhyay and S. M. Karisiddaiah) is on the coastal landforms and processes.

In Part III, 18 specific landscapes and landforms (Fig. 1) are described. This section of the book is organized into a sequence of chapters in a clockwise direction starting from the northern state of Jammu and Kashmir and ending in western India via central, northeastern, eastern and southern India.

The second longest glacier outside the polar regions, namely the Siachen Glacier is described by M. R. Bhutiyani, followed by chapters on the high-altitude cold desert of Ladakh (N. Juyal), the Vale of Kashmir (R. K. Ganjoo) and Sikkim-Darjeeling Himalaya (L. Starkel and S. Sarkar). Two special features of the Himalaya are the occurrence of longitudinal intermontane basins in the frontal zone, called 'Duns' (S. K. Tandon and V. Singh) and numerous megafans (R. Sinha). One of the most famous badlands is observed along the Chambal River in central India (V. U. Joshi) and the Brahmaputra River in Assam is one of the largest braided rivers in the world (J. N. Sarma). The landscape of Meghalaya (Shillong) Plateau is the focus of the next chapter (P. Prokop) and the characteristic features of the Sundarbans and the Ganga–Brahmaputra Delta are explained by K. G. Rogers and S. L. Goodbred Jr.

There are no classic karstic landscapes in India. However, several spectacular caves occur in eastern and northern parts of the subcontinent. A. C. Narayana et al. describe the special characteristics of two well-known karstic caves in eastern India, namely Belum and Borra Caves. Geographically, granite landforms cover vast areas of the Indian Shield. These are dealt with in the chapter by Y. Gunnell.

There are two archipelagos, a little away from the Indian mainland. One of them, the Andaman Archipelago, has been described by J. S. Ray. A prominent feature occurring in some parts of the east coast are the red sand dunes, presently dissected by gullies. The features of such a landscape occurring in the southern state of Tamil Nadu and known as Teri red sands are explained by R. Jayangondaperumal.

In the Deccan Volcanic Province, spectacular laterite-capped tablelands and mesas are present at Panchgani and form the focus of the chapter by V. S. Kale. The unique meteor impact crater in the basalts at Lonar is the main theme of the chapter by M. S. Bodas and B. Sen.

The surreal landscape of the Rann of Kachchh with an enormous stretch of salt-covered marsh is described by Navin Juyal and the special features of the largest playa in the Indian Desert, the Sambhar Lake, are discussed by R. Sinha. The concluding chapter

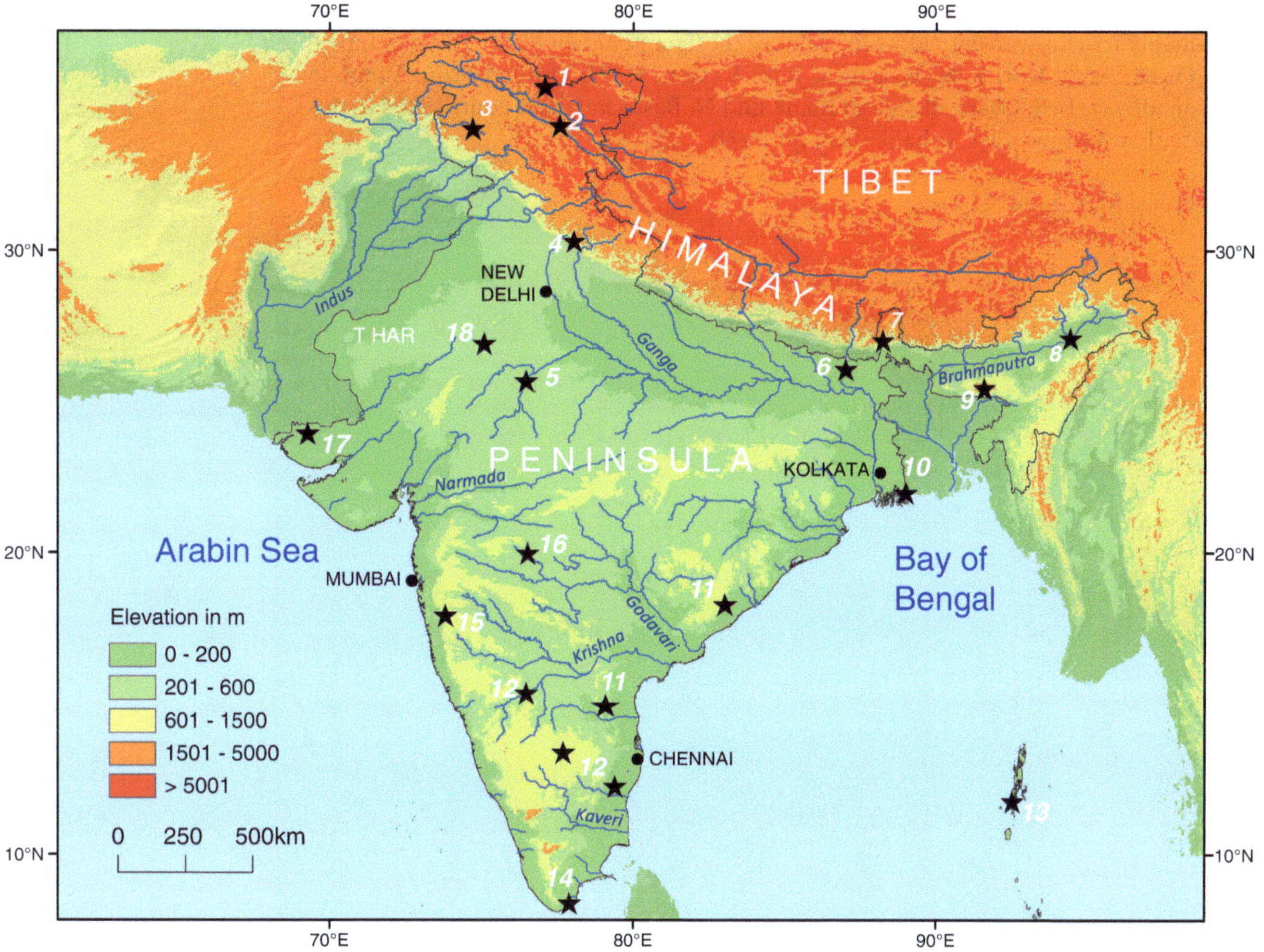

Fig. 1 Physical map of the Indian subcontinent. The map shows the location of 18 specific landscapes or landforms discussed in Part III of this book. 1 = The Siachen Glacier: The Second Longest Glacier Outside the Polar Regions, 2 = Ladakh—The High-altitude Indian Cold Desert, 3 = The Vale of Kashmir—Landform Evolution and Processes, 4 = Duns: Intermontane Basins in the Himalayan Frontal Zone, 5 = The Chambal Badlands, 6 = The Kosi Megafan: The Best known Himalayan Megafan, 7 = The Sikkim-Darjeeling Himalaya—Landforms, Evolutionary History and Present-day Processes, 8 = The Brahmaputra River in Assam: The Outsized Braided Himalayan River, 9 = The Meghalaya Plateau—Landscapes in the Abode of the Clouds, 10 = The Sundarbans and Bengal Delta: the World's Largest Tidal Mangrove and Delta System, 11 = The Spectacular Belum and Borra Caves of Eastern India, 12 = Granite Landforms of the Indian Cratons, 13 = The Andaman Archipelago, 14 = Teri Red Sands, Tamil Nadu, 15 = The Laterite-Capped Panchgani Tableland, Deccan Traps, 16 = The Lonar Crater—The Best Preserved Impact Crater in the Basaltic Terrain, 17 = The Great Rann of Kachchh: The Largest Saline Marshland in India, and 18 = The Sambhar Lake: The Largest Saline Lake in Northwestern India

of the monograph (V. S. Kale) constitutes Part IV and gives a list of four dozen potential geomorphosites in India.

This book could not have seen the light of the day without the help and support of several individuals. At the outset, I would like to express my profound thanks to Piotr Migoń, the Series Editor, for inviting me to edit this volume on India and for his guidance at every stage. He provided valuable comments that considerably improved the clarity and quality of the chapters. Special thanks are also due to Robert Doe and M. Agila of Springer for their constant support and guidance.

I would also like to express my sincere and heartfelt thanks to all the authors for their alacrity with which they responded to my invitation and for sending in their contributions for the monograph. My very special words of thanks are to R. Vaidyanadhan for reviewing some of the chapters and for his valuable inputs at every stage. I am also thankful to Sandip Pawar and Nilesh Susware for redrawing some of the figures. Navin Juyal, Yanni Gunnell, Lewis Owen, Amal Kar, S. N. Rajaguru, Bob Wasson, K. N. Prudhvi Raju, Sunando Bandyopadhyay, L. S. Chamyal, Nilesh Bhatt, and N. Basavaiah have contributed to this monograph in various ways. I thank them profusely.

Pune, India, May 5, 2014 Vishwas S. Kale

Contents

Part I

Physical Environment and Geomorphic History

Geological and Tectonic Framework of India: Providing Context to Geomorphologic Development

Sampat K. Tandon, Partha Pratim Chakraborty and Vimal Singh

Abstract

Surrounded by the Himalayan Orogen, Indian Ocean, Bay of Bengal and Arabian Sea on its north, south, east and west, respectively, the Indian subcontinent represents more than 3.5 Ga of geological history spanning from the Archean to Quaternary. The exposed Precambrian basement of India comprises four Archean cratonic nuclei welded together by several Proterozoic mobile belts. Proterozoic intracontinental basins that contain mostly undeformed sedimentary successions unconformably overlie the basement. The basement rocks are covered in the north along the Himalayan front by thick deposits of the Indo-Ganga alluvium, and in the west central region by the end-Cretaceous Deccan flood basalts. Basins related with intra- to inter-continental rifting of the Indian craton viz. the Gondwana Basins and those of the east and west coast record the Phanerozoic history of India from Permian to Cenozoic. Since $\sim$55 Ma, the continent-continent collision between the northerly drifting Indian Plate and the Eurasian Plate has guided the development of the structural design of the Himalayan Orogen. These tectono-geomorphic processes were responsible for the first order relief structure of the northern part of the Indian subcontinent including the development of the topography of the spectacular Himalayan Orogen and its foreland.

Keywords

Geology • Tectonic framework • India • Craton • Peninsula • Himalaya • Indo-Ganga plains

1 Introduction

The Indian region represents a peninsula in south Asia that is girdled by the Himalayan Orogen in the north, extends southward into the Indian Ocean, and is bordered by the Bay of Bengal and the Arabian Sea in its southeast and southwest, respectively (Fig. 1). It forms a part of the Indo-Australian Plate; one of the six major plates of the globe with an area of $\sim$12 million km^2. Broadly, three physiographic subdivisions viz. the Peninsula, the Indo-Ganga Basin, and the Himalaya constitute the Indian subcontinent. The geological structure of the Peninsula evolved through the welding of Archean cratonic nuclei by several Proterozoic mobile belts as a result of multiple orogenic cycles ranging in age from the Paleoproterozoic to early Palaeozoic; the Himalaya and the

S. K. Tandon (✉)
Department of Earth Sciences, Indian Institute of Technology, Kanpur 208016, India
e-mail: sktand@rediffmail.com

P. P. Chakraborty · V. Singh
Department of Geology University of Delhi, Delhi, 110007, India
e-mail: parthageology@gmail.com

V. Singh
e-mail: vimalgeo@gmail.com

V. S. Kale (ed.), *Landscapes and Landforms of India*, World Geomorphological Landscapes,
DOI: 10.1007/978-94-017-8029-2_1,

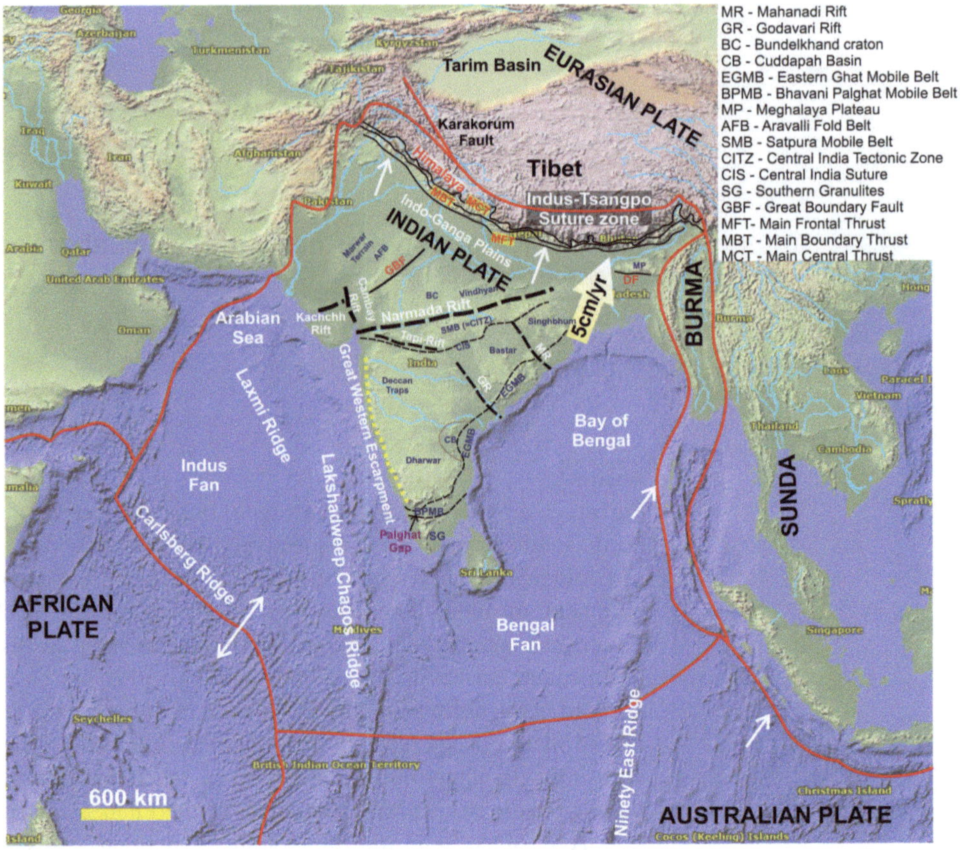

Fig. 1 Map showing tectonic framework of the Indian Plate with major regional physiographic features (base map taken from 3D world map)

Indo-Ganga (Indo-Gangetic) Basin evolved in response to the northward migration of the Indian Plate in the Cenozoic and collisional orogenesis involving the Indian Plate and the Eurasian Plate (Fig. 1).

The geological history of Peninsular India can be traced back to the Archean, a time when the early Earth was characterized by the formation of proto-continents of sialic composition. Proterozoic basins are well developed in various parts of the Peninsula, and some of these basins in the northern part of the Indian craton have been involved in the Himalayan Orogeny and now constitute an integral part of the Lesser Himalaya. Apart from some Early Cambrian strata, the earliest Phanerozoic geological record of Peninsular India is found in the rift-related Permian to Cretaceous coal bearing basins i.e., the Gondwana Basins. Because of the tectonic reactivation resulting from the Cenozoic uplift in the Himalaya and the Deccan Volcanism related plateau uplift, the landscapes in the Peninsula consist of several high-level geomorphic surfaces that show variable degrees of dissection.

In the Late Cretaceous, the Indian Plate split from Madagascar and began its northward journey at a rate of about 20 cm per year. A number of late Triassic to early Cretaceous basins located at the margin of the Indian Shield preserve the record of extension and subsidence of the Indian crust below the Neo-Tethys Sea in the north. Later, in the early Palaeogene the Indian Plate started its convergence below Asia; its leading edge dipped northwards and north-eastwards to be underthrusted beneath the Asian landmass. The collisional tectonics commenced ~50 to 55 Ma (million years) back and led to the initiation of the Himalayan Orogenic Belt. Southward propagating thrust sheets formed the Himalayan chain consisting of four main litho-tectonic units from the north to the south. Collisional tectonics took place through the Cenozoic in the Himalaya and is ongoing as evidenced by major seismicity all along the Himalayan arc in the Holocene and in the historical period. Continued underthrusting of the Indian Plate resulted in a thick (60–70 km) crust in the Himalayan region. Also, the continued loading of the lithosphere in the Himalayan thrust belt was accompanied by lithospheric flexure and the formation of a peripheral foreland basin—the Himalayan Foreland that extends along almost the entire southern flank of the orogen.

The Himalayan Foreland has acted as a loci of sediment accumulation from the Palaeogene, through the Neogene and into the Quaternary. This basinal fill has been deformed and uplifted into the Siwalik Hills in the outer Himalaya; the relatively distal and undeformed part of the basin constitutes the Indo-Ganga Basin whose surface expression corresponds with the major geomorphic units of the Indo-Ganga Plain. This Plain is a prominent feature on the surface of the globe, occupies an area of ~700,000 km^2, and

Table 1 A generalized stratigraphic framework for Peninsular and Extra-Peninsular India

		Peninsular India			Extra-Peninsular India
Phanerozoic	Quaternary	Indus, Ganga and Brahmaputra River Valleys, Coastal Tracts and Thar Desert			
	Cenozoic	Gujarat mainland, Saurashtra, Kachchh, Jaisalmer and Barmer Basins, Niniyur Formation, Baripada beds, Chhotanagpur Plateau	Northeast India Tipam Barail	Andaman-Nicobar basin	Siwaliks Kasauli Dagshai Subathus
	Mesozoic	*Rifting and formation of Coastline*			Collision and Himalayan uplift ($\sim$55 Ma)
		West Coast	*East Coast*		
		Kerala Basin, Konkan Basin	Bengal basin, Mahanadi basin		
		Mumbai Basin, Saurashtra Basin	Krishna-Godavari Basin, Palar Basin		
		Kachchh basin	Mesozoic Basins of Rajasthan (Kachchh and Jaisalmer Basins)		Abor volcanics Gondwanas in Bhutan and NE Himalaya
		Deccan Traps (68–60 Ma)			
		Rajmahal Trap Intracratonic rifting (Gondwanas) Pranhita-Godavari valley, Son-Mahanadi valley, Damodar-Koel valley			
	Palaeozoic				Tethyan sub-basins (i) Kashmir sub-basin (ii) Spiti-Zanskar sub-basin (iii) Garhwal-Kumaun sub-basin
Precambrian	Proterozoic	*Intracratonic to epicratonic Basins*	*Mobile belts*	*Rifts*	
		(i) Vindhyan (ii) Chhattisgarh (iii) Cuddapah (iv) Pranhita-Godavari (v) Kaladgi–Bhima (vi) Marwars etc. (1899 ± 20 Ma to $\sim$543 Ma)	CITZ (2.2–1.5 Ga) Eastern Ghat mobile belt (1.6–0.58 Ga) Pandyan mobile belt (0.75–0.55 Ma) Aravalli-Delhi mobile belt	Godavari Mahanadi Son-Narmada	Higher Himalaya Lesser Himalaya
	Archean	*Cratonic nuclei*			
		(i) Dharwar (3.4–2.5 Ga), (ii) Bastar (3–1.8 Ga), (iii) Singhbhum (3.4–1.6 Ga), (iv) Bundelkhand (3.3–1.8 Ga)			

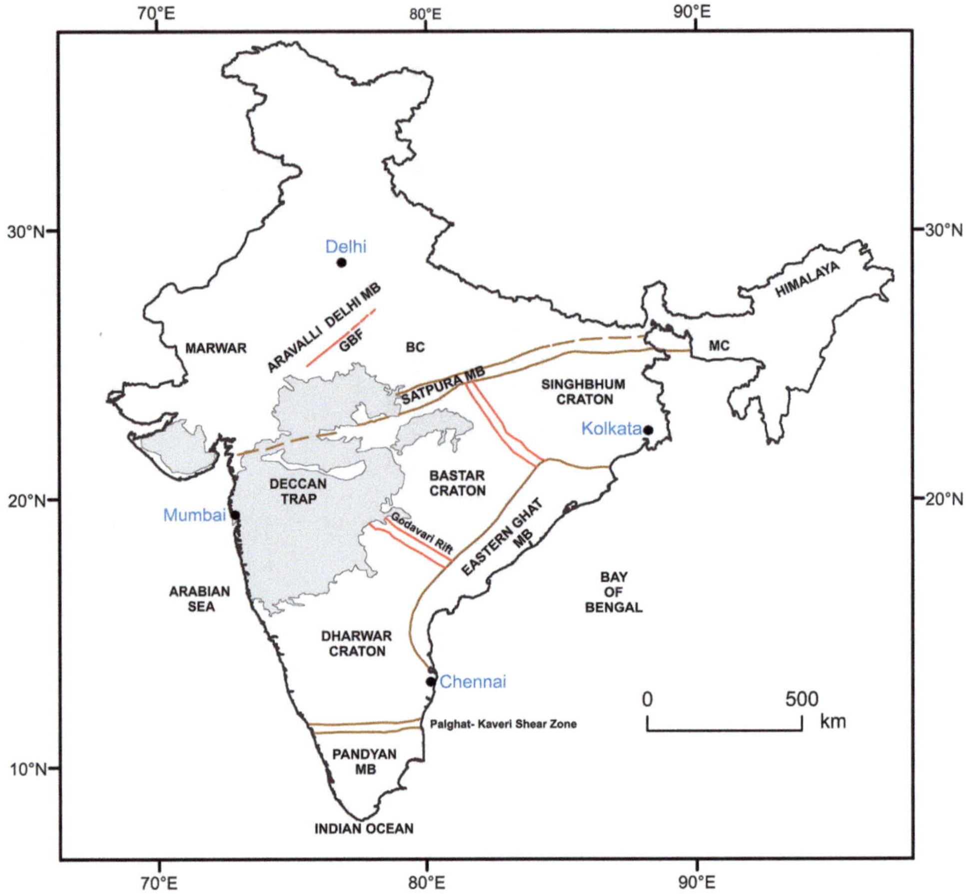

Fig. 2 Cratonic nuclei, marginal mobile belts and pericratonic rift systems of Peninsular India (modified after, Ramakrishnan and Vaidyanadhan 2010). *Dashed lines* suggest inferred extension of the faults. BC—Bundelkhand Craton; MC—Meghalaya Craton; GBF—Great Boundary Fault; MB—Mobile Belts

occurs in between the Indian Peninsula and the Extra-Peninsular or Himalayan Orogenic Belt. This alluvial plain slopes to the east-southeast and receives detritus both from the glacier-fed Himalayan rivers and those flowing from the positive cratonic areas in the south.

The Indo-Ganga Plain is flanked in its south by the craton-related positive areas of the Indian Peninsula, and to its west by the Thar Desert. Many of the orogenic belts of the Peninsula are morphologically expressed as hill ranges such as the Aravalli, Sahyadri, etc. Also, the Peninsular region is marked by plateaus, for example the Deccan Plateau covering an area of about half a million square km in the western and the central parts of the country, coastal lowlands in the eastern and western parts, and major deltas formed by west to east flowing major drainages whose origins have been connected to the plume-related uplift of the Deccan Volcanic Province (Cox 1989).

2 Broad Stratigraphic Framework of Peninsular India

Table 1 summarizes the general stratigraphic framework for Peninsular and Extra-Peninsular India. Tectonically, at a first order, Peninsular India consists of cratons and mobile belts that are curvilinear high-grade gneiss-granulite terrains surrounding the cratonic nuclei (Fig. 2).

The Indian Shield has been considered to be made up mainly of four cratons (1) Dharwar, (2) Bastar, (3) Singhbhum, and (4) Bundelkhand, and four mobile belts i.e. the Eastern Ghat mobile belt fringing the Dharwar, Bastar, and Singhbhum Cratons; the Pandyan Mobile Belt fringing the Dharwar Craton, the Satpura Mobile Belt fringing the Bastar, Singhbhum, Bundelkhand Cratons (Ramakrishnan and Vaidyanadhan 2010) and the Aravalli-Delhi Mobile Belt (ADMB) fringing the Bundelkhand Craton (Fig. 2). Radhakrishna and Ramakrishnan (1988) divided the Indian Shield along the Narmada-Son Lineament into a northern and southern block, each of which consisted of welded cratonic nuclei.

These old cratons and mobile belts that constitute the major part of the shield are succeeded by a younger cover, components of which are listed below (Ramakrishna and Vaidyanadhan 2010):

(1) Proterozoic sedimentary basins, also known as the Purana Basins, ranging in age from the mid-Proterozoic to the terminal Proterozoic
(2) Gondwana sedimentary basins in the Damodar, Satpura, Narmada-Tapti, Mahanadi, Krishna-Godavari rifts, ranging in age from the Permian to the Cretaceous,

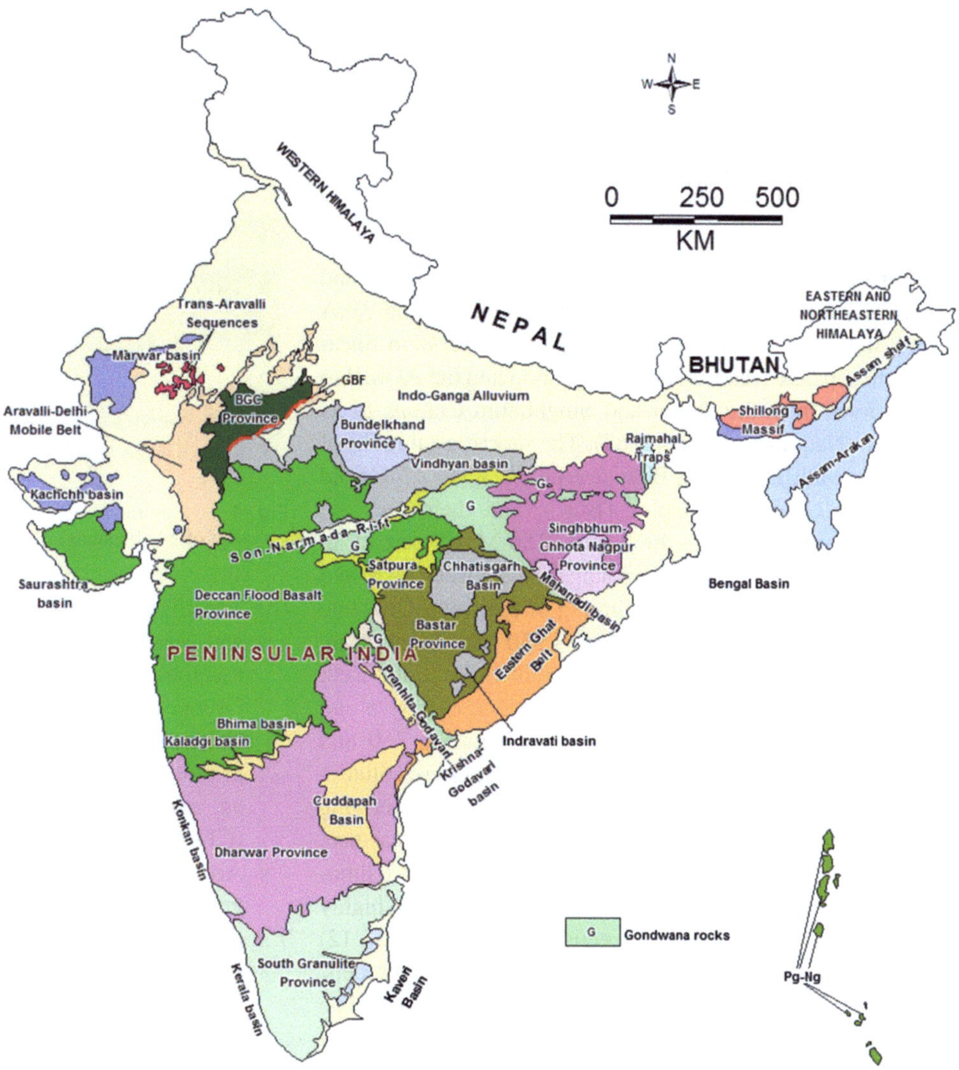

Fig. 3 Simplified map showing the geological provinces of Peninsular India (after Chakrabarti et al. 2006). BGC—Banded Gneissic Complex; GBF—Great Boundary Fault; Pg-Ng—Unclassified Paleogene-Neogene.

along with the Permian marine incursions marked by the occurrence of Umaria and Manendragarh marine beds

(3) Rajmahal Volcanic Province dating to ~115 Ma in eastern India
(4) Deccan Volcanic Province, a Late Cretaceous large igneous province in western and central India with the bulk of the volcanic activity being centered around the K-T (Cretaceous-Tertiary) boundary
(5) Cretaceous marine incursions in the Narmada region and in the Kaveri Delta, for example the well known Trichnopoly Beds
(6) Mesozoic-Cenozoic successions of Western Rajasthan, Kachchh and Saurashtra
(7) Late Quaternary alluvial sequences along the major river valleys of the Indian Peninsular, for example, Narmada, Son, Tapi, Ken, Betwa, Kaveri, Krishna-Godavari; as well as the Late Quaternary delta sequences along the east coast of India.

3 Precambrian Crustal Evolution of Peninsular India

The Precambrian geological history of the Peninsular India spans three billion years. The Precambrian shield comprises of Archean cratons, Proterozoic mobile belts and shear zones, and continental rifts such as Narmada-Son, Godavari and the Mahanadi Rifts (Figs. 2 and 3). The cratonisation of India was polyphase, and a stable configuration of the assembly was completed by ~2.5 Ga (Meert et al. 2010). Several Meso- to Neoproterozoic Purana Basins that include the Cuddapah, Chhattisgarh, Vindhyan, Pranhita-Godavari, Indravati, Bhima-Kaladgi, Kurnool, and the Marwar basins cover the Archean-early Proterozoic terrains, and are spread over many parts of the Indian landmass (Figs. 2 and 3).

3.1 Cratons, Purana Basins, and the Southern Granulite Terrain (SGT)

(a) The Bundelkhand Craton: The Great Boundary Fault (GBF) demarcates the boundary between the Bundelkhand protocontinent and ADMB in its west (Figs. 2 and 3). A crustal-scale lineament, namely the Narmada-Son lineament separates the Bundelkhand Craton and the ADMB to its north from the Bastar and Singhbhum Cratons to its south (Naqvi and Rogers 1987). The supracrustal rocks of the Aravalli-Delhi Fold Belts overlie the 3.3 to 2.5 Ga old Banded Gneissic Complex (BGC), and their ages are considered between ~1850 and 2150 Ma (Meert et al. 2010). Metamorphism in two stages i.e. between 1620 and 1725 Ma, related to the beginning of the Delhi Orogenic cycle (Roy et al. 2005), and at ~950 Ma (Buick et al. 2006) affected rocks of the Aravalli-Delhi Fold Belt. Following these events, the Malani felsic extrusive and intrusive igneous activity took place between 750 and 800 Ma (Gregory et al. 2009). These rocks constitute a part of the basement for the unconformably overlying Neoproterozoic Marwar Supergroup in western Rajasthan. The Bundelkhand Craton has been divided into three lithotectonic units (1) Archean enclaves constituted by highly deformed older gneissic-greenstone components, (2) undeformed granitoid plutons and associated quartz reefs, and (3) mafic dyke swarms and other intrusions. The Bundelkhand Granite has yielded an age of ~2492 Ma (Mondal et al. 2002). The mafic dykes that intrude the Bundelkhand Igneous Complex have yielded ages of 2.15 and 2.0 Ga, respectively (Mallikharjuna Rao 2004).
The intracratonic Marwar Basin of Neoproterozoic-Palaeozoic age occurs to the west of the Aravalli fold belt and the Vindhyan Basin of central Peninsular India to the west and south of the Bundelkhand Craton. The sickle-shaped Vindhyan Basin occurs between the Aravalli-Bundelkhand Province in the north and west and the Late Cretaceous Deccan Traps in the south; the Great Boundary Fault marks the western limit of this basin. Stratigraphically, the Vindhyan Basin comprises of a Lower Vindhyan Semri Group unconformably overlain by an Upper Vindhyan sequence consisting of the Kaimur, Rewa, and Bhander Groups. The geochronology of the Semri Group is based on a Pb-Pb isochron from the Lower Kajrahat Limestone of ~1721 Ma (Sarangi et al. 2004) and U-Pb ages from the porcellanites ranging from ~1630 to 1599 Ma. Most workers subscribe to a Paleo-Mesoproterozoic age for the Lower Vindhyan sedimentation.

Fig. 4 A panoramic view of the Dharwar granulite terrain (Photo courtesy of M. Jayananda)

(b) The Singhbhum Craton: This craton lies in the central and eastern part of India and consists of several assemblages that include the Older Metamorphic Group, the Singhbhum Granite, and the Iron Ore Group (IOG). It is bordered to the north by the Chhotanagpur granite-gneiss terrain (CGGT) which is considered to be an extension of the Central Indian Tectonic Zone (CITZ). The Singhbhum nucleus is composed of Archean granitoid batholiths, including the Singhbhum granite complex (Meert et al. 2010). The oldest enclaves within the batholiths are the Older Metamorphic Group (OMG) which has yielded U-Pb zircon ages of 3.5, 3.4, and 3.2 Ga. The OMG is intruded by the Singhbhum granite complex that consists of several domal magmatic bodies which range in age from 3100 to 3500 Ma. In the Singhbhum Craton, the Iron Ore Group marks the greenstone-gneiss terrain. Further, gneissic complex rocks of ~2.6 Ga, correlative with the OMG, recorded from Shillong-Mikir Hills Plateau in the northeastern part of India are also referred occasionally as 'Meghalaya Craton' (Sharma 2009).
The Palaeo-Mesoproterozoic sequences of the Singhbhum Craton include the Dhanjori Formation, the Chaibasa Formation, and the Dhalbhum, Dalma, and Chandil Formations. The Dhanjori Formation, representing the oldest (~2.5 to ~2.85 Ga) sedimentary basin in this craton consists of an assemblage of clastic, mafic-ultramafic volcanic, and volcaniclastic rocks (Mazumder 2005). The youngest of these Mesoproterozoic sequences, the Chandil Formation is probably older than ~1638 Ma (Mazumder 2005). In the southern part of this craton, the Kolhan Group, ~1.1 Ga in age, occurs as a transgressive sequence and has been related to the fragmentation of the Rodinia Supercontinent (Mukhopadhyay et al. 2006).

Fig. 5 A prominent escarpment in Pachmarhi sandstone of the Gondwana landscape, Satpura Belt. (Photo courtesy of Tapan Chakraborty)

(c) Bastar Craton: This craton, also referred to as the Central Indian Craton and the Bhandara Craton, consists of supracrustal sequences, mafic dyke swarms, and the Satpura Orogenic Belt (Naqvi and Rogers 1987). Ramakrishna and Vaidyanadhan (2010) pointed out that there are Tonalite-Trondjhemite Gneiss (TTG) assemblages dated to 2.5–2.6 Ga indicating a major period of crustal accretion. The Bastar Craton is intruded by several dyke swarms that cross cut the granitoids and the supracrustal sequences viz. the Dongargarh, the Sakoli, and the Sausar. The Bastar Craton is host to two major Mesoproterozoic Basins, the Chhattisgarh and the Indravati Basins.

(d) The Dharwar Craton: The Dharwar Craton (Fig. 4) is commonly subdivided into the Eastern and Western Dharwar Cratons. The Eastern Dharwar Craton (EDC) is composed of the Dharwar Batholith, greenstone belts, intrusive volcanics, Mesoproterozoic and younger sedimentary basins (Naqvi and Rogers 1987; Ramakrishnan and Vaidyanadhan 2010). The Western Dharwar Craton (WDC) consists of three generations of volcano-sedimentary greenstone-granite sequences i.e. the 3.1–3.3 Ga Sargur Group, the 2.6–2.9 Ga Dharwar Supergroup and the 2.5–2.6 Ga calc-alkaline to high potassic granitoids (Meert et al. 2010 and references therein).

Proterozoic sedimentation in the Dharwar Craton took place in the Cuddapah, Pranhita-Godavari, and Bhima Basins. The Cuddapah succession is estimated to be 12 km thick and is made up of two units—the Cuddapah Supergroup present throughout the basin and an unconformably overlying Kurnool Group limited to the western part of the basin. A thermal event at $\sim$1.9 Ga has been suggested for the initiation of the Cuddapah Basin. The Pranhita-Godavari Basin occurs between the Dharwar and the Bastar Cratons (Fig. 2), and consists of several unconformity-bounded packages adding up to $\sim$6 km in thickness. These range in age from $\sim$790 to 1330 Ma. The Bhima Basin is made up predominantly of calcareous facies.

(e) Southern Granulite Province: In addition to these cratons, a prominent feature of the geology of southern India is the Southern Granulite Province (SGP) which is considered to be made up of three late Archean to Neoproterozoic high grade metamorphic blocks that are joined together by a series of shear zones (Figs. 2 and 3). The Northern Block (NB) is separated from the Central Block (CB) by the Moyar-Bhavani Shear Zone. The Central Block is divided further into a western Nilgiri Block and an Eastern Madras Block. The Madras Block is bounded by the Palghat-Cauvery Shear Zone (PCSZ), and is followed southwards by the Madurai Block. The NE-SW trending Achankovil Shear Zone (ACSZ) separates the Madurai block from the southernmost Trivandrum Block (Naqvi and Rogers 1987; Ramakrishnan and Vaidyanadhan 2010).

Availability of well-constrained geochronological data, in the last two decades, allowed the delimitation of tectono-thermal events in craton-margin orogens and helped in understanding the forcing of such events in shaping the architecture of the Indian Peninsula. The most conspicuous early Paleoproterozoic (2.45–2.48 Ga) high grade metamorphism in the Indian crust is recorded from Nilgiri, Savoy Hill massifs and from the Palghat-Cauvery Shear Zone. The Aravalli Orogeny is considered to have closed around 1850 Ma. Unambiguous evidence for the Paleoproterozoic (1.6 Ga) ultrahigh temperature metamorphism at lower

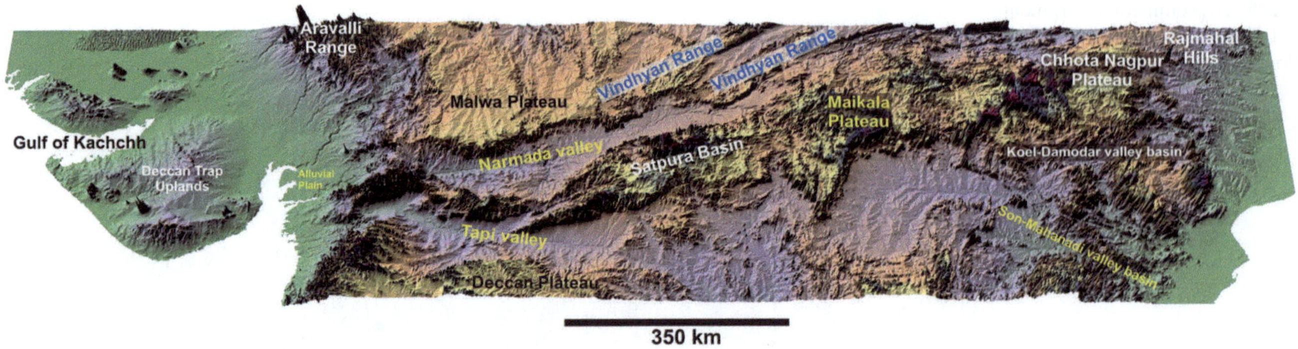

Fig. 6 Physiographic map of the central India showing distribution of Gondwana Basins

crustal depths is available from parts of the Central Indian Tectonic Zone (CITZ), Chhotanagpur Gneissic complex (CGC, 1.72 Ga) and within the southern part of the Eastern Ghat Belt (EGB; 1.76 Ga). It has been suggested that during the period 1.7–1.76 Ga, the EGB was behaving as an accretionary orogen with the development of fold-thrust belt at the craton margin. Also, there is a growing body of evidence in favour of the occurrence of major early Mesoproterozoic tectono-thermal and tectono-magmatic events in Peninsular India, which has resulted in speculations about a Pan-Indian Mesoproterozoic orogen, extending probably to North China, Australia and East Antarctica.

4 Rifting of Craton, Gondwana Sedimentation and Trap Volcanism

The Gondwana sediments (Fig. 5) accumulated in different areas of Peninsular India between Permo-Carboniferous and Triassic (290–208 Ma) and mark the resumption of sedimentation in the Peninsula in late-Carboniferous after a long hiatus (Veevers and Tewari 1995). Three well defined linear belts viz. the NW-SE trending Son-Mahanadi Valley and Pranhita-Godavari Valley Basins, and the east-west trending Damodar-Koel Basin demarcate the structural trends of the Gondwana Basins (Figs. 3 and 6) and are considered as relics of the original master basin that extends below the Cenozoic cover of the Bengal Basin to the coal belts of Bangladesh. The basins are typically bounded by faults that developed along Precambrian lineaments, and are also affected by intra-basinal faults indicating fault-controlled syn-sedimentary subsidence. The glaciogenic Talchir and the overlying coal-bearing Barakar Formations represent the lower part of the Gondwana succession showing uniform characteristics in all the basins. Overlying the pre-Gondwana metamorphics, the glaciogenic Talchir sedimentation comprises of glacial and glacial outwash (glacio-fluvial) deposits. A few beds containing marine fossils within the Talchir Formation indicate an early Permian marine incursion (Veevers and Tewari 1995). With the onset of humid climate two coal measures, the Barakar and Raniganj Formations, succeed the Talchir deposits. A fluvio-lacustrine sequence, devoid of coal and known as Iron Stone Shale (Barren Measure) separates the two coal measures. Rocks of the Panchet Formation, also barren of coal, overlie the Raniganj Formation and comprise of alluvial deposits formed in a semi-arid environment. The eruption of basaltic lava in Early Cretaceous (120–116 Ma) time over a vast stretch of land spanning from Rajmahal in the east to Sylhet region in north-eastern Bangladesh through Meghalaya in between marks the final phase of Gondwana sedimentation.

The Indian Shield, in particular in its central and western part, witnessed voluminous continental flood basalt eruptions in the form of Deccan Volcanics in Late Cretaceous time (Fig. 7). Present day areal extent of this province is 5×10^5 km^2 (Fig. 3) and its original areal extent is estimated at 1.5×10^6 km^2. Expressed as stepped hills, the Deccan lava pile represents its thickest development ($\sim$2000 m) along the Western Ghat region (*Sahyadri*) and thins progressively eastward and southeastward to a thickness of 200 m. Both plume-related and non-plume plate tectonic models are invoked for the interpretation of such a large igneous province like Deccan, which otherwise is considered to be related with rifting and breakup of the Seychelles Micro-continent from India in Late Cretaceous time (Sheth 2005). At places, the volcanic rocks are underlain by Cretaceous sediments that are described as infra-trappeans. In particular, the unique geomorphology of the west coast of India, consisting of elevated inland plateaus, erosionally controlled escarpment, coast parallel monoclinal flexure and the low lying coastal lowland, is interpreted as the response to the combined effects of several factors including shoulder uplift in a rift system, lateral scarp retreat, differential denudation of varying lithologies and flexural isostasy (Fig. 8; Gunnell and Fleitout 2000).

Fig. 7 A regional view of Deccan Traps exposed along the Western Ghat near Lonavala (Photo courtesy of Vishwas S. Kale)

5 Establishment of Indian Continental Margins

The continental breakup processes in East Gondwanaland involving India, East Antarctica, Madagascar and the Seychelles also have left their imprints on the evolution of the continental margins of India, which can be grouped under three different time windows: (a) 250–140 Ma breakup between Africa and the combined India-Madagascar, and its relation with the evolution of certain parts of the western continental margin of India (b) 140–100 Ma India-Antarctica breakup and the development of a passive margin along the east coast of India, and (c) 100–50 Ma India-Madagascar-Seychelles breakup and the development of some parts of the western continental margin of India. Segmented by sub-crustal tectonic elements, five major onshore pericratonic rift basins viz. the Kaveri, Palar, Krishna-Godavari, Mahanadi and Bengal basin, constitute the 2000 km long east coast of India, and are related with their corresponding thick offshore sedimentary depocenters. Towards southeast of the Indian mainland, the Andaman deepwater basin, involving Andaman-Nicobar (Island) accretionary complex in its forearc and the Andaman Sea at its back-arc, represent the arc-trench region in an active plate boundary between the Indian Plate and the Eurasian Plate. The records of rift-drift events related to the separation of Madagascar during the mid-Cretaceous and the Seychelles in Late Cretaceous from India are preserved in five major basins along the western margin of India viz. Kachchh, Saurashtra, Mumbai, Konkan and Kerala. The basins evolved through sequential rifting from north to south under the dominant influence of Dharwar trend (NW-SE to NNW-SSE), the Aravalli trend (NE-SW) and the Satpura trend (ENE-WSW to E-W). According to Biswas (1987), late Triassic rifting in the northern part of the west coast gave rise to three Mesozoic marginal rift basins in the onshore i.e., Kachchh, Narmada and Khambhat (Cambay) that preserve Mesozoic sediments of varying thickness.

Besides, the pericratonic basin of Rajasthan at the northwestern front of the Indian Shield preserves the record of Indus shelf sedimentation in its eastern extension. The basin hosts several sedimentary packages ranging in age from the Precambrian to Tertiary and is internally divided by structural elements into three sub-basins viz. Bikaner-Nagaur in the north, Jaisalmer in the middle, and Barmer-Sanchor in the south.

6 Cenozoic Flight of Indian Peninsula, Closure of Tethyan Sea and Uplift of the Himalayan Orogen

At ~130 Ma the Indian Plate was detached from Gondwanaland and started its northward journey resulting in subsequent closure of the Tethys Sea. The occurrence of island arc complexes in the suture zone between the Indian and the Eurasian Plates bears the imprint of island arc volcanism that was associated with the leading edge of the subducting Neo-Tethyan oceanic crust in the Late Jurassic and Early Cretaceous time. The volcanic arc, presently represented by a thick sequence of the Kohistan Complex in northern Pakistan, and Ladakh Batholith and related rocks to the east of Nanga Parbat, collided along the Indus Suture at the time of closure of the Neo-Tethys. Calc alkaline granite (e.g. Kohistan Batholith of 150 Ma and 85–26 Ma ages and equivalent Ladakh Granite) form a significant part of the island arc complex which also includes metamorphic, plutonic, volcanic and sedimentary rocks (Bouilhol et al. 2010). The structural evolution of the Himalayan Orogen initiated with continent-continent collision between the Indian and the Eurasian Plates in the early Eocene at ~55 Ma. The decrease in spreading rate in the Central Indian Basin from 8 cm/year to 4 cm/year, concomitant slowing down of the Indian Plate velocity from ~18–20 cm/year to ~5 cm/year (Kumar et al. 2006) and the change in the spreading direction from N-S to NE-SW constitute the evidence for this collision event. The formation of the Indus and Bengal Fans in the northern Indian Ocean also began sometime in the Oligocene (Curray and Moore 1971). It is generally agreed upon that the docking between the two plates is diachronous along the strike of the Himalayan Orogen. Evidence from the north-western edge of the Indian Plate suggests it occurred at ~65 Ma (Beck et al. 1995), whereas studies from the Zanskar and Ladakh regions constrain the collision ages to around 52 and 49 Ma. In spite of displaying first order similarities, the Himalaya varies along its strike in deformation styles, uplift-erosion characteristics, and in morpho-tectonic evolution. Additionally, at the eastern and western

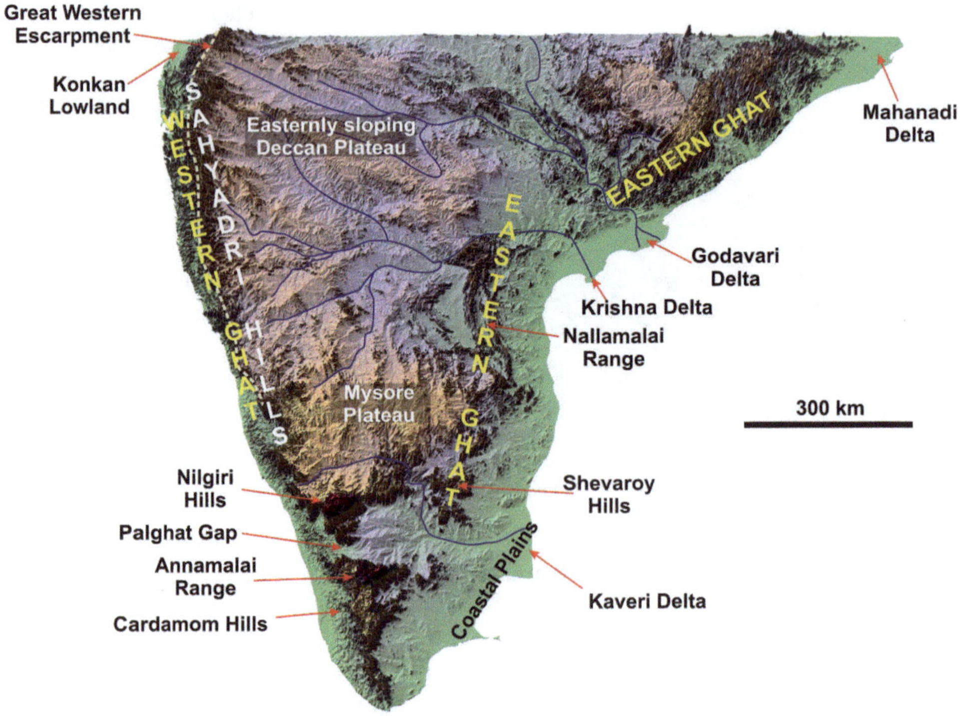

Fig. 8 Physiographic map of a part of Peninsular India showing first order relief structure viz. hills, plateaus, and deltas

Fig. 9 Panoramic view of a part of the Himalaya in Alaknanda Basin showing Lesser Himalaya (*LH*) with snow-covered Higher Himalayan (*HH*) mountains in the background

extremities of the Himalaya, syntaxial bends have been formed.

Stratigraphically, the Himalayan Orogen includes rocks spanning from the Precambrian to Holocene time. In the northwestern Himalaya, the orogen is subdivided into four litho-tectonic units juxtaposed along north-dipping thrust faults. From north to south, these are—(i) Tethyan Himalaya representing Proterozoic to Eocene sediments interbedded with Paleozoic and Mesozoic volcanics and batholiths, (ii) Greater (or Higher) Himalaya (Fig. 9) composed of crystallines, schists, gneisses and granites, (iii) Lesser Himalaya (Fig. 9) composed of Precambrian and minor Paleozoic strata, and orthogneisses, and (iv) Sub-Himalaya constituted of Cenozoic sequences including the Neogene continental Siwalik strata (Fig. 10). The thrust faults that demarcate the boundaries of these units are the South Tibetan Detachment (STD; separating Tethyan and Higher Himalaya), the Main Central Thrust (MCT; separating the Higher and Lesser Himalaya), the Main Boundary Thrust (MBT; separating the Lesser and Sub-Himalaya), and the Main Frontal Thrust (MFT; separating the Sub-Himalaya and Indo-Ganga Plain) (Fig. 10).

The poorly fossiliferous carbonate and siliciclastic sequences in both the Tethyan and Lesser Himalayan are assigned Proterozoic to early Cambrian age and are considered as the record of pre-rift sedimentation history in an epicratonic marine setting on the northern Indian margin (Valdiya 2010). Low to high grade metamorphosed

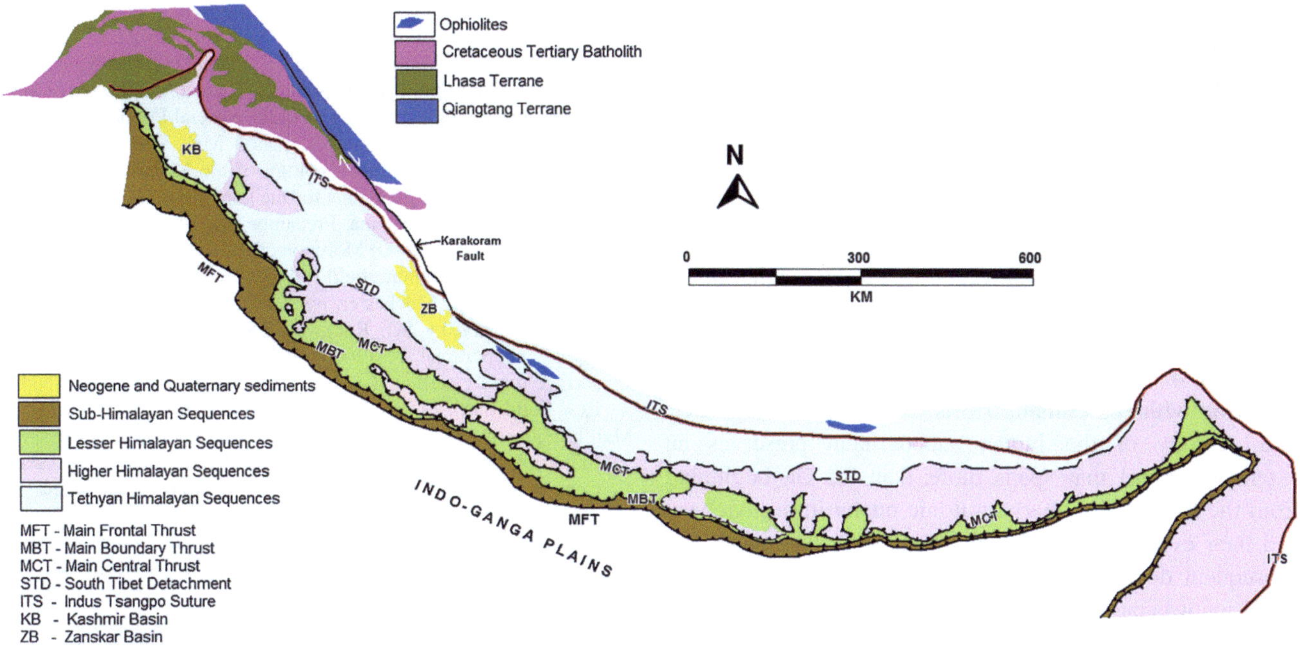

Fig. 10 Map showing geological subdivisions of the Himalaya (after Yin 2006; Valdiya 2010)

crystalline complex of the Higher Himalaya provides evidence of ductile deformation, intrusion and metamorphism at 2174, 1850, 500 Ma and in middle Miocene time. Between the Late Paleocene and the Early Oligocene, subduction of the Indian crust led to the formation of fold and thrust belt of the Tethyan Himalaya for which STD acted as the basal thrust. Continental sedimentation commenced only after mid-Oligocene time with the exhumation of the Higher Himalaya.

7 Cenozoic of Peninsular India

Restriction of Cenozoic records along the periphery of the Peninsula suggests that the craton remained elevated during this time period without any record of sedimentation. Onshore, basins in Gujarat mainland, Saurashtra, and Kachchh, Jaisalmer and Barmer basins of Rajasthan, carbonate succession of Niniyur Formation, Tamil Nadu and Baripada beds of Chhotanagpur Plateau constitute some of the best preserved Cenozoic records. In northeast India, the three important zones of Cenozoic sedimentation are (a) Shelf zone around Shillong-Mikir Plateau (b) Naga schuppen zone and (c) Kohima synclinorium and residual basins in Sylhet and Bengal. The fossiliferous Paleocene marine strata are recorded all along the West Indian shelf between Rajasthan and offshore Kerala and also in the East Indian shelf in the subsurface of West Bengal and outcrops in the South Shillong (Meghalaya) Front.

8 Quaternary

In India, several river valleys, coastal tracts and desert regions such as the Thar show well developed Quaternary sediments. The most important Quaternary fills occur in the valleys and extensive plains developed along the large river systems originating in the Himalaya i.e., the Indus, Ganga and Brahmaputra. These sediments have their provenance predominantly in the Himalaya.

Most of the river valleys in Peninsular India, both major and minor, contain late Quaternary strata with variable thicknesses. Significantly, a 74 Ka tephra layer commonly associated with the Youngest Toba Tephra (YTT) event has been identified in several river valleys in Peninsular India after being initially recorded in the Son-Narmada valleys. These river valley and floodplain sequences constitute an important archive for Quaternary palaeoclimatic studies and have served as recorders of variations in monsoonal strength and intensity that took place during the Late Quaternary in India. The wetter monsoonal phase corresponding to MIS 3 has been widely recognized in the river valley sediments.

The only Hominid fossil *Homo erectus narmadaensis* of the Indian subcontinent has been discovered from the conglomerates of the Narmada Valley (Sonakia 1984).

9 Epilogue and Summary

The trajectory of landscape and seascape evolution in the continents and oceans is to a large extent a function of geologic and tectonic evolution of the region. In continental landscapes, this trajectory is also influenced considerably by the nature of coupling between the geological and tectonic evolution with the climatic shifts at different time scales. The geology of the Indian subcontinent preserves an extensive record that spans more than 3.5 Ga beginning from the formation of early cratonic nuclei in the Archean and their evolution into a stable Peninsular landmass. The subsequent developments on which the geomorphological evolution was based are rooted in the rifting history of India that was initiated in the Late Palaeozoic and terminated in the end of Mesozoic. The rifting was also accompanied by two major episodes of volcanism, one of which, the 65 Ma Deccan volcanism, provided the basic relief structure on which geomorphological evolution proceeded from then onward in a large part of western and central India.

The final stages in the gross geomorphological evolution of the Indian region were largely determined by the processes connected with the northward movement of the Indian Plate and its subduction under the Eurasian Plate in the Cenozoic. These tectono-geomorphic processes resulted in the first order relief structure of the northern part of the Indian subcontinent with the progressive development of Himalayan topography from the Palaeocene onward as well as the isostasy driven evolution of a large subsiding area in front of the Himalaya due to lithospheric flexure.

References

Beck R, Burbank DW, Sercombe W, Riley GW, Barndt JK, Berry JR, Afjal J, Khan AM, Jurgen H, Metje J, Cheema A, Shafique NA, Lawrence RD, Khan MA (1995) Stratigraphic evidence for an early collision between Northwest India and Asia. Nature 373:55–58

Biswas SK (1987) Regional tectonic framework, structure and evolution of the Western Marginal basins of India. Tectonophysics 135:302–327

Bouilhol P, Schaltegger U, Chiaradia M, Ovtcharova M, Stracke A, Burg JP, Dawood H (2010) Timing of juvenile arc crust formation and evolution in the Sapat complex (Kohistan-Pakistan). Chem Geol 280:243–256

Buick IS, Allen C, Pandit M, Rubatto D, Hermann J (2006) The Proterozoic magmatic and metamorphic history of the banded gneiss complex, Central Rajasthan, India; LA-ICP-MS U/Pb zircon constraints. Precambr Res 151:119–142

Cox KG (1989) The role of mantle plumes in the development of continental drainage patterns. Nature 342:873–877

Curray JR, Moore DG (1971) Growth of the Bengal deep-sea fan and denudation in the Himalayas. Bull Geol Soc Am 82:563–572

Chakrabarti C, Basu Mallick S, Pyne TK, Guha D (2006) A manual of the geology of India, vol 1. Precambrian, part I: Southern part of the Peninsula, Geological Survey of India, Special Publication, No. 77

Gregory LC, Meert JG, Bingen B, Pandit MK, Torsvik TH (2009) Paleomagnetism and geochronology of the Malani igneous suite, Northwest India: Implications for the configuration of Rodinia and the assembly of Gondwana. Precambr Res 170:13–26

Gunnell Y, Fleitout L (2000) Morphotectonic evolution of the Western Ghats, India. In: Summerfield MA(ed) Geomorphology and global tectonics. Wiley, Chichester, pp 321–336

Kumar S, Wesnousky SG, Rockwell TK, Briggs RW, Thakur VC, Jayangondaperumal R (2006) Paleoseismic evidence of great surface rupture earthquakes along the Indian Himalaya. J Geophys Res: Solid Earth 111:B03304. doi:10.1029/2004JB003309

Mallikharjuna Rao J (2004) The wide spread 2 Ga dyke activity in the Indian shield: evidences from Bundelkhand Mafic dyke swarm, Central India and their tectonic implications. Gondwana Res 7:1228–1291

Mazumder R (2005) Proterozoic sedimentation and volcanism in the Singhbhum crustal province, India and their implications. Sed Geol 176:167–193

Meert JG, Pandit MK, Pradhan VR, Banks J, Sirianni R, Stroud M, Newstead B, Gifford J (2010) Precambrian crustal evolution of Peninsular India: a 3.0 billion year odyssey. J Asian Earth Sci 39:483–515

Mondal MEA, Goswami JN, Deomurari MP, Sharma KK (2002) Ion microprobe $^{207}Pb/^{206}Pb$ ages of zircons from the Bundelkhand massif, Northern India: implications for crustal evolution of the Bundelkhand–Aravalli protocontinent. Precambr Res 117:413–419

Mukhopadhyay J, Ghosh G, Nandi AK, Chaudhuri AK (2006) Depositional setting of the Kolhan group: its implications for the development of a Meso to Neoproterozoic deep-water basin on the South Indian craton. S Afr J Geol 109:183–192

Naqvi SM, Rogers JJW (1987) Precambrian geology of India. Oxford monographs on Geology and Geophysics, vol 6. Oxford University Press, Oxford

Ramakrishnan M, Vaidyanadhan R (2010) Geology of India, vol 1. Geological Society of India, Bangalore

Radhakrishna BP, Ramakrishnan M (1988) Archean-Proterozoic boundary in India. J Geol Soc India 32:263–278

Roy AB, Kröner A, Bhattacharya PK, Rathore S (2005) Metamorphic evolution and zircon geochronology of early Proterozoic granulites in the Aravalli mountains of Northwestern India. Geol Mag 142:287–302

Sarangi S, Gopalan K, Kumar S (2004) Pb–Pb age of earliest megascopic, eukaryotic alga bearing Rohtas formation, Vindhyan supergroup, India: implications for Precambrian atmospheric oxygen evolution. Precambr Res 132:107–121

Sharma R (2009) Cratons and fold belts of India. In: Lecture notes in earth sciences, vol 127. Springer, Berlin. doi:10.1007/978-3-642-01459-8_1

Sheth HC (2005) From Deccan to reunion: no trace of a mantle plume. In: Foulger GR, Natland JH, Presnall DC, Anderson DC (eds) Plates, plumes and paradigms. Geological Society of America Special Paper, vol 388, pp 477–501

Sonakia A (1984) Skull cap of an early man from the Narmada valley alluvium (Pleistocene) of central India. Am Anthropol 87:612–615

Valdiya KS (2010) The making of India: geodynamic evolution. Macmillan, New Delhi

Veevers JJ, Tewari RC (1995) Permian-Carboniferous and Permian-Triassic magmatism in the rift zone bordering the Tethyan margin of southern Pangea. Geology 23:467–470

Yin A (2006) Cenozoic tectonic evolution of the Himalayan orogen as constrained by along-strike variation in structural geometry, exhumation history, and foreland sedimentation. Earth-Sci Rev 76:1–131

Past and the Present Climate of India

Ashok K. Singhvi and R. Krishnan

Abstract

The Indian monsoon, which comprises the seasonal reversal of winds and implies rainfall on land, is a complex system. Given that the national economy depends critically on monsoon, its rigorous understanding it warranted. We trace here the geological history of monsoon and the present trends of its performance. The monsoon as it is now is about 10–8 Ma old. It has fluctuated around a mean value at all timescales due to the response of causative factors to various global forcings—the dominant being the Sun-Earth geometry and consequent asymmetric heating of land versus oceans. Whereas most regions show analogous responses to global forcings, significant spatial heterogeneities are seen on shorter timescales. Instrumental data suggest general weakening of monsoon system and at the same time an increase in extreme events. A broad brush scenario of the monsoon and its variability through geological time is presented.

Keywords

Monsoon • Droughts • Floods • El Niño • Geological records • Proxies • Marine records • Long-term trends

1 Introduction

The climate of the Indian subcontinent is dominated by the monsoons which are pronounced seasonal reversals of winds and transitions from drier to wetter regimes. During the summers, the southwesterly winds pick moisture from the northern Indian Ocean and drop this on the landmass, providing the summer rainfall that is critical to the survival of the Indian society and its economy. The winter winds blow from the northeast regions of the Asian continent and travel towards the Indian Ocean. This annual cycle of atmospheric circulation arises from interactions of the seasonal changes in solar insolation with the land-ocean distribution such that variations in the thermal fluxes produce differential warming between the land and the sea, leading to development of convective cells and consequent moisture transport and deposition (i.e. rain), inland.

The food grain production of India has a proportional relationship to the monsoon, with its critical dependence on the onset, duration and distribution of rainfall and the periods of break monsoon conditions. While the present monsoonal conditions are tracked using variety of instruments, its past history is studied using geological archives. Studies on the monsoon using such geological archives and instrumental record have now established the following (Singhvi and Kale 2009; Singhvi et al. 2010, 2012):

A. K. Singhvi (✉)
Geosciences Division, Physical Research Laboratory
Navrangpura, Ahmedabad 380009, India
e-mail: 2aksprl11@gmail.com

R. Krishnan
Indian Institute of Tropical Meteorology, Pashan 411008, Pune, India
e-mail: krish@tropmet.res.in

V. S. Kale (ed.), *Landscapes and Landforms of India*, World Geomorphological Landscapes,
DOI: 10.1007/978-94-017-8029-2_2, © Springer Science+Business Media Dordrecht 2014

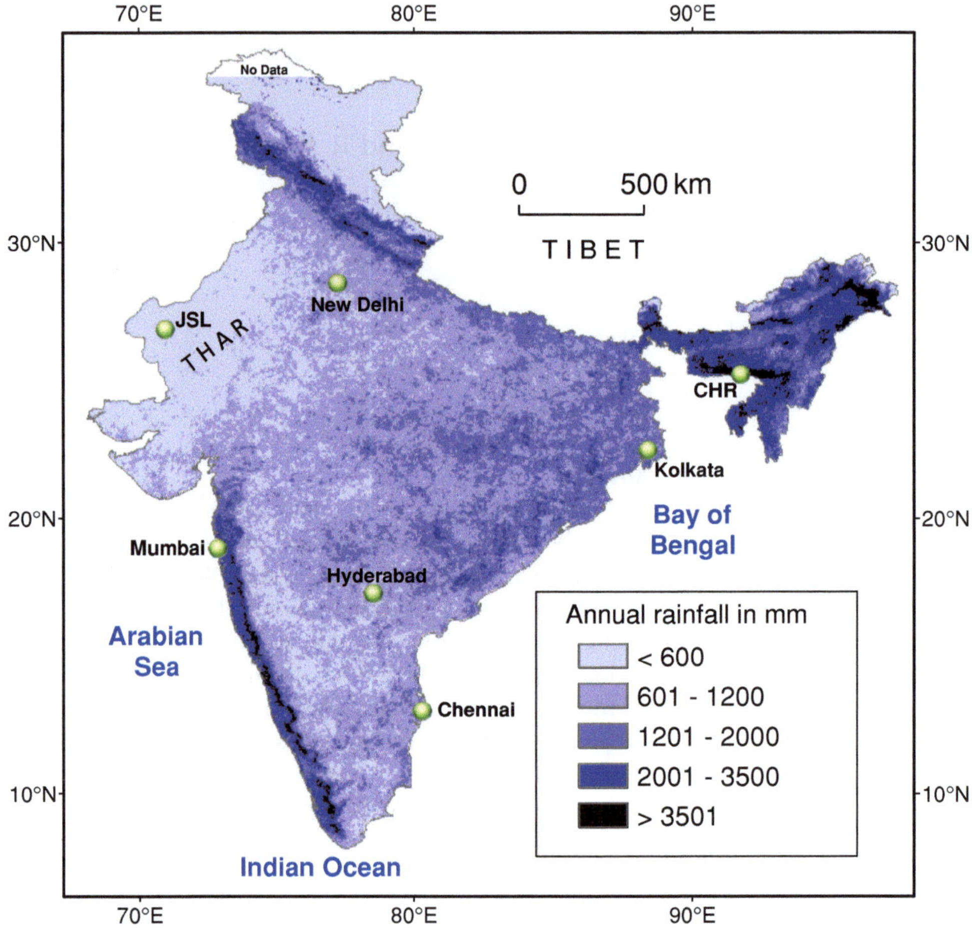

Fig. 1 Spatial distribution of annual rainfall over India. Two sites with extreme rainfall are Jaisalmer (JSL) and Mawsynram, near Cherrapunji (CHR). Average annual precipitation from TRMM satellite (*Data source* http://www.geog.ucsb.edu/ ~ bodo/TRMM/)

1. As per the Climate Profile of India, published by the India Meteorological Department, the mean southwest monsoon (June, July, August and September) rainfall is 877.2 mm, which is about 74.2 % of annual rainfall (1,182.8 mm).
2. Monsoon is a stable atmospheric system, has never failed totally and has fluctuated around a mean southwest monsoon value of ~ 877 mm by about 20–30 %.
3. Significant spatial and temporal variability in monsoon rainfall exists (Figs. 1 and 2). However, the amount, intensity and distribution of rain through a season are highly variable. It is this variability that causes droughts and floods. Thus, in a region, 20–30 % lower rainfall implies drought and as much higher rainfall may imply floods. The long-term average of rainfall, however, remains nearly constant (Fig. 2).
4. The rainfall is spatially variable with the lowest average annual rainfall being ~ 130 mm at Jaisalmer and highest average annual rainfall being 11,410 mm at Mawsynrum (near Cherrapunji) in Meghalaya (Fig. 1). Despite this variability, it is expected that over a long term, the monsoon across the regions should have fluctuated analogously around their local means. Much of the moisture is derived from the oceans, particularly the Bay of Bengal, and is due to its thermal stratification with critical sea surface temperature for monsoon to establish being ~ 28 °C. The Arabian Sea being well mixed by

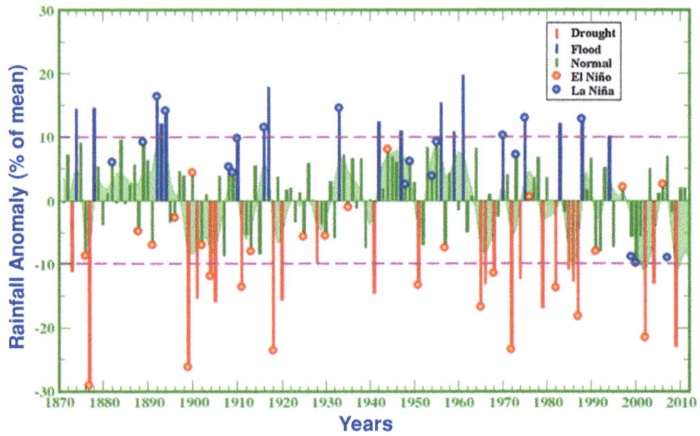

Fig. 2 Time-series of all-India summer monsoon rainfall anomalies expressed as percentage departure from normal for the period 1871–2011, based on IITM homogeneous Indian monthly rainfall dataset (*source* http://www.tropmet.res.in). Note that several instances of monsoon droughts have co-occurred with El Niño events; and conversely several excess monsoon rainfall years have co-occurred with La Niña events in the Pacific

wind fields has cooler waters and, hence, with the exception of western regions, it does not contribute much to the precipitation over the entire subcontinent.

5. Several factors, such as the Sun-Earth geometry, the condition in the North Atlantic Ocean, extent of snow cover in Eurasia, vegetation change, El Niño Southern Oscillation (ENSO), etc. contribute to rainfall in a given season, but an estimation of their exact influence and its long-term stability is difficult. The current rainfall forecast models use the following parameters; (i) NW Europe land surface temperature, (ii) Equatorial Pacific warm water volume, (iii) north Atlantic sea surface temperature, (iv) east Asian mean sea level atmospheric pressure, (v) central Pacific sea surface temperature, (vi) north Atlantic mean sea level atmospheric pressure, and (vii) north central Pacific winds, with statistically determined weightage for each of these (Kelkar 2009). Dependence of monsoon system on these parameters is not constant through time. Statistical considerations suggest that the predictability of monsoon system is 50 %, at the most. Thus, although good predictability is needed for water resources and agriculture, the prediction of key parameters like onset of rainfall, duration and intensity cannot be predicted with desired accuracy (Kelkar 2009).

This chapter reviews the present understanding of monsoon through geological time including the instrumental record. Direct measurement of rain and temperatures using instruments is ~150 year old. The pre-instrumental record of monsoon is deduced from geological archives such as tree rings, corals, speleothems, peat deposits, sediments from lakes, dunes, sedimentological character and style of deposition of sediments by rivers, sediments in lagoons and mangrove and their pollens, phytoliths, diatoms, geochemical changes, changes in isotopic ratios in specific fractions in them and, foraminifera from marine sediments. Being reconstructions, the geological records of monsoon are generally qualitative. Some efforts towards more quantitative reconstruction using annually laid archives, such as tree rings, have been made using statistical correlations of measured rainfall or temperature records and tree ring parameters. Each of the geological proxies records signatures of climate and monsoon signal in its distinctive manner and rate and this aspect has to be understood before interpreting them reliably. Sediment deposition on the land itself betrays a complex relationship of the rainfall and winds with endogenic processes like tectonics and, a proper care is needed in their interpretation. The response time to any given climate forcing is proxy dependent and the response time of proxies may range from being near instantaneous to millennia or even longer. This makes the interpretation and correlation of proxy records a non-trivial exercise (Singhvi et al. 2010).

2 Monsoon from Geological Records

Reconstruction of monsoon has been based on both the ocean and land based records. Preservation and issues arising from post-depositional diagenesis have implied that older Miocene to Quaternary records of climate are mostly available from ocean cores, whilst the younger records from the Quaternary are available both from the oceans and from land.

Changes in foraminiferal assemblages and their isotopic ratios in ocean bottom sediments, sedimentology and other attributes, provide a proxy for the monsoon winds and consequent changes in water masses due to wind induced upwelling. The land records on the other hand present a direct record of rainfall and associated changes in sediment production, transport, delivery and preservation. Such reconstructions use response of organic or inorganic system to rainfall. Sediment cores from the Bay of Bengal, the Indian Ocean and the Arabian Sea have provided records of changes in monsoon winds and water masses including sediment fluxes from rivers.

On the land, a considerable amount of long-term, chronometrically (age) constrained database comes from the Thar Desert and its margins (Singhvi 2004), the Ganga Plains (Singh 2005), Siwaliks and high altitude lakes. Such studies also used sedimentology, pedology, isotopic analysis and field correlations of sedimentary sequences besides isotope or relative chronology. Millennial scale to shorter term records have been reconstructed using sediments from rivers and lakes, chemical precipitates and tree rings across India and from ocean bottom sediments from Arabian Sea and the Bay of Bengal (Singhvi et al. 2012 and references therein).

2.1 Early History: Onset and Till 500 ka

Studies on sequences from the Vastan lignite mine near Surat in Gujarat, provided isotopic and pollen evidence of high precipitation in the tropical land mass during 56–52 Ma, the period of Paleocene-Eocene Thermal Maximum (PETM). Similar precipitation on the land must have existed even during earlier geological times. The only driving force would have been, accentuated greenhouse conditions and the temperature contrast between land and oceans. At this time period, India was far south than its present location and Himalaya was almost not even in existence (Samanta et al. 2013).

Establishment/intensification of present day Indian monsoon has been variously placed between 23 and 7 Ma, based on the type of evidence being examined and the age resolution of the archive. The exact origin of present monsoon is debated and suggestions range from its association to the rise of Tibet and Himalaya (to a height sufficient enough to block moisture laden winds), to its being a natural consequence of land-sea temperature contrasts. Differential heating of Tibetan Plateau and Bay of Bengal is considered as one of the driving mechanisms of the Indian Monsoon. A variety of marine records ranging from changes in sediment fluxes into the Bengal and the Indus Fans; foraminiferal abundances, and others indicate that the erosion of the Himalaya due to monsoon rain possibly began at around mid to late Miocene i.e. 10–8 Ma. The sediment records from Siwalik foreland basin and carbon isotope ratios provide a record of changes from tree dominated to grass dominated ecology around 7.5 Ma, which could have happened with the initiation of monsoon. Records from Arabian Sea using the foraminifer *G. Bulloides* suggest the initiation of upwelling around 8.5 Ma and this is also seen in the atmospheric modeling studies which place initiation of monsoon sometime during 10–8 Ma. The sediment transported by Luni River in western India varied from gravel bedload to sandbed stream to sheet flow, interspersed by deposition of aeolian sands and reflected changes in the river flow regime which in turn was related to monsoon rainfall. The gravel deposits had an estimated age bracket of late Miocene-Pliocene ($\sim$7–3 Ma). Thus, multiple evidences ranging from increased upwelling to rain induced erosion of the Himalaya, etc., converge to the inference that monsoon was initiated during the period 10–8 Ma. The monsoon was also intense and this intensification has been variously ascribed to the Tibet and Himalaya attaining a critical height or to the expansion of Antarctic ice sheets. It has also been suggested that vegetation change was in response to global changes in CO_2 budget. Though the rise of Himalaya began some 40–50 Ma ago, recent studies provide evidence that Tibet attained a height of around 1 km at 3.4 Ma, and was uplifted to 3 km at around 1.67 Ma. This implies that the height of Tibet and Himalaya possibly played only a subordinate role in the strengthening of the monsoon. Thus, the jury is still out on the possible causes, but the timing of initiation seems to be reasonably well bracketed.

Post 8.5 Ma, reduced upwelling phases around 5.5 and 1 Ma, indicating reduced monsoon strength, have been reported. The intensification of winter monsoon is placed at $\sim$2.8–2.5 Ma based on micro-paleontological studies in the eastern Indian Ocean sediments. Results from a lake in China indicate monsoon was weak during 2.78–2.27 Ma, 2.61–1.44 Ma and 0.34–0.12 Ma and strong during 2.27–2.61 Ma and 1.44–0.34 Ma. This suggested that the monsoon fluctuated on millennium timescales. We refer to Molnar et al. (1993), Quade et al. (1989), Chao and Chen (2001), Gupta et al. (2004), and Clift and Plumb (2008) for more detailed account on monsoon, its initiation and later evolution.

In the Quaternary Period, most evidence on climate/monsoon is provided by calcretes, and fluvial and aeolian sediments of the Thar Desert. Radiometric ages and isotope geochemical analysis on these deposits indicate that the desert was well watered but had high seasonality with semi-aridity between 1.5 and 0.5 Ma and then for the past 0.5 Ma, the region has experienced increasing desiccation with interludes of phases with conditions comparable to the present. The sedimentation style and later cementation of fluvial sediments of Luni River also provide clues to changes in the monsoon and evidence of increasing desiccation is seen from 500 ka onwards till the present (Singhvi 2004).

2.2 Monsoon: 500–10 ka

Very few records for the earlier part of this period are available and the density of records increases with their proximity to the present. The sedimentation style and the sedimentology of river sediments have been interpreted in terms of river discharge such that normal monsoon implies conditions similar to the present, more intense monsoon conditions and drier conditions are then defined relative to the present. The records indicate periods of fluctuating monsoon in gross alignment with global climatic changes. Thus for example, records from the sediments of Luni, Ganga and Mahi Rivers and from the accreted dunes of the Thar Desert indicate that the period 150–100 ka was a period of monsoon, similar to the present with its later part being wetter (Fig. 3). The period 100–70 ka witnessed fluctuating condition followed by drier conditions from 70–60 ka. It was in this period, that the glaciers expanded and came down to $\sim$2600 m a.s.l. as compared to the present elevation of $\sim$3400 m a.s.l. in Himalaya. This was due largely to higher precipitation and lower temperature. The location of moraine deposits and their radiometric ages are the records of such changes.

A general phase of more humid condition is seen during 50–30 ka in a variety of records across the region. This is seen as red soils in Gujarat and Rajasthan, presence of elephants tusk and reddened horizon in Kalpi along Yamuna River in central India, the isotopic ratios of calcretes in Thar Desert, and the foraminiferal assemblages and salinity changes in the sediments from Bay of Bengal. The Bay of Bengal records suggest rapid fluctuation in monsoon during 75–15 ka and in this period, a phase of stronger monsoon during 60–40 ka. This has been linked to an internal feedback between snow and dust accumulation over Himalaya via albedo changes.

The monsoon was significantly weaker during the last ice age that peaked at $\sim$21.4 ka (Last Glacial Maximum or LGM), and it was re-strengthened only around 13 ka attaining its full vigor $\sim$11 ka. This evidence accrues from ocean upwelling records using foraminiferal assemblages,

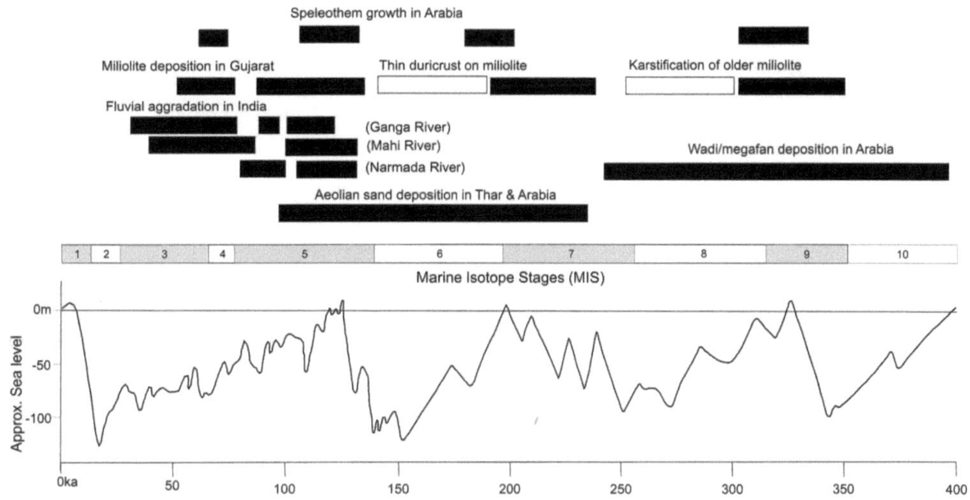

Fig. 3 Schematic of the major monsoon-controlled events during the past 500 ka. Global sea level changes and marine isotopic stages are also depicted. *Dark* and *white shades* respectively represent higher than and lower than normal monsoon. (Reproduced from Singhvi et al. 2012, with permission from Blackwell-Wiley, UK)

the trace metal records from the Arabian Sea and general absence of any geomorphic activity in the Thar Desert. During the LGM, the glaciers descended to lower elevation but the extent of descent was smaller compared to that during 70–60 ka. The descent during both time periods was due to lower temperature but at LGM still lower moisture availability implies lower amplitude of descent. Thus, despite favorable temperatures, the descent of glaciers was limited by the avalibility of precipitation. Several lines of evidence clearly suggest that the monsoon was weaker during the ice age that peaked at ~21.4 ka, the reduction in rainfall or associated winds is not yet established quantitatively. The difficulties in establishing this is compounded by a variety of factors including the presence of thresholds in each proxy. Interestingly, the desert sand from the margins of Thar Desert provides a good record of specific monsoon conditions that facilitate movement, deposition and preservation of sands over specific time windows. These time windows occur during transitional climate—from cooler to warmer. The sand accretion in Thar Desert is controlled by both wind and rain (i.e. vegetation limited) and its accretion followed by preservation occurs only over a short time window that occurs during a transitional regime from dry to wetter. The most recent window occurred around ~13 ka, when as per the oceanic records, the monsoon was re-establishing and the dune accretion and preservation across Thar was optimum. Summer winds (linked to southwest monsoon) were strong enough to move the sands and the moisture/vegetation was sufficient to trap the sand resulting in accretion. Prior to this window, the winds were weaker and moisture was absent to trap the sands, and after this window the vegetation was denser so that despite stronger winds associated with monsoon, sand migration did not occur. Extending this analogy to other episodes of regionally extended dune accretion in Thar, it is suggested that conditions of monsoon winds and rain that existed at around 13 ka should have also existed at 30, 60, 100 and 150 ka (Singhvi 2004). Beyond this time range, the dune accretion record is not accessible.

An external element of climate change in India was introduced by the explosive Toba Super Eruption at ~74 ka. Analysis of pollen from a core in Bay of Bengal led to a suggestion that the eruption was followed by cooling and prolonged desiccation which led to a decline in tree cover such that forests with trees were replaced by grasses and this resulted in a large-scale extinction of mammals (Williams et al. 2009). This indicates that even short-term abrupt events, like volcanic eruptions, could perturb monsoon dependent eco-system for an extended period to cause mammalian extinctions.

2.3 Past 10 ka

A large volume of data is available for this time period. Ocean bottom sediment records from the Arabian Sea suggest that the monsoon was fully established after the LGM, around 13 ka onwards. Sediment discharge from Ganga and Brahmaputra Rivers during 11–7 ka was twofold higher compared to the present flux, indicating that the monsoon and sediment availability in this time interval was stronger than the present. Pioneering work on the lakes of Thar Desert using the sediment character, evaporite mineralogy, and pollens indicated that the saline lakes fluctuated between—saline and freshwater condition during 12.8–6.5 ka, mostly fresh water during 6.3–4.4 ka and then experienced a extended dry phase at around 4 ka and then at 1 ka (Fig. 4) (Singhvi et al. 2012). This reconstruction is also generally borne by cave speleothem records from northwest Himalaya, record from rivers from Peninsular India and peat deposits in Himalaya and other studies (Fig. 4). An event of extreme dryness around 4 ka is seen in other records, which might possibly represent an extended period of droughts. Farther away but also dominated by the

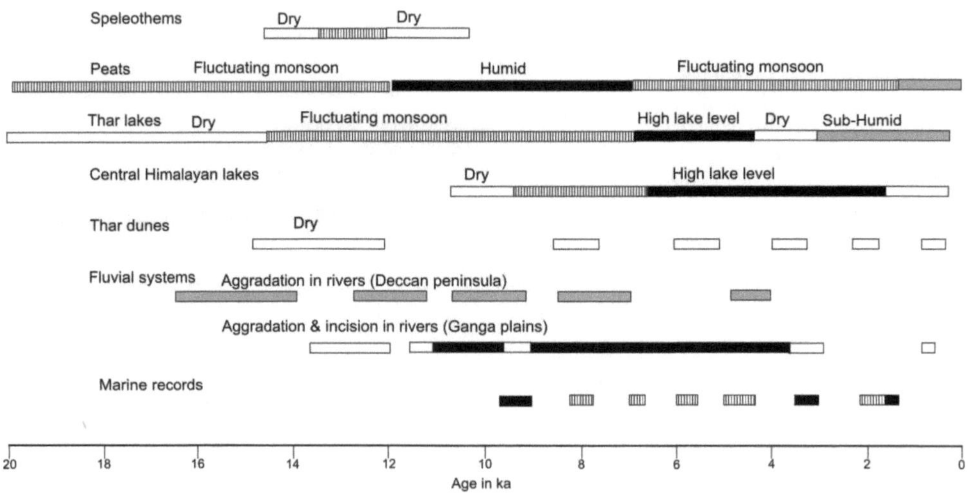

Fig. 4 Summary of geological data of changes in climate/monsoon over Indian subcontinent. (Reproduced from Singhvi et al. 2012, with permission from Blackwell-Wiley, UK)

southwest monsoon, the speleothem records from Oman indicate a phase of higher monsoon during 9–6 ka followed by a period of dryness 3–1 ka. A short lived drier/cooler event at 8.3 ka is also suggested.

An important lesson that accrues from the lakes in Thar Desert, which though indicate a similar patterns of change in their hydrology, yet show a time lag from the east to the western margins of Thar Desert, such that the lakes in western Thar Desert desiccated a thousand or more annum earlier than those in the east. This observation for the first time indicates the presence of spatial heterogeneity in geomorphic response for a given climate forcing and this to some measure reflects the changes in monsoon gradient through time. The dune record from Thar Desert also shows a similar response with geomorphic gradients, both from the east to west and from the south to north.

On a more recent timescale, reconstruction of palaeoflood records from slackwater deposits of rivers in central and western India, have provided evidence of changes in the frequency of large floods during past two thousand years (Kale 2012). Floods occur under specific rainfall conditions and are somewhat correlated to excess rainfall though floods occur in periods of overall low monsoon as well. Analysis of palaeoflood record indicated absence of large floods during 700–150 annum ago suggesting a period of low rainfall. Using oceanic records it was recently suggested that the monsoon winds strengthened during the past 400 years, however both the instrumental records and those from a first order Pennar River in Peninsular India, suggest that no increase in flood frequency is documented. This suggests the need to examine the relationship of the results from ocean upwelling and river floods in some detail.

Record from peats in Himalaya suggests a dry phase around 2 and 1.2 ka and a high rainfall phase around 600 years ago (Singhvi and Kale 2009). Around 2 ka, the dry phase is seen in dune migration rate in the Thar that was high at 0.9 cm/year around 2 ka reducing to 0.25 mm/ka during 1.5–0.6 ka and high during the past 200 years, this time due to human interventions. Slackwater deposits in Himalaya show 25 large floods during the past 1,000 years with 14 events between 1,000 and 700 years ago and 8 events between the last 700 and 200 years ago. In the recent flood in Kedarnath, palaeoflood studies along with sediment provenance using isotopes indicated the occurrence of large-magnitude floods over a few hundred year timescale—an aspect that was totally ignored in the economic development of the region (Wasson et al. 2008). Studies on tree rings have also provided useful information such as; no significant change in temperature during the past 500 years, decadal scale fluctuation with decades of cool and warm periods. Thus, for examples, 1801–1810 was cooler with a temperature lowering of 0.31 °C and warming of 0.25 °C during 1978–87. Rainfall reconstruction also showed a similar decadal scale change. We refer to Singhvi and Kale (2009) and to the original papers cited therein for more details. Gupta et al. (2006) have discussed the relationship of proxy records and human adaptations.

2.4 Past 200 years: The instrumental records

(a) Observed changes during the 20th Century
Unlike the long-term warming signal in surface temperature data (Fig. 5), the summer monsoon precipitation on an all-India scale is characterized by significant inter-annual variations, but still does not exhibit any pronounced long-term trend (Fig. 2). However, on smaller spatial scales, some notable trends in the monsoon rainfall emerge. In particular, significant declining trends in the monsoon rains have been observed over the areas of Chattisgarh and Jharkhand in north-central India and in Kerala and parts of Western Ghat (Guhatakurtha and Rajeevan 2006; Krishnan et al. 2012 and references

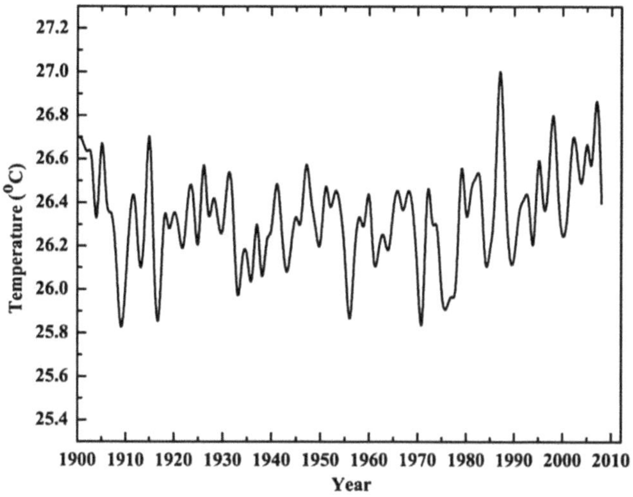

Fig. 5 Time-series of surface temperature (°C) averaged over India for the June–September season. The data are based on the global gridded surface temperature dataset from the University of Delaware (http://www.esrl.noaa.gov/psd/data/gridded/data.UDel_AirT_Precip.html)

therein) since 1950s (Fig. 6). In the recent three decades, the seasonal monsoon rainfall over Kerala has decreased by ~6 % and is supported by a consistent weakening trend of the summer monsoon circulation. Further, the period 1940–1960 and 1980–1990 experienced large floods when in general the monsoon system was inferred to be weakening (Kale 2012). Such an anti-correlation is intriguing and a possible reason could be the increased occurrence of extreme rainfall events that whilst have kept the average value nearly intact, nonetheless caused floods.

Studies on changes in the statistics of extreme precipitation events over central India during the last five decades suggest a significant increase in the frequency of very heavy rainfall events (>150 mm/day) and a reduction in the frequency of low and moderate monsoon rainfall days (Goswami et al. 2006; Rajeevan et al. 2008). Nandargi and Dhar (2012) analyzed rainstorms over northwestern Himalaya and noted that some of the most severe storms occurred during 1951–1975 and 1976–2000, but not during the recent decade (2001–2010). Overall these analyses present a somewhat counter intuitive-inferences, suggesting that we still do not completely understand all the internal feedbacks in monsoon system and their temporal evolution.

Two extreme episodes of abnormal precipitation in the recent times are the 26 July 2005 heavy rainfall event in Mumbai, and the 14–17 June 2013 multi-day event in Kedarnath, Uttarakhand. The Mumbai incident, which recorded 944 mm of rainfall in 24 hours, was associated with vigorous interactions of the large-scale monsoon circulation with embedded synoptic and meso-scale weather systems. The June 2013 Uttarakhand heavy-precipitation event was dominated by deep convection and repeated occurrence of cloud-bursts. Analysis of daily weather charts showed that the heavy rains in the Uttarakhand region were sustained by persistent interactions between the moist monsoonal circulation and anomalous southward intruding mid-latitude westerly troughs over northwest Himalayan region.

(b) Changing summer monsoon circulation

Using atmospheric wind data since 1950s, Joseph and Simon (2005) suggested a weakening trend in the summer monsoon low-level southwesterly flow. Long-term observed sea-level pressure records also indicate a weakening trend of the large-scale meridional pressure difference between the Indian monsoon trough and the subtropical southern Indian Ocean. It turns out that much of the decrease in monsoon orographic precipitation over the Western Ghat during the last 50+ years can be attributed to the weakening trend of the large-scale southwesterly monsoonal winds.

3 Understanding of the Monsoonal Trends in Recent Decades

In addition to changes in the seasonal mean monsoon, consistent long-term trends in the activity of monsoon synoptic systems and the intra-seasonal variations are seen. It has been reported that the frequency and intensity of prolonged monsoon "breaks" over India have increased during recent decades (Ramesh Kumar et al. 2009). Further, the frequency of the monsoon depressions, which are an important rain producing synoptic system forming over Bay-of-Bengal and propagating west-northwest into the Indian region, have significantly decreased during the last few decades (Dash et al. 2004). This decrease in the frequency of monsoon depressions is consistent with the weakening of the large-scale monsoonal flow, which is known to be important for the development of sheared instabilities in the summer monsoon environment (see Krishnakumar and Lau 1997).

A possible explanation for the recent weakening trend of the Indian monsoon has been the increasing concentration of anthropogenic aerosols (e.g., sulphates, black carbon, soot, etc.). Climate model simulations suggest that the radiative effects due to anthropogenic aerosols can modulate the tropical atmosphere-ocean coupled system and cause weakening of the South Asian monsoon circulation, leading to a decrease in precipitation (Bollasina et al. 2011 and references therein). Contrastingly, other climate modeling studies show that large-scale tropical atmospheric circulations can actually weaken in response to global warming alone i.e., purely from enhanced concentration of greenhouse gases,

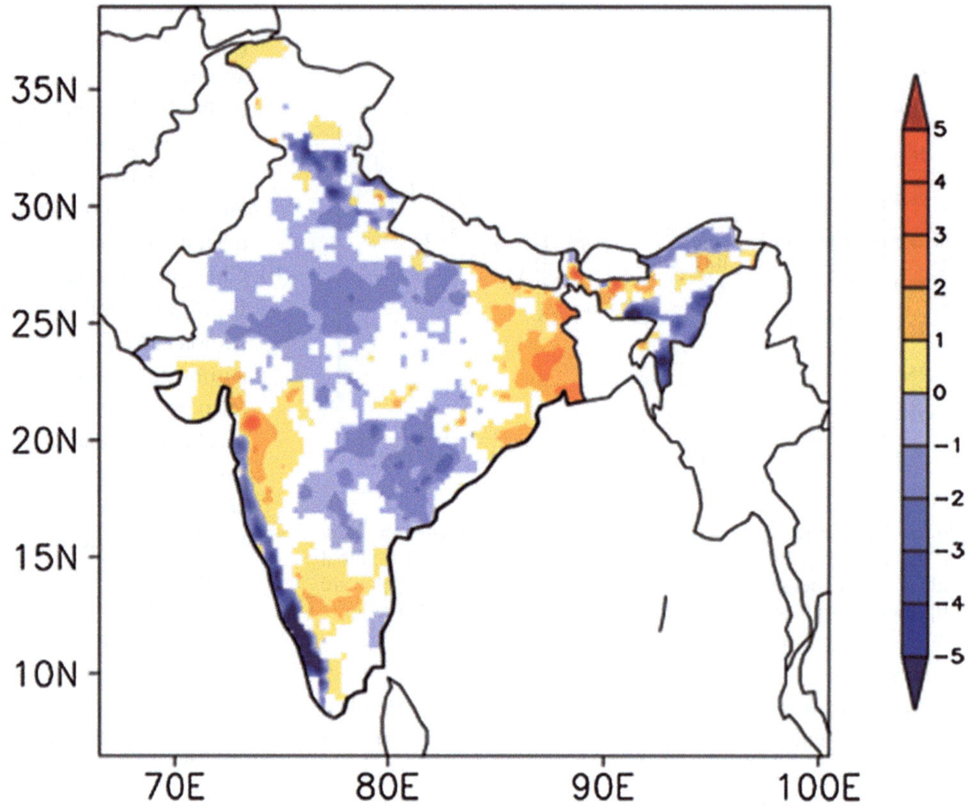

Fig. 6 Spatial map of linear trend of rainfall rate for monsoon season (JJAS) based on the APHRODITE rainfall dataset (1951–2007). The units are mm/day for the entire 57-year period. Values exceeding the 5 % level of significance have been shaded

even in the absence of anthropogenic aerosol forcing (Veechi et al. 2006). Very high resolution climate model simulation experiments show that global warming can lead to weakening of southwesterly monsoon flow and decrease of monsoon precipitation over the Western Ghat (see Krishnan et al. 2012).

Besides an increase in greenhouse gases and anthropogenic aerosols, the tropical Indian Ocean (TIO) has experienced rapid warming of sea surface temperature (SST) at a rate of 0.5–1.0 °C during the past five decades with the warming being strongest during the June-September (JJAS) summer monsoon season. The SST warming in the TIO is apparently related to the weakening of southwesterly monsoon winds (Swapna et al. 2013). Comprehensive modeling and observational studies will be necessary to unravel the details of the coupled interactions and quantify the influences of natural and anthropogenic forcing on the monsoon precipitation variations over the Indian subcontinent both in the historical past and the future.

4 Conclusions

This chapter has attempted to provide a general overview of the Indian monsoon and its variability through time. The monsoon has fluctuated on all timescales, yet it has been a stable system that has never totally failed. Given that historical records suggest that societal changes were largely driven by rainfall changes and not by temperature changes, it becomes imperative that the relationship of monsoon rainfall with temperature changes—both global and regional—need proper elucidation. The monsoon dynamics is complex and with a limit on its predictability makes it imperative to reconstruct robust palaeo records not only to further constrain the models but also provide palaeo-analogs of pristine Earth and its ecosystem so as to deduce the human impact aspects. We refer to Singhvi and Kale (2009) for more detailed discussion on these aspects and to Gupta et al. (2006) on human adaptation through times of climatic shifts.

References

Bollasina MA, Ming Y, Ramaswamy V (2011) Anthropogenic aerosols and the weakening of the South Asian monsoon. Science 334(6055):502–505

Chao WC, Chen B (2001) The origin of monsoons. J Atmos Sci 58:3497–3507

Clift PD, Plumb AR (2008) The Asian monsoon: causes, history and effects. Cambridge University Press, London

Dash SK, Jenamani RK, Shekhar MS (2004) On the decreasing frequency of monsoon depressions over the Indian region. Curr Sci 86:1404–1411

Goswami BN, Venugopal V, Sengupta D, Madhusoodanan S, Xavier PK (2006) Increasing trend of extreme rain events over India in a warming environment. Science 314:1442–1445
Guhatakurtha P, Rajeevan M (2006) Trends in the rainfall pattern over India. National climate centre (NCC). India Meteorol Dept Res Rep 2:1–23
Gupta AK, Singh RK, Joseph S, Thomas E (2004) Indian Ocean high-productivity event (10–8 Ma): linked to global cooling or to the initiation of the Indian monsoons ? Geology 32(9):753–756
Gupta AK, Anderson DM, Pandey DN, Singhvi AK (2006) Adaptation and human migration, and evidence of agriculture coincident with changes in the Indian summer monsoon during the Holocene. Curr Sci 90:1082–1090
Joseph PV, Simon A (2005) Weakening trend of the southwest monsoon current through peninsular India from 1950 to the present. Curr Sci 89:687–694
Kale VS (2012) On the link between extreme floods and excess monsoon epochs in South Asia. Clim Dyn 39:1107–1122
Kelkar RR (2009) Monsoon prediction. BS Publications, Hyderabad
Krishnakumar V, Lau K-M (1997) Symmetric instability of monsoon flows. Tellus 49A:228–245
Krishnan R, Sabin T, Ayantika DC, Kitoh A, Sugi M, Murakami M, Turner AG, Slingo GM, Rajendran R (2012) Will the South Asian monsoon overturning circulation stabilize any further? Clim Dyn 40:187–211
Molnar P, England P, Maritond J (1993) Mantle dynamics, uplift of the Tibetan Plateau, and the Indian Monsoon. Rev Geophys 31:357–396
Nandargi S, Dhar ON (2012) Extreme rainstorm events over the northwest Himalayas during 1875–2010. J Hydrometeorol 13:1383–1388
Quade J, Cerling TE, Brownman JR (1989) Dramatic ecologic shift in the late Miocene of northern Pakistan and its significance to the development of Asian Monsoon. Nature 342:163–166
Rajeevan M, Bhate J, Jaswal AK (2008) Analysis of variability and trends of extreme rainfall events over India using 104 years of gridded daily rainfall data. Geophys Res Lett 35 L18707. doi:10.1029/2008GL035143
Ramesh Kumar MR, Krishnan R, Syam S, Unnikrishnan AS, Pai DS (2009) Increasing trend of "break-monsoon" conditions over India -Role of ocean-atmosphere processes in the Indian Ocean. IEEE Geosci Remote Sens Lett 6:332–336
Samanta A, Bera MK, Ghosh R, Bera S, Fillet T, Pande K, Rathore SS, Rai J, Sarkar A (2013) Do the large carbon isotopic excursions in terrestrial organic matter across Paleocene: Eocene boundary in India indicate intensification of tropical precipitation ? Palaeogeogr Palaeoclimatol Palaeoecol 387:91–103
Singh IB (2005) Climate change and human history in Ganga Plain during Late Pleistocene: Holocene. Palaeobotanist 54:1–12
Singhvi AK (ed) (2004) Quaternary history and palaeoenvironmental record of the Thar Desert in India. Proc Indian Acad Sci (Earth Planet. Sci.) 113(3):367–368
Singhvi AK and Kale VS (2009) Paleoclimate studies in India: last ice age to the present. Indian National Science Academy ICRP-WCRP-SCOPE Report Ser. 4, New Delhi
Singhvi AK, Bhatt N, Glennie KW, Srivastava P (2012) India, Arabia and adjacent regions. In: Metcalfe SE, Nash DJ (eds) Quaternary environmental change in the tropics. Wiley, New York, pp 151–206
Singhvi AK, Rupakumar K, Thamban M, Gupta AK, Kale VS, Yadav RR, Bhattacharya A, Phadtare NN, Roy PD, Chauhan MS, Chauhan OS, Chakravorty S, Sheikh MM, Manzoor N, Adnan M, Ashraf J, Khan AM, Quadir DA, Devkota LP, Shreshtha AB (2010) Instrumental, terrestrial and marine records of the climate of South Asia during the Holocene: present status, unresolved problem and societal aspects. In: Mitra AP, Sharma C (eds) Global environmental change in South Asia: a regional perspective. Capital Publications, New Delhi, pp 54–124
Swapna P, Krishnan R, Wallace JM (2013) Indian Ocean and monsoon coupled interactions in a warming environment. Clim Dyn. doi:10.1007/s00382-013-1787-8
Veechi GA, Soden BJ, Wittenberg AT, Held IM, Leetma A, Harrison MJ (2006) Weakening of tropical Pacific atmospheric circulation due to anthropogenic forcing. Nature 411:73–76
Wasson RJ, Juyal N, Jaiswal M, Culloch M, Sarin MM, Jain V, Srivastava P, Singhvi AK (2008) Deforestation and sediment transport in the Upper Ganges catchment. Environ Manage 88:53–61
Williams MAJ, Ambrose SJ, van der Kars S, Ruehlemann C, Chattopadhyay U, Pal J, Chauhan PR (2009) Environmental impact of the 73 ka Toba super-eruption in South Asia. Palaeogeogr Palaeoclimatol Palaeoecol 284:295–324

Geomorphic History and Landscapes of India

Vishwas S. Kale

Abstract

The Indian subcontinent is a land of great geomorphic diversity and grand scenery. Two major tectonic influences have affected the Indian subcontinent since the Mesozoic times: fragmentation of the Gondwanaland and the Himalayan Orogenesis. These events, along with Deccan Volcanism, differential uplift and reactivation of faults in the Peninsula, and the onset of monsoon climate over the subcontinent have primarily shaped the megascale architecture and scenery of the Indian landmass. The erosional landscape of the Peninsula has inherited substantial elements from and throughout the Cenozoic. The foundations of the modern drainage systems, the coastal margins, and the inland terrains of the Peninsula were laid around the mid Mesozoic. In comparison, the Himalayan landscape and the foreland basin (Indus-Ganga-Brahmaputra Plains) formed and evolved during Neogene and Quaternary. The glacial-interglacial cycles of the Quaternary and associated oscillations in the sea level (erosional base-level) and the monsoon strength have left an indelible impression on the drainage systems and landforms of the subcontinent in a variety of ways. The net effect is the present magnificent scenery of the subcontinent.

Keywords

Gondwanaland • Drift • Differential erosion • Peninsular India • Himalaya • Ganga Plains • Brahmaputra Plains • Cenozoic history • Quaternary • Uplift • Climate change

1 Introduction

The Indian subcontinent is a land of great geomorphic diversity and grand scenery. The subcontinent encompasses virtually every geomorphic landscape present on the Earth surface, from spectacular glacier-carved valleys and frozen ice peaks to vast mangrove swamps, chains of coral islands and a scorched rocky-sandy desert, though landforms resulting from continental glaciations so common in the northern latitudes of North America and Europe have not been reported in the Indian subcontinent. The fundamental reasons for the development of the grand scenery are, rifting along the continental margins, the northward drift of India, collision with Eurasia, and formation of towering Himalaya out of the Tethys Sea and the adjacent foreland basin. The Indian region also has the distinction of having some of the oldest rocks on the Earth surface. Rocks ranging in age from Archean to Quaternary are found in the subcontinent. Long timescale is, thus, involved in landscape evolution of the Indian landmass.

The geomorphic history of the diverse landscapes that we see today in different parts of the subcontinent, in general, and the Indian Peninsula in particular, can be traced back to at least the last ~170 Ma (mid-Mesozoic), when

V. S. Kale (✉)
Department of Geography, University of Pune, Pune 411007, India
e-mail: vskale.unipune@gmail.com

V. S. Kale (ed.), *Landscapes and Landforms of India*, World Geomorphological Landscapes,
DOI: 10.1007/978-94-017-8029-2_3,

Fig. 1 Summary of the major tectonic and geomorphic events during the long history of the Indian subcontinent (1 to ~250 Ma)

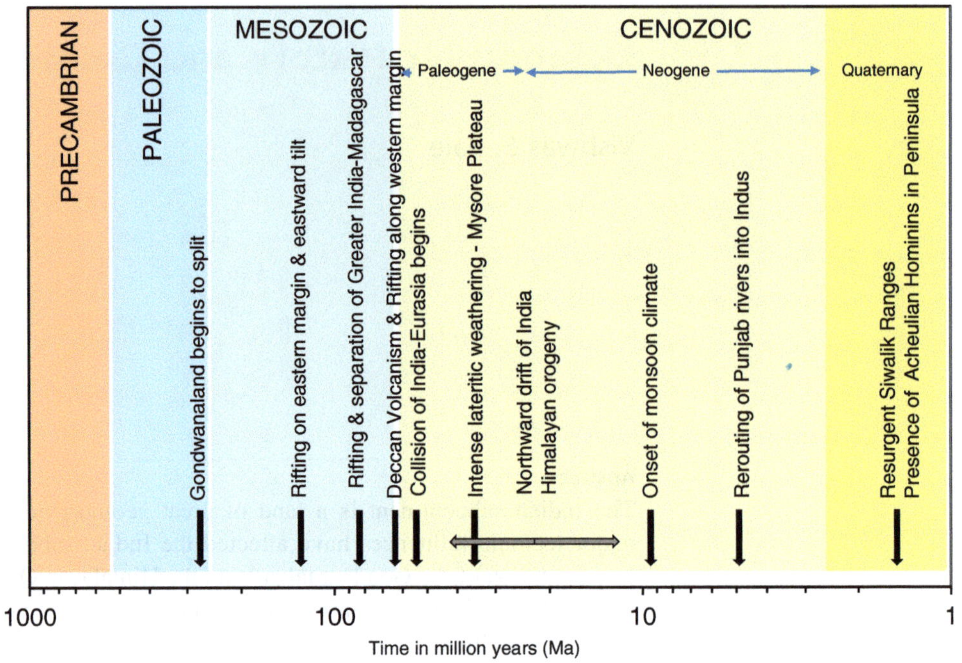

India started detaching itself from the Gondwanaland (Fig. 1). Evidence indicates that the Indian Peninsula was a part of the Gondwanaland in the early Mesozoic that broke away from the supercontinent and drifted northwards and collided with the Eurasian Plate to form the lofty Himalaya Mountains. The Himalayan Orogeny had a decisive role to play in the onset of monsoon climate over the subcontinent ~8–10 Ma and formation of the vast, fertile alluvial plains adjacent to and south of the elevated mountains. These major geological, climatic and tectonic events (Fig. 1) have largely shaped the architecture and scenery of this vast landmass.

2 Geomorphic Provinces

On the basis of common geologic and geomorphic attributes, the Indian region can be broadly divided into three spatial entities or provinces and several sub-provinces (Table 1). The three distinct geomorphic provinces are (Figs. 2 and 3):

(a) The Indian Peninsula (Geologic Province—Indian Shield), including the Deccan Traps Region (Geologic Province—Large Igneous Province)
(b) The Himalaya Mountains (Geologic Province—Indian Orogenic Belt)
(c) The Indus-Ganga-Brahmaputra Plains (Geologic Province—Indian Foreland Basin), including the Thar Desert.

2.1 The Indian Peninsula

The Indian Peninsula constitutes the oldest and the largest (~2.1 million km^2) geomorphic province of India. The Peninsula, chiefly made up of Precambrian cratonic blocks and Proterozoic fold belts, is the most geologically exposed part of the Gondwanaland and largely displays an erosional landscape. By and large, bedrock landforms (Fig. 4) and partially to deeply weathered rocks dominate the scenery of this ancient landmass.

The Deccan Plateau, with an easterly tilt, is the principal sub-province of the Indian Peninsula. The nearly 1,500-km long, Western Ghat (*Sahyadri*) Escarpment forms the western edge of the plateau. In the east, the plateau is flanked by discontinuous hill ranges of the Eastern Ghat (e.g. Shevaroy, Nallamalai and Mahendragiri Hills). The Eastern Ghat is breached by large peninsular rivers. The two Ghats converge in the south at Nilgiri Hills, with Doda Betta (2,637 m a.s.l.) as the 2nd highest peak. Anai Mudi (2,695 m a.s.l.), south of the Palghat Gap, is the highest point in the Western Ghat and the Peninsula. The Deccan Plateau is bounded on each side by coastal lowlands of variable width and morphology. Whereas the west coast lowland is generally rocky, relatively narrow (except along the Gujarat coast) and dominated by hills and laterite plateaux, the eastern seaboard is featured by large deltas, deltaic plains and predominantly aggradational littoral features (Vaidyanadhan 2002). There are two archipelagos in the adjoining seas, namely the Andaman and

Table 1 Principal geomorphic provinces of India

Abbreviation	Geomorphic province (area within India)	Geologic province	Sub-provinces and their abbreviations	Geological characteristics	Geomorphic characteristics
IP	The Indian Peninsula (~2.1 million km^2)	Indian Shield	A. The Deccan Plateau (DP) B. The Western Ghat Escarpment (WGE) C. The Deccan Traps Region (DT) D. The coastal lowland (CL) E. The Archipelagos—Andaman and Nicobar Islands and Lakshadweep Islands (AR)	The shield composed of large areas of exposed Precambrian crystalline igneous and high-grade metamorphic rocks Four cratons bordered by fold belts and rifts Tectonically relatively stable	Oldest geomorphic province with ancient rocks, surfaces (erosional/planar) and rivers. Dominantly erosional landscape with wide open valleys. Two conspicuous landscapes—granitic and basaltic Great Escarpment of India on the western margin of Deccan Plateau—the Western Ghat
HM	The Himalaya Mountains (~0.5 million km^2)	Indian Orogenic Belt	A. The Siwalik Ranges (SR) B. The Lower Himalaya (LH) C. The Higher Himalaya (HH) D. Tethys Himalaya (TH) E. Indo-Myanmar Ranges (IM)	An active mountain belt in a collisional setting Long tracts of highly deformed rocks Four tectonic zones separated by thrust faults	Multiple, parallel, mountain ranges having rugged topography with prominent peaks and ridges reaching 7–8 km above sea level Dominated by fluvial erosion, landslides and glacial erosion. Impressive deep gorges, strath and fill terraces and glacial landforms. Massive landslides
IGB	The Indus-Ganga-Brahmaputra Plains (~0.7 million km^2)	Indian Foreland Basin	A. The Indus or Sindhu Plains (SP) B. The Ganga Plains (GP) C. The Brahmaputra Plains (BP) D. The Thar Desert (TD)	A foreland basin. A large geological and sedimentary depression developed adjacent and parallel to the Himalayan mountain belt An aggregation of sediments ranging for boulders to clay, that filled up a depression or accumulated in an area	An extensive, low-relief alluvial plain ~100–500 km wide and ~3,000 km long, between Himalaya and the Indian Peninsula Sediments deposited almost continuously since Eocene-Oligocene times (~55 million years ago). Constitutes fan, floodplain, channel and deltaic deposits

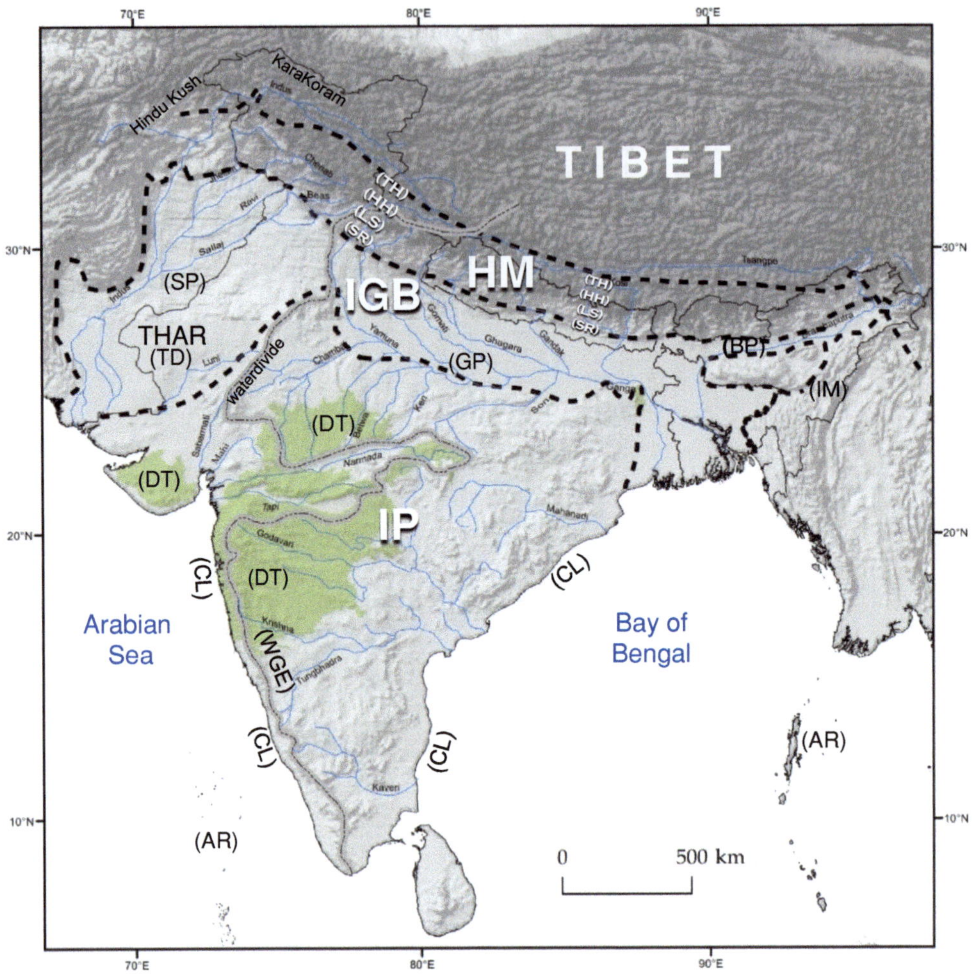

Fig. 2 Map showing the principal geomorphic provinces and sub-provinces of the Indian subcontinent. Refer Table 1 for explanation of the abbreviations

Nicobar Islands in the Bay of Bengal and Lakshadweep Islands in southern Arabian Sea (Fig. 2).

In the west-central part of the Peninsula, the Precambrian cratonic blocks and Vindhyan rocks are concealed under thick piles of late Cretaceous-Eocene Deccan basalts. This vast igneous province (5×10^5 km^2), partitioned by Narmada-Son fault-trough (Sheth 2007), is the youngest geological unit of the Indian shield. Consequently, the drainage and landforms are also much younger, albeit older than the Himalayan drainage.

A line joining the Western Ghat and Aravalli Hills via Gawilghar-Maikal-Vindhyan Ranges constitutes the major waterdivide of the Indian Peninsula (Fig. 2) that separates the Arabian Sea drainage from the Bay of Bengal drainage. Four large rivers, namely Mahanadi, Godavari, Krishna and Kaveri drain into the Bay of Bengal and form large deltas. Whereas the Narmada, Tapi, Sabarmati and Mahi debouch into the Arabian Sea via estuaries; the Chambal, Betwa, Ken, Tons and Son form part of the Yamuna-Ganga drainage system. The courses of Narmada, Tapi and Son are strongly controlled by a regional geofracture that transects the Indian shield in the middle. All the Peninsular rivers are monsoon-fed and are presently incised in bedrock (Fig. 5) or late Quaternary alluvium. By and large, the channels are stable and floodplains are narrow, discontinuous or absent, except in the deltaic portions.

Although landforms developed within sedimentary rocks are present in certain pockets, the peninsular landscape is overwhelmingly dominated by landforms of the granite-gneissic terrain and basaltic terrain.

Many of the major landforms (valleys, high-level surfaces, inland plateaux, duricrusted landforms, etc.) on the face of Indian Peninsula are the legacy of its long geomorphic history that extends back to the Cenozoic or pre-Cenozoic times (Vaidyanadhan 2002). It is the Gondwana breakup that initiated the development of the current macroscale morphology of the Indian Peninsula. Much of the geomorphology of the subcontinent relates to tectonic and

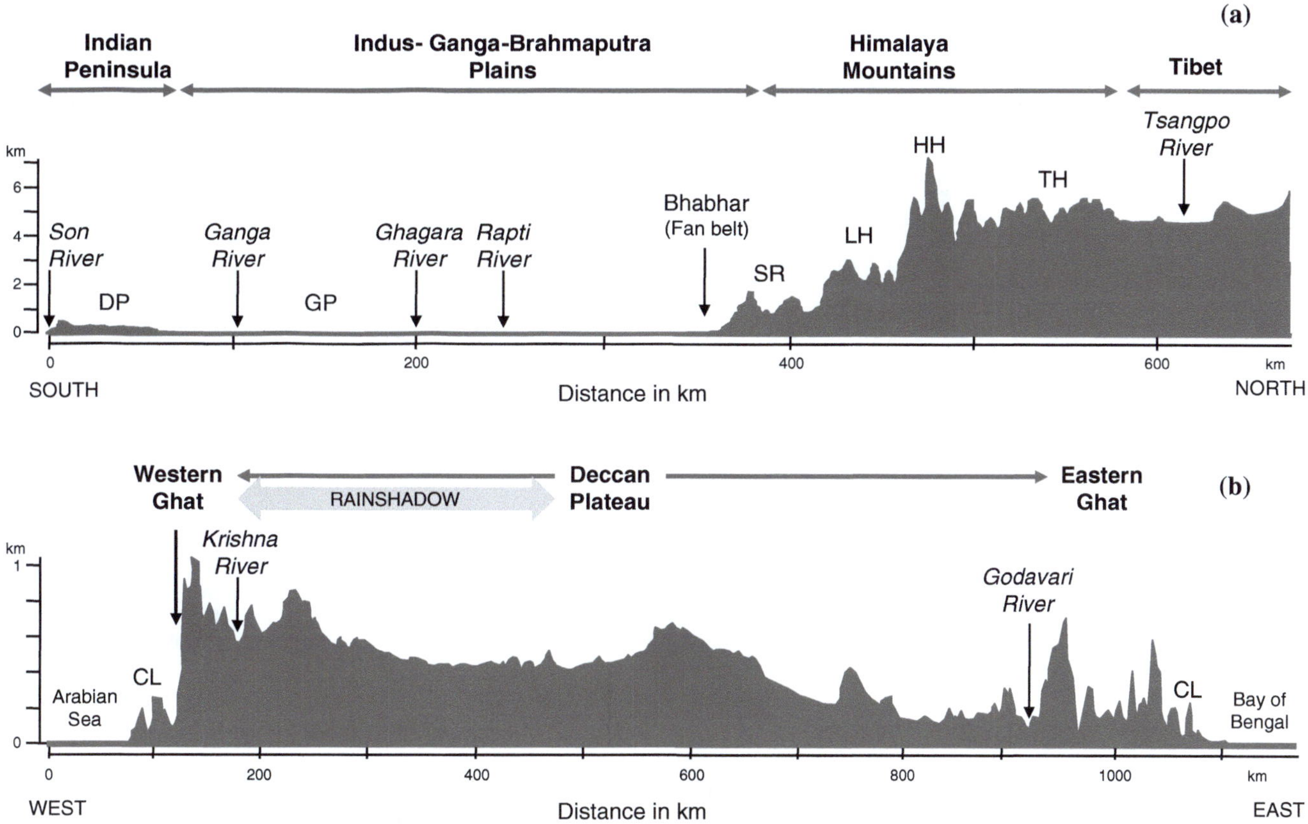

Fig. 3 Transverse profiles showing major geomorphic provinces and sub-provinces of India. The *upper panel* **a** shows a *south-north profile* (approximately along 83.3° E longitude) across the *northern part* of the Indian Peninsula to Tibet via Ganga Plains and Himalaya Mountains. The *lower panel* **b** depicts a *west-east profile* (approximately along 17.3° N latitude) across the Indian Peninsula. The *cross profile* shows the coastal lowlands, the Western Ghat Escarpment, the Deccan Plateau and the Eastern Ghat. Refer Table 1 for abbreviations

Fig. 4 Granitic hills near Mandya on the Mysore Plateau. The influence of jointing patterns on the form and development of these hills is clearly seen

Fig. 5 Turbulent waters of the Kaveri River through a bedrock gorge near Mekedatu, Karnataka

climatic events, and base level changes that have taken place since the mid-Jurassic ($\sim$170 Ma). By and large, divergent weathering and differential erosion during its long geological history have fashioned the peninsular landscape.

2.2 The Himalaya Mountains

Along the northern rim of the Indian subcontinent exists a grand and immense mountain range—the Himalaya-Karakoram Mountains. The ranges include the world's highest peak, the Sagarmatha or Mt. Everest (8,848 m a.s.l.) and over a dozen peaks rising above 7,500 m a.s.l. (for example, Mt. Kanchenjunga, Dhaulagiri, Nanga Parbat, and Nanda Devi). These highest mountain ranges in the world constitute an enormous geophysical-geomorphologial system and display a whole range of spectacular landforms, such as alpine mountains, long glaciers, glacial troughs, steep gorges, intermountain basins (called *Duns*) and multiple strath and fill terraces. The fundamental reason for the development of such rugged relief and grand scenery is the tectonic uplift and upthrusting caused by the collision of the Indian and Eurasian Plates that started $\sim$55 Ma and is continuing to the present. The Himalayan Orogen is one of the most impressive examples of an active mountain belt in a collisional setting.

Himalaya, *the abode of snow*, is one of the youngest mountain ranges in the world. The nearly 2,500-km long, arc-shaped mountain range (Fig. 2) extends from Afghanistan-Pakistan in the west to Indo-Myanmar Ranges in the east, and varies in width from $\sim$250 to 300 km. Syntaxial bends are present at the western (Nanga Parbat) and eastern (Namcha Barwa) ends.

The Himalayan province is sub-divided into four principal morpho-tectonic zones or sub-provinces (Figs. 2 and 3). Each province is separated by a major thrust fault. From south to north these longitudinal sub-provinces are: the Siwalik Ranges, the Lesser Himalaya, the Higher (or Greater) Himalaya and the Tethys Himalaya (Valdiya 1998; Yin 2006). The Siwalik Ranges (elevation <1,000 m a.s.l.) rise abruptly above the Indus-Ganga-Brahmaputra Plains along the Main Frontal Thrust (MFT). Neogene and Quaternary sediments dominate these sub-Himalayan ranges. The ranges occur in the form of rows of hills, separated by elongated piggyback basins (called *Duns*). The Main Boundary Thrust (MBT) defines the boundary between Siwaliks and Lesser Himalaya. The Lesser Himalaya (including Pir Panjal, Dhaulagiri, Dhauladhar, Mahabharat and other ranges) consists of Precambrian and Cambrian sequences (Fig. 6). Here summits can rise above 3500 m a.s.l. (Fig. 7). Well known intermountain valleys and hill resorts, such as Srinagar, Kathmandu, Kangra, Shimla, Mussoorie and Darjeeling are located in this sub-province. The Higher Himalayan crystalline is separated from the Lesser Himalaya by the Main Central Thrust (MCT). In this zone, snow and glacier capped peaks range in elevation from 3,000 to >8,000 m a.s.l. The South Tibetan Detachment (STD) separates the Higher Himalaya from the Tethys Himalaya. This northern-most morpho-tectonic zone consists of sedimentary sequences that

Fig. 6 Near-vertical slopes developed in Lesser Himalayan granite/banded gneiss, Upper Beas Valley, Himachal Pradesh

Fig. 7 Snow covered Himalayan peaks and small glaciers as seen from Manali, Himachal Pradesh

range in age from late Precambrian (>600 Ma) to Cretaceous-Eocene (95–45 Ma) (Valdiya 1998). Whereas south of the Siwalik Ranges lies the Duar and Terai Belt, on the Tibetan side lie the Karakoram, Zanskar, Ladakh and Kailas Ranges. On the extreme eastern flank of the subcontinent are, the roughly north-south trending Indo-Myanmar Ranges.

Tectonic uplift, rapid valley incision, landslide erosion and glacial erosion are the fundamental processes responsible for the spectacular rugged relief of the Himalayan landscape. The continuing rapid uplift of the Himalaya, great differences in altitude over short distances, and orography-induced monsoon precipitation have all combined to provide

Fig. 8 The Ganga River upstream of Rishikesh. The river has incised into Lesser Himalayan metasediments

geomorphic conditions favourable for highly elevated rates of denudation and sediment transport. The river systems are supply-limited (sediment transport capacity exceeds sediment supply), and hence, they are forced to cut down into rising anticlines. The contribution of glacial erosion to the overall denudation in Himalaya is also noteworthy. Cryogenic weathering is conspicuously observed at higher elevations.

Despite rapid uplift (tectonic or isostatic) and increased stream power and incision, the mountain slopes are maintained at threshold angles by high rates of landsliding (Larsen and Montgomery 2012). As a result, formation of landslide dams and generation of large floods due to the failure of such dams is nothing extraordinary in this terrain. Such extreme floods, in turn, play a key role in promoting incision and coarse sediment transport.

Major Himalayan rivers, such as, Indus, Brahmaputra, Satluj, Karnali, Kali-Gandaki, and Kosi (Arun) flow across the geologic structures (thrust faults) and the orientation of the mountain ranges, right from the edge of the Tibetan Plateau. There is a broad consensus that these trans-Himalayan rivers are antecedent and predate the Himalayan Orogeny. Wherever these rivers breach the transverse ranges or anticlines, spectacular gorges have formed (Fig. 8). The deeply incised river valleys of the trans-Himalayan rivers are related to high rates of erosion, favoured by rapid uplift and late Cenozoic changes in monsoon strength. Some of the transverse rivers are laterally diverted at Siwalik anticlines, giving rise to gridiron drainage pattern (Gupta 1997). Such rivers emerge through the mountain front and form large alluvial fans (megafans).

Due to its high elevation, there are over 15,000 glaciers in the Himalaya. As a result, snowmelt constitutes an important component of the hydrological budget of the Himalayan rivers. Of the total annual discharge, snowmelt contributes up to 50 % in the Indus catchments, ~25 % in Tsangpo-Brahmaputra catchments, and <20 % in other basins (Bookhagen and Burbank 2010). The Himalayan glaciers (continental and maritime) are more extensive in the west due to higher average altitude. On account of the presence of large number of glaciers, there is abundance of glacial erosional and depositional features. The Himalayan landscape is also modified by glacial lake outburst floods (GLOFs) from time to time.

2.3 The Indus-Ganga-Brahmaputra Plains

To the south of the Himalayan front lies a vast alluvial plain (~1.17 million km^2) created by three large Himalayan rivers and their tributaries, namely Indus, Ganga and Brahmaputra. These highly fertile Indus-Ganga-Brahmaputra (IGB) Plains are primarily composed of fan, floodplain, channel and deltaic deposits, and are intensively cultivated and densely settled. The IGB Plains are built by sediments derived from the Himalayan Orogen to its north as well as from the

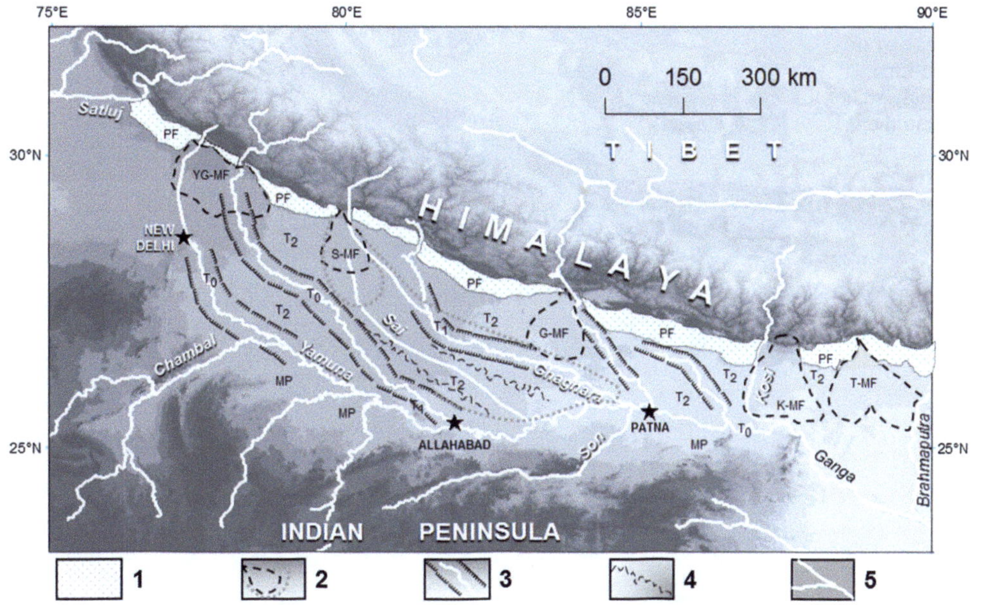

Fig. 9 Generalized geomorphic map of the Ganga Alluvial Plains. *1* fan belt at the foot of the Siwalik Ranges, *2* megafans; *3* incised valleys, *4* abandoned channel belt, *5* rivers. *PF* piedmont fan belt, *MP* marginal plain upland surface, *K-MF* Kosi Megafan, *G-MF* Gandak Megafan, *S-MF* Sarda (Ghaghara) Megafan, *YG-MF* Yamuna-Ganga Megafan, *T-MG* Tista Megafan, T_0 active floodplain surface, T_1 river valley terrace surface, T_2 upland interfluve surface. (Modified after Singh 1996)

cratonic hinterland in the south. The plains generally have an altitude of less than 150 m a.s.l. and slope gently towards the Bay of Bengal and the Arabian Sea.

The vast foreland basin is a large geological and sedimentary depression, adjacent and parallel to the Himalayan Mountain belt. The alluvial plains were largely formed by net alluviation and filling up of a foredeep basin by sediments derived from the rising Himalaya as well as Peninsular India since about Paleocene. The foredeep was initially formed due to the flexing down of the Indian Plate following India-Eurasia collision and later deepened due to sinking of the basin floor as the Indian Plate slid under the Himalaya (Valdiya 2010). Over the last few million years the foreland basin has been gradually filled by sediments brought in by rivers. The effect of this deposition was to level the landscape producing a flat terrain. The thickness of the sediments under the plains varies from a few hundred meters up to $\sim$7 km in structural depressions.

Rivers emerging from Himalaya Mountains onto the alluvial plains rapidly lose their stream power and carrying capacity. As a result, they dump their excess sediment load at the mountain-plain interface, giving rise to alluvial fans. All Himalayan rivers having sizable catchments have developed large alluvial fans (Fig. 9). A continuous belt of coalescing gravel fans (*Bhabhar*) exists at the foot of the Siwalik Ranges. Some of the rivers, such as Kosi and Gandak, bring enormous amounts of sediment load and hence form gigantic fans at the mountain front, called megafans (Singh 2007). The gigantic fans have shoved the Ganga drainage southward (Fig. 9). Over the fan surface, the channels are unstable; hence, they frequently change their path by avulsion. The Kosi River provides one of the best examples.

Traditionally, two types of alluvial deposits are recognized in the Ganga Plains (also along the Indus), namely *Bhangar* (older) and *Khadar* (recent) (Singh 2007). The former represents more clayey and calcareous alluvium, and occurs in the form of a high, extensive terrace or interfluves. The younger alluvial deposits are confined to the modern floodplains of the Ganga and its tributaries.

Although the channel forms vary according to sediment load, slope and discharge variability, meandering, braided and anabranching channel patterns and alluvial styles predominate the IGB Plains. Amongst the principal rivers, only the Yamuna River displays a distinct meandering channel pattern. Ganga, Brahmaputra, Satluj, Kosi, etc. exhibit either braided or anabranching pattern. Examples of palaeochannels (Fig. 9), natural levees, ox-bow lakes and meander scars are found in abundance throughout the IGB plains (Sarma 2005; Inam et al. 2007; Singh 2007). Due to the alluvial nature of the rivers and large monsoon floods, changing of river courses, avulsion and bankline migration have been going on since times immemorial. One of the noteworthy characteristics of the Ganga drainage is the occurrence of groundwater-fed streams such as Gomati, Sai, etc. These rivers originate on the plains and are perennial with low sediment load to discharge ratio.

Rivers have played a decisive role in the distribution of settlements from pre-historic times. The oldest civilization in the subcontinent is the Harappan (*aka* Indus Valley),

Fig. 10 The Luni River downstream of Tilwara, *east* of Barmer, Rajasthan. The Luni (meaning salty) is the only well integrated river system in the Thar Desert

which first flourished on the banks of the Indus River (~2–4 millennium BCE) and then extended into other rivers of northwest India (e.g. Ghaggar, Satluj, Ravi, Saraswati, etc.).

The interfluve area between the Ganga and the Indus drainage is a dryland area known as the Great Indian Desert or the Thar Desert. Spread over an area of ~285,000 km^2, the Thar is the only region in India, where there is evidence of strong aeolian activity during the late Quaternary. Shallowness of the monsoon current, subsiding air and anticyclonic circulation are the fundamental reasons for aridity over this area. Though the exact age of the Thar Desert is not known, some workers believe that aridity has prevailed over the area for the last few million years.

Apart from sand dunes (parabolic, barchans, longitudinal and transverse), rocky landforms (hammada) and desert pavements (reg), the Thar Desert is featured by numerous saline lakes or playas. The Sambhar Lake is the largest saline lake (playa) on the eastern fringe of the desert. Blockage of rivers and streams by advancing sand dunes, deflation, and tectonic subsidence have been suggested as the possible causes for the formation of these lakes. Their subsequent desiccation has been attributed to weakening of the monsoon.

Today, the Luni River (Fig. 10) is the only well-integrated drainage system in the Indian Desert. It is an ephemeral river and in some years it is completely dry. It receives tributaries from the Aravalli Hills, but none from the sandy desert in the west. At least one major river, sometimes referred to as the 'Saraswati' or 'Vedic Saraswati', is believed to have flowed through the Harappan-sites dominated section of the Thar Desert (Valdiya 2010). There is difference of opinion about whether the former river was a large glacier-fed Himalayan river or monsoon-fed river system during the Holocene.

3 Macroscale Geomorphic History of India

3.1 Pre-Cenozoic Geology and Geomorphology

There is little doubt that some of the large-scale landforms occurring in the Indian Peninsula are the legacy of the geological and geomorphic history that extends back to the Mesozoic and or even earlier. Even though the Gondwana successions provide evidence of the existence of glacial, fluvial and marine regimes in central and eastern India from late Permian to Cretaceous times, and the palaeocurrent trends from various Gondwana Basins suggest drainage alignment towards the northwestern or northeastern direction (Valdiya 2010), there is not much tangible evidence of the pre-Gondwana-breakup landscapes. This is primarily due to the fact that the Gondwana sediments in India and elsewhere are confined to only rifted grabens.

The megascale geomorphic history of India approximately starts with the breakup of eastern Gondwana when the Indian Peninsula came into existence (Fig. 1). The eastern margin of the Peninsula was created by the breakup of India from Antarctica around early Cretaceous (~130 Ma). This was followed by the breakup of Greater India from Madagascar/Malagasy around 80–90 Ma (Storey et al. 1995), as India started drifting northwards. The initial breakup resulted in the eastward tilt of the Peninsula and establishment of the easterly drainage (Murthy et al. 2010). During its northward journey the western continental passive margin of India was impacted by two major hotspots (Marion and Réunion) and has experienced two episodes of separation—Madagascar-Greater India ~80–90 Ma and Seychelles-India ~65 Ma (Norton and Sclater 1979; Storey et al. 1995). As a result of the formation of the eastern and western margins and tilt of the Peninsula, the pre-Cenozoic rivers had to adjust and work headward from the new coastlines.

During late Cretaceous-Eocene, western India experienced intense explosive volcanism of Deccan Traps basalts. The extensive outpourings of basalts not only buried the low-relief, pre-Trappean landscape over an area of ~1.5 million km^2 but also created a completely new topography over which sub-aerial processes started operating and new drainage network was established. The extrusion of Deccan Traps lavas was also associated with major rifting along the western margin of India, and failed rifting along the Khambhat (Cambay) and Kachchh grabens. Erosion commenced on the newly formed rifted margin and operated to the then existing erosional base level. It is the continuance of this headward-working erosion along the western passive margin that gave rise to the most prominent physiographic feature of the Peninsula: the ~1,500 km long Great Escarpment of India, the Western Ghat (*Sahyadri*) and consequently the Deccan Plateau.

The massive Deccan volcanism was perhaps associated with a major global change in climate and mass extinction. This, along with the fact that the northward-drifting Indian Plate had already entered the warmer low-latitudes by the early Cretaceous, suggests that the period from late Jurassic to Eocene was characterized by warmer and more humid conditions (Gunnell 1998). These conditions facilitated deep weathering of the rocks underlying the valleys, erosional surfaces and plateaux and formation of deep weathering fronts. The formation of laterites over the youngest lava flows in the Deccan Traps Region is an indicator of intense and deep weathering immediately after the cessation of the Deccan volcanism. Some of the oldest high level surfaces, capped in places by laterite duricursts and inferred to be late Cretaceous-Paleocene in age (Gunnell 1998), are observed over the Nilgiri and Palni Hills (~2,200 m a.s.l.) in southern Peninsula.

3.2 Cenozoic Geomorphology: The Paleogene-Neogene Period

After the breakup and separation of India, much of the geomorphology of the Indian subcontinent relates to geological, tectonic and climatic events that have taken place during Cenozoic, since the end of Deccan Volcanism (Fig. 1). These events were linked to the northward migration of the Indian Plate through different latitudinal zones, the Himalayan Orogeny and the establishment of the monsoon system over the subcontinent. There is conspicuous absence of Paleogene-Neogene continental records (except in Tamil Nadu, Kachchh and Rajasthan) and this has inhibited the erection of the detailed Cenozoic geomorphic history of the Indian Peninsula.

At the beginning of the Paleogene the subcontinent existed as a huge island, surrounded by the Tethys Sea in the north and the Indian Ocean on the other three sides. The Earth's climate was extremely warm during the Paleocene-Eocene thermal maximum (PETM). The beginning of the Cenozoic also witnessed significantly higher eustatic sea-level (Gunnell 1998) and submergence of large areas along the margins of the Indian Peninsula. The elevated temperatures and high sea-stands had a significant impact on the weathering fluxes and denudation rates. Development of duricrust armouring over the deep weathered profiles on multiple surfaces of the Peninsula was one of the consequences of warmer climate. $^{40}Ar/^{39}Ar$ dates suggest intensive lateritic weathering over Mysore Plateau between ~26 and 36 Ma under warm and wet climate similar to that prevailing in tropical humid forests (Nicolas et al. 2014).

Even though the foundations of the modern drainage system and fluvial landscape were laid in the middle Mesozoic, there is enough evidence to suggest that the landscape of Peninsula was sculptured predominantly by fluvial systems throughout Cenozoic. The major landscape elements, the Godavari, the Krishna, the Kaveri, the Narmada, the Mahanadi, and some other rivers very likely existed throughout the Cenozoic, although complete integration of younger Deccan Traps drainage with the older Peninsular drainage must have taken some time. Apart from changes in the catchment hydrology, the long-term downward trend in the eustatic sea-level during mid to late Cenozoic not only lowered the erosional base level but also subaerially exposed previously submerged areas along the margins. The rivers most likely responded to the long-term base-level lowering by downcutting their channels, dissecting inland plateaux and high-level surfaces, exposing the weathering fronts (giving rise to bornhardts and boulder inselbergs) and by increasing their sediment output. It appears that most of the present relief in the Indian Peninsula was acquired during Neogene (Nicolas et al. 2014).

Phases of uplift of Himalaya Mountains in the Mio-Pliocene had a profound effect on the climate of the subcontinent as well as on the regional drainage system. This in turn, controlled the sediment supply to the foreland basin (IGB Plains), coastal areas as well as the adjoining seas. Since their formation the Himalaya Mountains have been attacked and broken down by weathering and erosion (fluvial and glacial). The erosion of the Himalaya was remarkably high between 15 and 10 Ma and since late Plio-Pleistocene (Clift et al. 2008). This was also the time when the monsoon climate was firmly established over the subcontinent. Tectonic movements provided the template for changes in the fluvial system. Although the Himalayan drainage configuration is believed to have remained, more or less, unchanged during the India-Eurasia collision (Brookfield 1998), rerouting of the major Punjab rivers into the Indus after 5 Ma (Clift and Blusztajn 2005), and capture of Tsangpo-Irrawaddy by Brahmaputra a few million years ago (Brookfield 1998), suggest major changes in the Himalayan drainage geometry at the end of Neogene.

In the Indian Peninsula, there is every reason to believe that periods of tectonic stability (indicated by widespread high-level erosion surfaces) were interspersed with periods of tectonic/isostatic uplift during the Cenozoic. Unlike Australia, which is equally old but has been worn down by subaerial erosion to give rise to the flattest continent, the Indian Peninsula shows considerable relief (1000–2000 m). This itself is a strong evidence of the presence of differential tectonic movements during Cenozoic Era.

3.3 Cenozoic Geomorphology: The Quaternary Period

The Quaternary is known for moderate to extreme fluctuations in the climatic conditions over the subcontinent in response to glacial-interglacial cycles. As a result, throughout the Quaternary, the river systems underwent repeated adjustments. Quaternary glacial-interglacial cycles and associated sea-level changes led to intermittent periods of aggradation and degradation resulting in the formation of the alluvial plains or river terraces. In the tectonically active Himalaya, there is ample evidence of multiple glaciations during the Quaternary. There is no reason to suppose that differential uplift did not occur in Peninsular India during Quaternary. However, there are difficulties in isolating the impact of tectonic movements from climatic changes.

All the available climate proxies of the last glacial cycle in general and the Last Glacial Maximum (LGM) in particular, indicate a cooler and drier period. During the LGM, a low sea level (ca. –120 m), increased seasonality and weakening of monsoon rainfall are indicated. Fluvial activity was generally subdued (Kale et al. 2003). In comparison, the early Holocene climatic optimum was associated with revival of fluvial activity (Kale 2007 and references therein). Thus, by and large, all Quaternary glacial cycles were characterized by lowering of the snowline, aggradation in river valleys, recession of the sea, and expansion of sandy desert. In contrast, the warmer interglacial periods were dominated by coeval strengthening of the monsoon and revival in the fluvial activity, and marine transgression. There were, of course, deviations from this general pattern. For instance, the glacial advances in Himalaya and strong aeolian activity in the Thar Desert were associated with stronger monsoons and did not coincide with the LGM as expected.

The Quaternary Period also witnessed numerous changes in the drainage in dynamic settings. One of the noteworthy events was the capture of the Yamuna River by Ganga drainage via a tributary of the Chambal River (Valdiya 1996).

4 Conclusions

The three distinct geomorphic provinces of the Indian subcontinent—(a) The Indian Peninsula, (b) the Himalaya Mountains, and (c) the Indus-Ganga-Brahmaputra Plains—display distinct landform assemblages and have different evolutionary history. The Indian Peninsula constitutes the largest geomorphic province and is the most geologically exposed part of the Gondwanaland. Bedrock landforms and partially to deeply weathered rocks dominate the scenery of this ancient landmass. The Himalayan landscape and the foreland basin (Indus-Ganga-Brahmaputra Plains) formed and evolved during Neogene and Quaternary. Tectonic uplift, rapid valley incision, landslide erosion and glacial erosion are the fundamental processes responsible for the spectacular rugged relief of the Himalayan landscape. The vast fertile Indus-Ganga-Brahmaputra Plains are primarily composed of fan, floodplain, channel and deltaic deposits.

Rifting along the continental margins, the northward drift of India, the Deccan Volcanism, the Himalayan orogenesis, onset of monsoon climate over the subcontinent, differential uplift, and the glacial-interglacial cycles of the Quaternary have primarily shaped the megascale architecture and scenery of the Indian landmass. The net effect is the present magnificent scenery of the subcontinent.

References

Bookhagen B, Burbank DW (2010) Toward a complete Himalayan hydrological budget: spatiotemporal distribution of snowmelt and rainfall and their impact on river discharge. J Geophys Res, 115, F03019. doi:10.1029/2009JF001426

Brookfield ME (1998) The evolution of the great river systems of southern Asia during the Cenozoic India-Asia collision: rivers draining southwards. Geomorphology 22:285–312

Clift PD, Blusztajn J (2005) Reorganization of the western Himalayan river system after five million years ago. Nature 438:1001–1003, doi:10.1038/nature04379

Clift PD, Hodges KV, Heslop D, Hannigan R, Long HV, Calves G (2008) Correlation of Himalayan exhumation rates and Asian monsoon intensity. Nat Geosci 1:875–880

Gupta S (1997) Himalayan drainage patterns and the origin of fluvial megafans in the Ganges foreland basin. Geology 25:11–14

Gunnell Y (1998) The interaction between geological structure and global tectonics in multistoreyed landscape development: a denudation chronology of the south Indian shield. Basin Res 10: 281–310

Inam A, Clift PD, Giosan L, Tabrez AR, Tahir M, Rabbani MM, Danish M (2007) The geographic, geological and oceanographic setting of the Indus river. In: Gupta A (ed) Large rivers: geomorphology and management. Wiley, New York, pp 333–346

Kale VS (2007) Fluvio–sedimentary response of the monsoon-fed Indian rivers to Late Pleistocene-Holocene changes in monsoon strength: reconstruction based on existing ^{14}C dates. Quatern Sci Rev 26:1610–1620

Kale VS, Gupta A, Singhvi AK (2003) Late Pleistocene-Holocene palaeohydrology of Monsoon Asia. In: Gregory KJ, Benito G (eds) Palaeohydrology: understanding Global Change. Wiley, Chichester, UK, pp 213–232

Larsen I, Montgomery DR (2012) Landslide erosion coupled to tectonics and river incision. Nat Geosci 5:468–473

Murthy KSR, Subrahmanyam V, Subrahmanyam AS, Murty GPS, Sarma KVLNS (2010) Land–ocean tectonics (LOTs) and the associated seismic hazard over the eastern continental margin of India (ECMI). Nat Hazards 55:167–175

Nicolas J, Bonnet NJ, Beauvais A, Arnaud N, Chardon D, Jayananda M (2014) First $^{40}Ar/^{39}Ar$ dating of intense late Palaeogene lateritic weathering in Peninsular India. Earth Planet Sci Lett 386:126–137

Norton IO, Sclater JG (1979) A model for the evolution of the Indian Ocean and the breakup of Gondwanaland. J Geophys Res 84:6803–6830

Sarma JN (2005) Fluvial process and morphology of the Brahmaputra River in Assam, India. Geomorphology 70:226–256

Sheth HC (2007) Plume-related regional pre-volcanic uplift in the Deccan Traps: absence of evidence, evidence of absence. Geol Soc Am Spec Pap 430:785–813

Singh IB (1996) Geological evolution of Ganga Plain—an overview. J Palaeontol Soc India 41:99–137

Singh IB (2007) The Ganga River. In: Gupta A (ed) Large rivers: geomorphology and management. Wiley, New York, pp 347–371

Storey M, Mahoney JJ, Saunders AD, Duncan RA, Kelley SP, Coffin MF (1995) Timing of hotspot-related volcanism and the breakup of Madagascar and India. Science 267:852–855

Vaidyanadhan R (2002) Geomorphology of the Indian sub-continent. Monograph published by the Indian Society of Remote Sensing, Dehra Dun

Valdiya KS (1996) River piracy. Saraswati that disappeared. Resonance 1:19–28

Valdiya KS (1998) Himalaya: emergence and evolution. Universities Press, Hyderabad

Valdiya KS (2010) The making of India: geodynamic evolution. Macmillan, New Delhi

Yin A (2006) Cenozoic tectonic evolution of the Himalayan orogen as constrained by along-strike variation of structural geometry, exhumation history, and foreland sedimentation. Earth-Sci Rev 76:1–131

Part II

Main Geomorphic Provinces

Himalayan Landscapes of India

Lewis A. Owen

Abstract

The Himalaya of northern India comprise a series of mountain ranges, including the Siwaliks, Lesser and Greater Himalaya and Transhimalaya. The region is amongst the most geomorphically dynamic on the planet. The mountain ranges that constitute the Himalaya are the consequence of the continued collision of the Indian and Eurasian continental lithospheric plates. In addition to the tectonic processes that help create the mountains, glaciers and their associated processes, vast rivers that drain the mountains, mass movements, aeolian processes and weathering, are actively shaping them. Human-influences are also becoming important in modifying Himalayan landscapes, but their magnitude is yet to be fully assessed. The relative importance of tectonic and surface processes varies across the Himalaya, and their magnitude and frequency is greatly influenced by the climate gradients and topography. The spatial and temporal variation in processes has led to a great diversity in landscapes throughout the Himalaya of northern India, making it one of the most fascinating and challenging places to study geomorphology and landscape evolution.

Keywords

Siwaliks • Transhimalaya • Glaciers • Rivers • Landslides • Tectonics • Lakes • Dunes

1 Introduction

The northern regions of India are home to the high mountains of the Himalaya. The Himalaya comprises of a series of mountain ranges that stretch east-west, rising from the Indo-Ganga (Indo-Gangetic) Plains in the south to the Tibetan Plateau in the north. The main ranges, from south to north, include the Siwaliks, the Lesser Himalaya, Greater (or Higher) Himalaya and Transhimalaya (Fig. 1). These mountain ranges stretch for about 2,500 km from Pakistan through NW India, Nepal, Bhutan, NE India, and Myanmar (Fig. 1). Within each of these mountain ranges distinct regions are delimited based on their geology, topography, drainage systems and climate, and include in northwest India, for example, the Kulu, Garhwal, Lahul-Spiti Himalaya, and the Pir Panjal, Zanskar and Ladakh Ranges (Fig. 2).

The name Himalaya is derived from the Sanskrit *hima* meaning 'snow' and *alaya* being 'dwelling, abode' implying that snow and glaciers cover much of the mountainous region. Indeed much of the region is glaciated and during winters extensive areas are covered by snow. However, this is highly simplistic and much of the region is unglaciated and only experiences a thin winter snow cover.

Arguably one of the greatest characteristics of the Himalaya is its diversity of environments, landscapes and landform types. This is partially because of the strong north-south and east-west precipitation gradients (Fig. 1b), which

L. A. Owen (✉)
Department of Geology, University of Cincinnati,
PO Box 210013 Cincinnati, OH 45221-0013, USA
e-mail: Lewis.Owen@uc.edu

V. S. Kale (ed.), *Landscapes and Landforms of India*, World Geomorphological Landscapes,
DOI: 10.1007/978-94-017-8029-2_4,

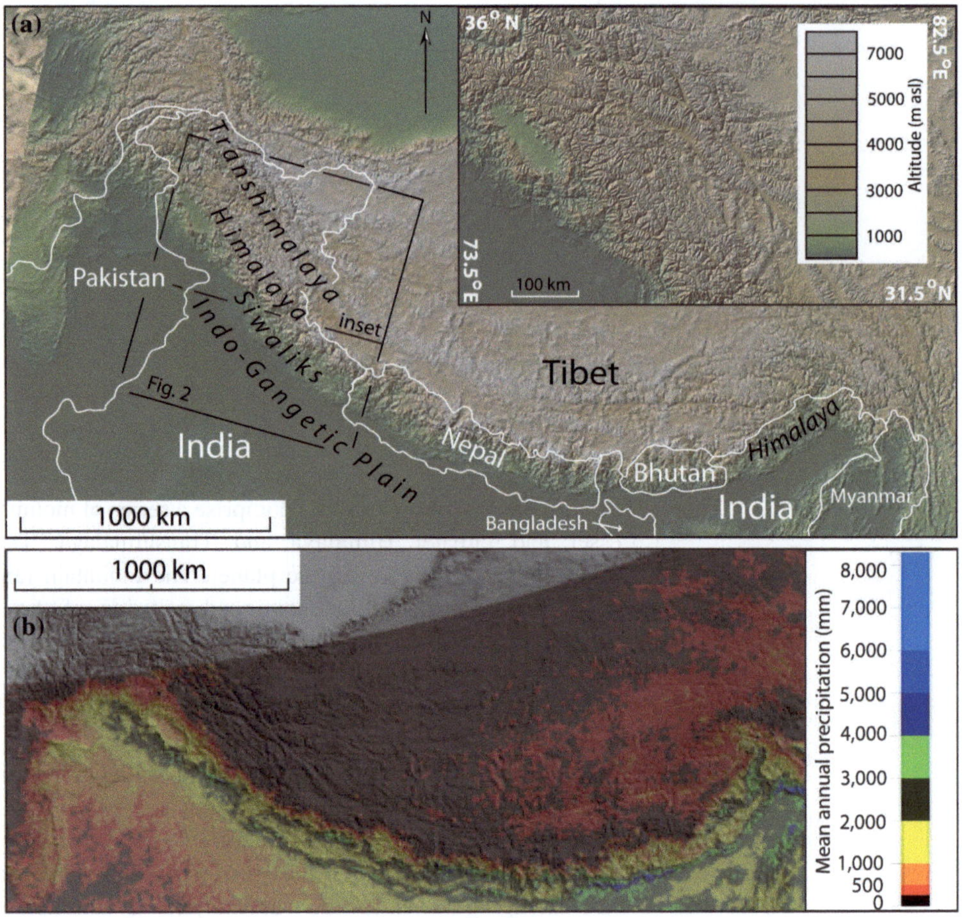

Fig. 1 Digital elevation models for **a** the Himalaya showing regional setting and with the inset showing detailed DEM of the Himalaya in NW India produced using GeoMapApp (http://www.geomapapp.org). The political boundaries are according to the Survey of India. **b** Annual precipitation averaged for 1998–2009 from TRMM data (http://www.geog.ucsb.edu/~bodo/TRMM/#ASCII) draped over the DEM of the Himalaya and adjacent Tibet showing the strong north-south and east-west precipitation gradients. Note the double precipitation maxima across the Himalaya that adds to the complexity of the climate gradients

have a profound influence on the dominant type and magnitude of Earth surface processes and in particular the style and extent of glaciation. Environments range from semi-arid mountain areas in the northernmost regions to some of the wettest places on Earth in the south (Figs. 1 and 3). The topography also has a great influence on surface processes and how landscapes evolve. Topography is highly variable, with some of the greatest relative relief on Earth in the Transhimalaya and lesser slopes in the foothills of the Himalaya, to the flatlands of the Indo-Ganga Plain (Fig. 1). Most importantly, the relative relief and steepness of slopes has a major influence on the type of surface processes that operate in the Himalaya and which create the mountain landscapes. The great rivers that drain the Himalaya, for example, the Indus, Ganga and Brahmaputra act as conduits to transfer the eroded sediment from the mountains. The environments along these rivers are also diverse, ranging from torrent headwaters, to long stretches of wide valleys most notably along longitudinal drainages, and to deep gorges along the transverse rivers.

To fully describe the geomorphology of the Indian Himalaya is challenging because of the diversity of landscapes and landforms. But in an attempt to do so concisely, several major themes can be considered. Firstly, the Himalaya are the results of tectonics, the consequence of the continued collision of the Indian and Eurasian continental lithospheric plates since ~50 Ma (Yin and Harrison 2000). This has resulted in ~2000 km of crustal shortening, which has produced the Himalaya and Tibet mountain mass with an average elevation of ~5000 m a.s.l. The mountains are a complex assembly of different rock types representing much of the whole geologic record and juxtaposed by major continental scale faults and crustal sutures (Fig. 3a). The dynamic tectonism that is producing the Himalaya is expressed periodically by large to great earthquakes that shake the region, resulting in uplift and also denudation by such processes as earthquake-triggered landsliding.

The resultant high topography creates abodes for glaciers. The glacier system has a profound influence on shaping landscapes, sediment transfer and the hydrological system. The role of glaciation has oscillated over time due to Quaternary climate change. The mighty rivers of the Himalaya constitute the next major theme. They erode, transport and deposit vast quantities of sediment, and are also the lifelines for millions of people in the mountains and the Himalayan forelands. Mass movement is also a pervasive process throughout the Himalaya, helping to shape the slopes and pose a severe threat to the people who live in the

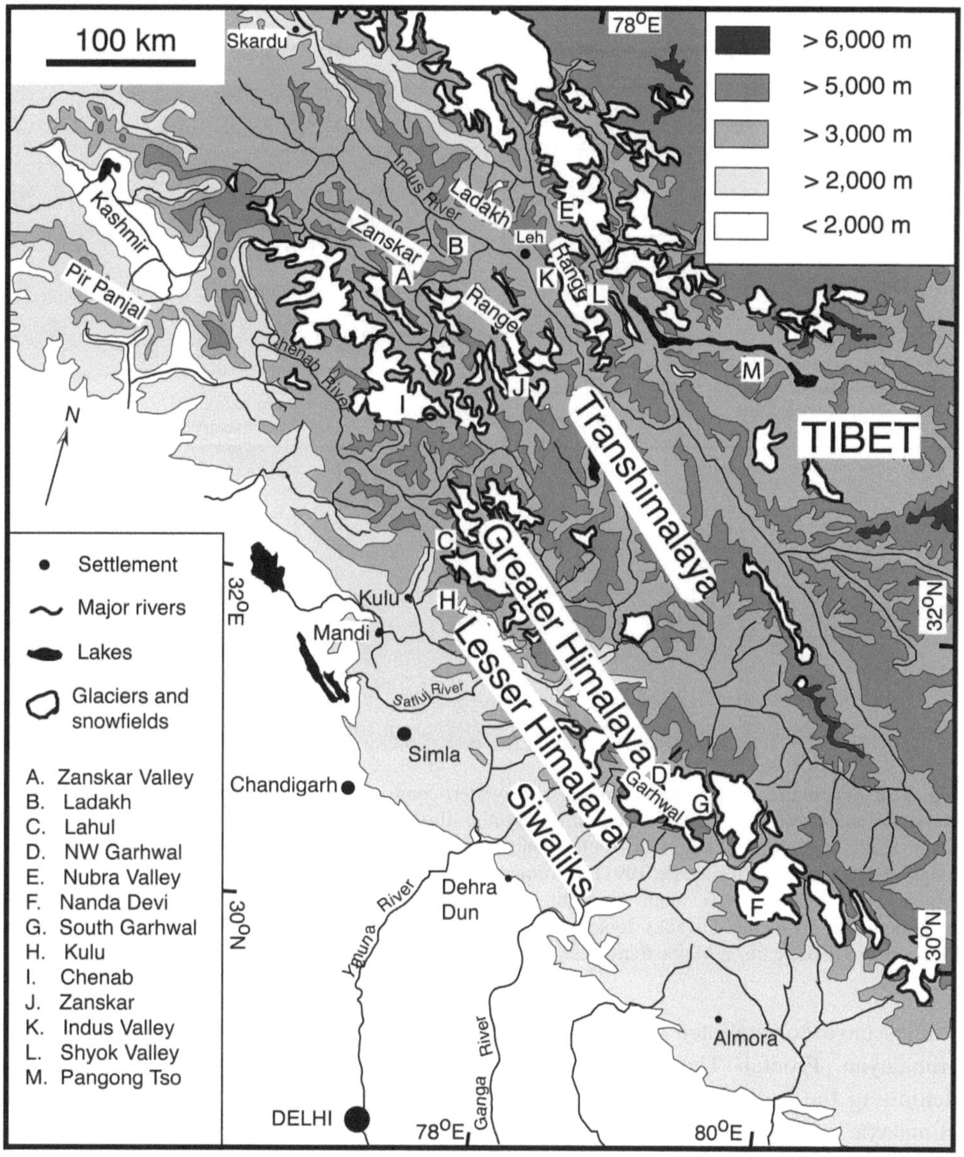

Fig. 2 The NW Indian Himalaya showing major locations mentioned in the text. This is the most studied region of the Indian Himalaya. There is a paucity of geomorphic studies in the Eastern Himalaya of northern India, and hopefully researchers will focus more of their efforts in those regions in future years

region. Less obvious are the roles of wind and weathering in the Himalaya. Weathering ranges from deep chemical types in the warm monsoonal areas to intense physical types in the periglacial and glacial regions of the Himalaya. Wind processes are particularly evident and important around glaciated regions, and also influence deposition of sediment such as loess in the foreland of the Himalaya. In recent times, human influences have begun to strongly impact landscape development through the Himalaya, most notably accelerating erosion and increasing landsliding.

To help examine the variety of landscapes and landforms in the Indian Himalaya the geomorphology will be considered under the following themes: (i) tectonic geomorphology; (ii) Himalayan glacier systems; (iii) hydrological systems; (iv) mass movements; (v) aeolian system; (vi) weathering; and (vii) human influences.

2 Tectonic Geomorphology

Numerous studies have been undertaken on the geology and tectonics of the Himalaya (Yin and Harrison 2000 and references therein). Research has included studies to try to define the rate of slip on active faults, paleoseismic studies, and projects to examine the relationship between tectonics and erosion, and work on the influence of earthquakes on landscape evolution. These studies are helping to quantify the dynamic nature of the Himalaya and the relative role of tectonics on landscape evolution.

Particularly impressive are the paleoseismic studies being undertaken along the Himalayan front, which are showing that great earthquakes are common phenomena and pose a real and major threat to the vast populations of the Himalaya and its foreland (Kumar et al. 2006). These

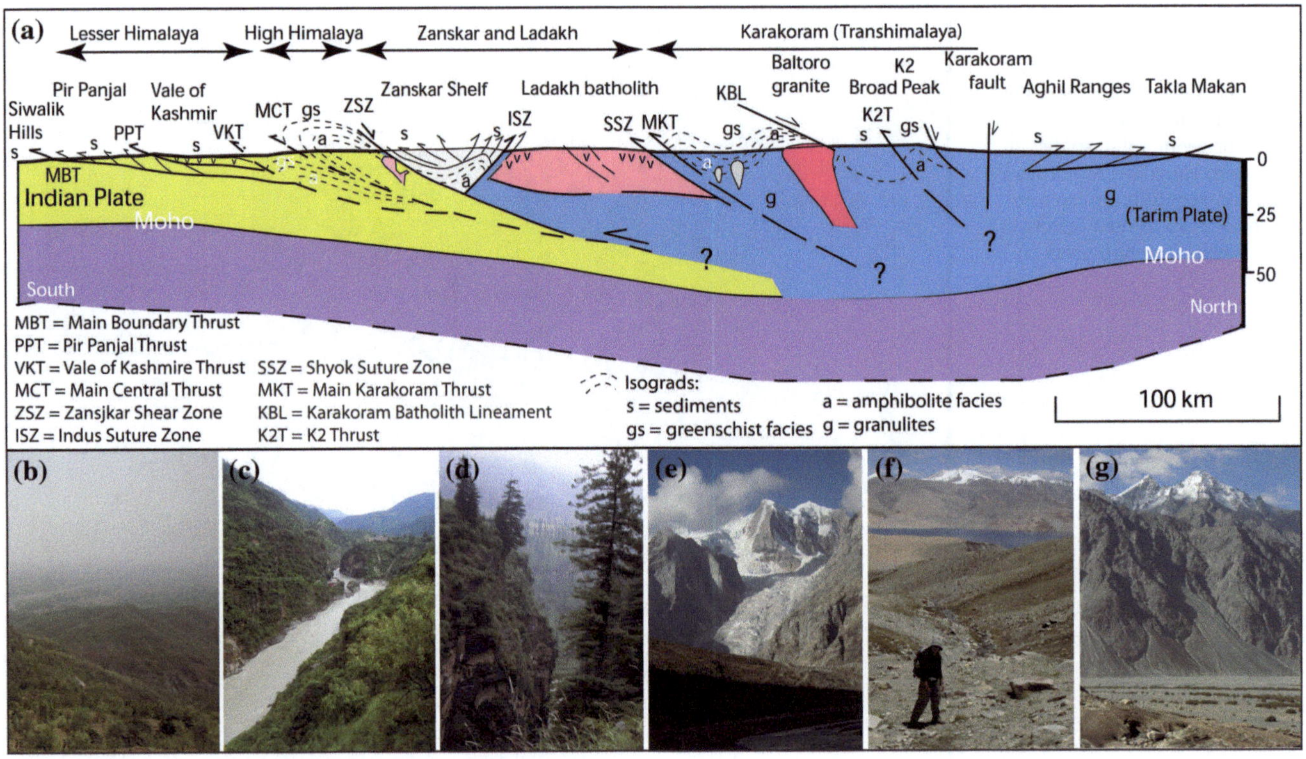

Fig. 3 **a** Schematic cross-section through the western end of the Himalaya across NE India and the Central Karakoram illustrating the major tectonic-geologic features that helps contribute to the diverse range of landscapes (after Searle 1991). **b** through **g** are example of typical environments from south to north across northern India. **b** View south from the Siwaliks looking towards the Indo-Ganga Plains near Chandigarh, **c** Major transverse drainage of the Ganga in the monsoon-influenced Lesser Himalaya, **d** Deeply incised monsoon-influenced tributary valley of the Beas River on the southern slopes of the Pir Panjal near Manali, **e** Bara Shugri Glacier in the Greater Himalaya in the rain shadow of the Indian monsoon, **f** View of Tso Morari in the mountain desert of Zanskar, **g** View across the Nubra Desert Valley looking at snow covered Karakoram in the distance

studies have concentrated on the major thrust faults (e.g. Himalayan Frontal Thrust) that bound the Himalaya delimiting the Indo-Ganga Plains and the foothills of the Himalaya.

Rates of slip on major faults such as the Karakoram Fault in Ladakh are being defined by mapping and dating offset landforms (e.g. Brown et al. 2002; Fig. 4) and are contributing greatly to our understanding of the partitioning of deformation across the Himalayan-Tibetan orogen. The Karakoram Fault, for example, is major right-lateral strike-slip fault, which is considered to be important in contributing to the eastward lateral crustal extrusion of Tibet. The fault is expressed as an impressive topographic feature helping to define the Nubra Valley and Pangong Tso in Ladakh.

Defining rates of surface uplift is more challenging, but recent work on exhumation and erosion are beginning to illustrate strong contrasts in rates of erosion, uplift and exhumation across mountain ranges, such as in Lahul Himalaya (Adams et al. 2009) and Ladakh Range (Dortch et al. 2011a). Exhumation, uplift and incision rates are generally in the range of a few mm/year, but may exceed several tens of mm/year in some regions (Dortch et al. 2011a).

Particularly interesting are the new geomorphic paradigms that are being developed based on tectonics and geomorphic research in the Himalaya. These include the extrusive flow/extrusion model, which theorizes southward flow of ductile middle and lower crust that continuously replenishes the Himalayan range front between the Main Frontal Thrust and the South Tibetan Fault System as the Himalayan front is actively eroded away under a monsoon climate (Hodges 2006).

3 Himalayan Glacial Systems

The glacial systems throughout the Indian Himalaya are complex, ranging from sub-polar continental type glaciers in the semi-arid high mountains of Ladakh and Zanskar to maritime type glaciers in the monsoon-influenced regions of the Greater (or Higher) Himalaya such as in Garhwal (Fig. 5). Some of the glaciers, such as Siachen and Gangotri Glaciers, are among the longest outside of the polar regions. The valleys that many of the large maritime type glaciers occupy are deeply eroded with impressive relative reliefs. In

Fig. 4 The Karakoram Fault. **a** View of the linear Nubra Valley along the trace of the Karakoram Fault in Ladakh. **b** View of a fault scarp (highlighted by the hands) along the Karakoram fault in a debris flow fan in the upper Nubra Valley. The *arrows* show the sense of displacement

contrast, the smaller, notably sub-polar type glaciers, generally reside in small v-shaped valleys that show little evidence of glacial erosion. Many of the glaciers have impressive debris mantles and deposit vast successions of sediments (Benn and Owen 2002).

Throughout Himalaya valleys there is much evidence for former glaciations. In some regions, for example, in the Indus Valley near Leh, moraines of great antiquity are preserved (dating back to >400 ka). There are marked contrasts in the extent and timing of glaciation between mountain ranges in northern India (Owen 2011). In the Lahul Himalaya, glaciation was very extensive during the Late Glacial with an extensive valley glacier system filling the main trunk valleys for >100 km (Owen 2011). But in contrast, glaciers in Zanskar and Ladakh only advanced a few kilometers from their present positions during the Late Glacial. These different patterns likely reflect temporal and spatial variability in the two major climate systems, the Indian monsoon and mid-latitude westerlies, that influence the region, and their consequent resultant regional precipitation gradients. Dortch et al. (2013) examined the evidence for time of glaciation in the semi-arid regions of India and the adjacent regions, and identified 16 regional glaciations over the past ~500 ka. In addition, Murari et al. (2014) identified 27 regional glaciations for the monsoon-influenced areas of the Himalaya and Tibet, many of which Owen and Dortch (2014) showed correlated with glaciations in the semi-arid regions of the India and adjacent regions.

Himalayan landscapes in northern India have therefore been greatly influenced by numerous glacial oscillations throughout the Quaternary. This includes erosion as glaciers advance during glacial times; and landscape readjustment during deglaciation, which has become known as paraglaciation. Many researchers have argued paraglacial processes are amongst the most important in shaping Himalayan landscapes. Barnard et al. (2004) working in the Garhwal Himalaya, for example, showed major resedimentation of moraines to form extensive alluvial fans, river terraces and valley fills.

4 Hydrological Systems

The hydrological system is dominated by two major influences in the Indian Himalaya: (i) the monsoon; and (ii) glacier melting and snowmelt. As such, river discharge varies considerably throughout the year in the Himalaya, being greatest in the summer as the glaciers and snowfields melt at high altitudes and as the monsoon dumps its heavy rains. The waters drain into some of the largest rivers on our planet, transporting vital waters and transporting vast amounts of sediment to the Indus-Ganga-Brahmaputra Plains and the Arabian Sea and the Bay of Bengal. Armed with vast quantities of debris the rivers are powerful and important erosional agents, eroding deep gorges, such as the Bhagirathi and Brahmaputra Gorges, which are among the most impressive on our planet.

Impressive strath terraces are present along many of the valleys of the Himalaya in northern India (Fig. 6a). These are a testament to the erosive power of the river and can be dated to determine the rates of fluvial incision. Dortch et al. (2011a) provide a summary of all strath terraces throughout the Himalaya that have been dated, some of which date

Fig. 5 Views of glaciers in the Indian Himalaya. **a** Sub-polar continental type glaciers in Zanskar, **b** Batal Glacier in the Lahul Himalaya illustrating the thin debris cover that exists on many Himalaya glaciers, **c** Maritime type glacier in the Garhwal Himalaya ~10 km NW of Kedarnath

Fig. 6 **a** Strath terraces in granite along the Chandra River in the Lahul Himalaya. These were dated using ^{10}Be cosmogenic nuclide surface exposure dating to a few thousand years before present by Dortch et al. (2011b). **b** Views of valley fills in at Sachu in Zanskar comprising fluvial gravels

back to >700 ka. The data show that incision rates vary considerably, both temporally and spatially. Holocene incision rates, for example, range from 0.02 to ~26 mm/year, as compared to Pleistocene incision rates that are ≥5 mm/year. Dortch et al. (2011b) suggest that the temporal pattern for rates of fluvial incision is probably controlled by episodic incision linked to significant precipitation changes throughout the Quaternary. Dortch et al. (2011a) illustrated spatial differences by examining straths on the north side of the Ladakh Range (~1 mm/year) with those on the northern side of the Zanskar Range that are considerably less (<0.06 mm/year), likely reflecting different climatic and tectonics regimes.

Catastrophic flooding is an important formative process in the Himalaya. This was well illustrated by the horrific floods in Garhwal during the summer of June 2013 where 1,000s of people lost their lives, in settlements such as Kedarnath, during a series of exceptional cloud bursts that resulted in torrential floodwaters and debris flows. Flooding has also been recognized as the consequence of glacial lake outburst floods (GLOFs), such as in the Lahul Himalaya related to deglaciation at the end of the last glacial (Coxon

Fig. 7 Alluvial fans in **a** the Indus Valley emerging from the Zanskar Ranges, and **b** in the Nubra Valley. These alluvial fans were the first to be described in the literature (Drew 1873)

et al. 1996). Dortch et al. (2011b) identified flood deposits in Ladakh related to catastrophic partial drainage of Pangong Tso.

Many of these vast rivers that drain the Himalaya leave behind thick valley fills, especially in intermontane basins/Duns of the Himalaya (Fig. 6b). These thick sediment fills record a long history of hydrological change. River terraces are abundant; they are formed not only from fluvial deposits, but also include landslide, glacial and lacustrine deposits, and provide a rich history of landscape evolution. Studies by Barnard et al. (2004) and Ali et al. (2013) in the Goriganga Valley illustrate the richness of terrace deposits.

Lakes are abundant throughout the Indian Himalaya. As Owen (1996) highlighted they can form by tectonic processes and occur in schuppenstruktur (thrust bounded basins) along the southern margin of the Himalaya (e.g. Kashmir Basin), along strike-slip faults (e.g. Pangong Tso), and in areas of crustal subsidence (e.g. Tso Morari). Lakes also occur in glacial settings behind ice dams, moraines (e.g. Tapovan along Gangotri Glacier) and within eroded bedrock depressions (e.g. Chandra Tal). In addition, lake may from behind blockages caused by debris flows, rock falls and landslides.

Alluvial fans are common landforms in the semi-arid regions of the Himalaya, especially in the Indus and Nubra Valleys (Fig. 7). These form as rivers and debris flows deposit their loads when they exit confined valleys; but many alluvial fans also form as a result of resedimentation of glacial deposits during deglaciation as in the upper Bhagirathi Valley in Garhwal (Owen and Sharma 1998). Goswami et al. (2013) describe examples of other impressive alluvial fans in the eastern Himalayan foothills of West Bengal and discuss the tectonic controls on their development.

5 Mass Movements

Mass movement is pervasive throughout the Himalaya of northern India, ranging from small-scale soil creep to mega-landslides involving more than a million m^3 of debris. Landsliding is very common during the heavy rainfalls of the monsoon season, especially along highways where fresh road cuts that are devoid of vegetation are very susceptible to failure. Most of these landslides are shallow (<a few meters deep) and they often develop into debris flows. Once formed the landslides will commonly continue to develop, excavating their way up slopes and persisting for many years. Larsen and Montgomery (2012) using a study of the Eastern Himalaya highlight the likely importance of landsliding in helping to limit topography and underscore the significance of mass movement in the landscape evolution of the Himalaya.

Snow avalanching is also a common process throughout the Himalaya, occurring mostly in spring and throughout the summer months. Snow avalanches often incorporate a considerable amount of rock and soil debris, helping to shape slopes, and in glacial catchments are important in supplying ice and debris to glaciers.

Landslides are also triggered during earthquakes, as was well demonstrated during the October 20, 1991 and March 28, 1999 earthquakes in the Garhwal Himalaya (Fig. 8a). Owen et al. (1996) and Barnard et al. (2001) describe the characteristics and distribution of the earthquake-triggered landslides during these earthquakes and compared them to monsoon- and human-triggered landslides. Most were shallow failures of rock avalanche type and surprisingly the earthquake-triggered landslides were less significant in modifying the landscape as compared to the monsoon- and human-triggered landslides.

There is abundant evidence throughout the Himalaya of India for catastrophic long-runout landslides (Fig. 8). Dortch et al. (2009) used ^{10}Be cosmogenic nuclides to date four of these landslides to the early Holocene and compared their ages with other dated mega-landslides throughout the Himalaya. The ages of the mega-landslides clustered during times of enhanced monsoon activities, which caused Dortch et al. (2009) to suggest that they may have been earthquake-triggered during times of increased monsoonal rainfall.

Rock glaciers are another form of mass movement and are very common landforms in the drier permafrost areas of northern India, especially in northern most Lahul and Zanskar (Owen and England 1998; Fig. 9a).

6 Aeolian System

The aeolian is one of the least well-studied geomorphic systems in the Himalaya of northern India. This is probably because aeolian deposits are not particularly abundant. But where they exist they can be quite impressive such as the large dune fields at the confluence of the Nubra and Shyok Rivers in Ladakh (Fig. 9b). Impressive sand ramps are also present. Notable ones occur east of Leh along the Indus Valley in Ladakh. Colluviated aeolian deposits are present on many landforms, sometimes containing buried soils such in the Suru Valley on the north side of the Nun Kun Massif (Lee et al. 2014). Many of the aeolian deposits are ephemeral, easily blown and/or washed away.

7 Weathering

Few quantitative studies have been undertaken on weathering processes in the Indian Himalaya and it is therefore difficult to assess its importance in landscape evolution. However, the landforms produced by weathering are abundant and testify to its importance in the geomorphic evolution of the Himalaya in northern India. In the Siwaliks and Lesser Himalaya, chemical weathering has decomposed bedrock to depths of many meters as can be seen in the numerous road cuttings along the main highways. Core stones are left on the surfaces where grus, the weathered rock product, has been washed away. In contrast, deep chemical weathering is less common in the Greater (or Higher) Himalaya and Transhimalaya, but abundant talus slopes are a proof of abundant cryogenic weathering at high altitudes.

8 Human Influences

Human influences on Himalaya landscapes have become more evident in recent decades, especially with the construction of numerous highways that traverse the region and connect isolated villages. The construction of highways has resulted in an increase in landslides, as noted by Barnard et al. (2001) in the Garhwal Himalaya. The management of water has also had an influence on the landscape, including the construction of dams and canals that has influenced the hydrological character of stretches of many rivers. Over the past decades much debate has focused on the nature and role of deforestation on landscape development in the Himalaya (Ives and Messerli 1989).

Defining the extent and degree of human influence and comparing it to the role of natural processes is challenging, particularly because of the diversity of environments throughout the Indian Himalaya. Nevertheless, as regions throughout the Indian Himalaya continue to develop it is likely that human activity will lead to accelerated erosion and significant environmental changes. Knowledge of the natural processes, their spatial and temporal variability, is needed to fully assess the human impact on landscapes in the Indian Himalaya.

Fig. 8 **a** Reactivated ancient mega-landslide triggered by the March 28, 1999 Garhwal earthquake at Dear in the Alaknanda River Valley. Barnard et al. (2001) dated the ancient landslide to ~8 ka using ^{10}Be terrestrial cosmogenic nuclides surface exposure dating, **b** Long-runout mega-landslides comprising >million m^3 of debris at Dacha (note the highway for scale). Dortch et al. (2009) dated this landslide to ~8 ka using ^{10}Be terrestrial cosmogenic nuclides surface exposure dating

9 Endogenetic Versus Exogenetic Processes

The Indian Himalaya is one of the best natural laboratories to assess the relative roles of endogenetic (Earth's internal) and exogenetic (Earth's external/surface) processes in the evolution and development of active mountain landscapes. In recent decades there has been much debate on what processes control the height and erosion of mountains, and defining the interactions and feedbacks between the different sets of exogenetic and endogenetic processes (Molnar and England 1990; Whipple et al. 1999). Tectonic uplift, for example, produces high topography that allows glaciers to develop. The glaciers then erode the landscape leading to

Fig. 9 **a** Protalus rock glaciers in the upper Bhaga Valley in the Lahul Himalaya. **b** Large barchan sand dune at the Shyok-Nubra confluence in Ladakh

denudational unloading, which in turn leads to more uplift and possibly an increase in the altitude of peaks and interfluves between glaciers.

The challenge for future geomorphic research in the Himalaya, and especially for hazard mitigation and environmental risk assessment, is to quantify the various different exogenic and endogenetic processes and their links, and to understand their spatial and temporal variability. Only then can accurate predictions be made about future geomorphic and landscape changes.

10 Conclusion

The Himalaya of northern India is arguably one of the most dynamic geomorphic environments on our planet. The landforms and landscape are impressive and are a consequence of both exogenetic and endogenic processes. These sets of processes interact with one another having complex feedbacks, which are only now beginning to be understood in any detail. The spatial and temporal variations in the relative roles of these processes have resulted in a very diverse range of geomorphic environments throughout the Himalaya of northern India. New methods including remote sensing, computer modeling, geochronology, thermochronology and real time monitoring of processes will shed new light on the geomorphic evolution of the Himalaya.

References

Adams B, Dietsch C, Owen LA, Caffee M, Spotila J, Haneberg B (2009) Exhumation and incision history of the Lahul Himalaya, northern India, based on (U-Th)/He thermochronometry and terrestrial cosmogenic nuclide dating techniques. Geomorphology 107:285–299

Ali SN, Biswas RH, Shukla AD, Juyal N (2013) Chronology and climatic implications of late Quaternary glaciations in the Goriganga valley, Central Himalaya, India. Quatern Sci Rev 73:59–76

Barnard PL, Owen LA, Sharma MC, Finkel RC (2001) Natural and human-induced landsliding in the Garhwal Himalaya of Northern India. Geomorphology 40:21–35

Barnard PL, Owen LA, Sharma MC, Finkel RC (2004) Late Quaternary (Holocene) landscape evolution of a monsoon-influenced high Himalayan valley, Gori Ganga, Nanda Devi, NE Garhwal. Geomorphology 61:91–110

Benn DI, Owen LA (2002) Himalayan glacial sedimentary environments: a framework for reconstructing and dating former glacial extents in high mountain regions. Quatern Int 97–98:3–26

Brown ET, Bendick R, Bourles DL, Gaur V, Molnar P, Raisbeck GM, and Yiou F (2002) Slip rates of the Karakoram fault, Ladakh, India, determined using cosmic ray exposure dating of debris flows and moraines. J Geophys Res 107(B9, 2192):7-1–7-8

Coxon P, Owen LA, Mitchell WA (1996) A late Quaternary catastrophic flood in the Lahul Himalayas. J Quat Sci 11: 495–510

Dortch JM, Owen LA, Haneberg WC, Caffee MW, Dietsch C, Kamp U (2009) Nature and timing of mega-landslides in northern India. Quatern Sci Rev 28:1037–1056

Dortch JM, Owen LA, Schoenbohm LM, Caffee MW (2011a) Asymmetrical erosion and morphological development of the Ladakh Range, Northern India. Geomorphology 135:167–180

Dortch JM, Owen LA, Caffee MW, Kamp U (2011b) Catastrophic partial drainage of Pangong Tso, Northern India and Tibet. Geomorphology 125:109–121

Dortch JM, Owen LA, Caffee MW (2013) Timing and climatic drivers for glaciation across semi-arid western Himalayan-Tibetan orogen. Q Sci Rev 78:188–208

Drew F (1873) Alluvial and lacustrine deposits and glacial records of the upper Indus basin; Part 1, Alluvial deposits. Geol Soc London Q J 29:449–471

Goswami CC, Mukhopadhyay D, Poddar BC (2013) Geomorphology in relation to tectonics: a case study from the eastern Himalayan foothills of West Bengal, India. Quatern Int 298:80–92

Hodges K (2006) Climate and the evolution of mountains. Sci Am August:72–79

Ives JD, Messerli B (1989) The Himalayan dilemma: reconciling development and conservation. Routledge, London 324 p

Kumar S, Wesnousky SG, Rockwell TK, Briggs RW, Thakur VC, Jayangondaperumal R (2006) Paleoseismic evidence of great surface-rupture earthquakes along the Indian Himalaya. J Geophys Res 111:B003309

Larsen I, Montgomery DR (2012) Landslide erosion coupled to tectonics and river incision. Nat Geosci 5:468–473

Lee SY, Seong YB, Owen LA, Murari MK, Lim HS, Yoon HI, You K-C (2014) Late Quaternary glaciation in the Nun-Kun massif, Northwestern India. Boreas 43: 67–89

Molnar P, England P (1990) Late Cenozoic uplift of mountain ranges and global climate change: Chicken or egg? Nature 346:29–34

Murari MK, Owen LA, Dortch JM, Caffee MW, Dietsch C, Fuchs M, Haneberg WC, Sharma MC, Townsend-Small A (2014) Timing and climate drivers for glaciation across monsoon-influenced regions of the Himalayan-Tibetan orogan. Quatern Sci Rev 88:159–182

Owen LA (1996) Quaternary lacustrine deposits in a high-energy semi-arid mountain environment, Karakoram Mountains, Northern Pakistan. J Quat Sci 11:461–483

Owen LA (2011) Quaternary glaciation of Northern India. In: Elhers J, Gibbard P, Hughes PD (eds) Quaternary glaciations: extent and chronology: a closer look. Developments in Quaternary science, vol 15, 2nd edn. Elsevier, Amsterdam, pp 929–942

Owen LA, Dortch JM (2014) Quaternary glaciation of the HImalayan Tibetan orogen. Quatern Sci Rev 88:14–54

Owen LA, England J (1998) Observations on rock glaciers in the Himalayas and Karakoram Mountains of northern Pakistan and India. Geomorphology 26:199–213

Owen LA, Sharma MC (1998) Rates and magnitudes of paraglacial fan formation in the Garhwal Himalaya: implications for landscape evolution. Geomorphology 26:171–184

Owen LA, Sharma M, Bigwood R (1996) Landscape modification and geomorphological consequences of the 20 October 1991 earthquake and the July–August 1992 monsoon in the Garhwal Himalaya. Z Geomorphologie 103:359–372

Searle MP (1991) Geology and tectonics of the Karakoram mountains. Wiley, Chichester

Whipple K, Kirby E, Brocklehurst S (1999) Geomorphic limits to climatically induced increases in topographic relief. Nature 401:39–43

Yin A, Harrison TM (2000) Geologic evolution of the Himalayan-Tibetan orogen. Ann Rev Earth Planet Sci 28:211–280

Indus-Ganga-Brahmaputra Plains: The Alluvial Landscape

Rajiv Sinha and Sampat K. Tandon

Abstract

The vast alluvial plains of the Indus-Ganga-Brahmaputra river systems form one of the three major geomorphic provinces of the landmass of the Indian subcontinent. Drained by these three large rivers, these plains show significant variations in terms of landform development due to significant hydrological differences and variable tectonic-geomorphic regimes of their hinterlands. These plains have accumulated several kilometers of alluvial sediments over Quaternary timescales and are underlain by Siwaliks or older basement. Distinctive river processes in these three basins have often resulted in severe fluvial hazards such as floods, bank erosion, and rapid migration affecting millions of people.

Keywords

Large rivers • Alluvial plains • Himalayan foreland • Geomorphic diversity • Fluvial hazards

1 Introduction

The Indus-Ganga–Brahmaputra (IGB) Plains form an integral part of one of the largest sediment routing systems of the globe (Fig. 1), the Himalayan-Bengal-Nicobar and the Indus submarine fan systems, which was initiated following the commencement of Himalayan orogenesis in the Eocene. The India-Eurasia collision led to a thickening of the crust in the Himalayan region, which in turn resulted in lithosphere flexure and the formation of a peripheral foreland basin that extends in the south along the length of the orogen. This foreland basin has acted as a depocenter through most of the Cenozoic and continues to receive detritus from the Himalayan as well as the cratonic highlands. The proximal part of the foreland has been deformed as a consequence of the southward migration of the Himalayan fold thrust belts, and forms the outermost part of the Himalaya. To the south of the sub-Himalaya, the foredeep is covered by vast alluvial plains that are drained by the Indus, Ganga and Brahmaputra Rivers (Table 1). Much of the sediment brought from the highlands is stored in these alluvial tracts in the sub-surface, and in places may attain thicknesses of several kilometers. At a synoptic scale, these plains have a low relief and are marked by the development of piedmont zones, channel belts, floodplains and large interfluves.

From the west to east, these plains are divisible into the Indus Plains, the Ganga Plains which are further divisible into an eastern Ganga Plain dominated by fan-interfan setting and a western Ganga Plain dominated by a valley-interfluve setting, and the Brahmaputra Plains which are characterized by large widths of channel belts and the development of large alluvial islands and wetlands.

The Indus-Ganga-Brahmaputra (IGB) Plains constitute one of the three major geomorphic provinces of the Indian subcontinent with the Himalaya occurring along the

R. Sinha (✉) · S. K. Tandon
Department of Earth Sciences, Indian Institute of Technology, Kanpur, 208016, India
e-mail: rsinha@iitk.ac.in

S. K. Tandon
e-mail: sktand@rediffmail.com

V. S. Kale (ed.), *Landscapes and Landforms of India*, World Geomorphological Landscapes,
DOI: 10.1007/978-94-017-8029-2_5, © Springer Science+Business Media Dordrecht 2014

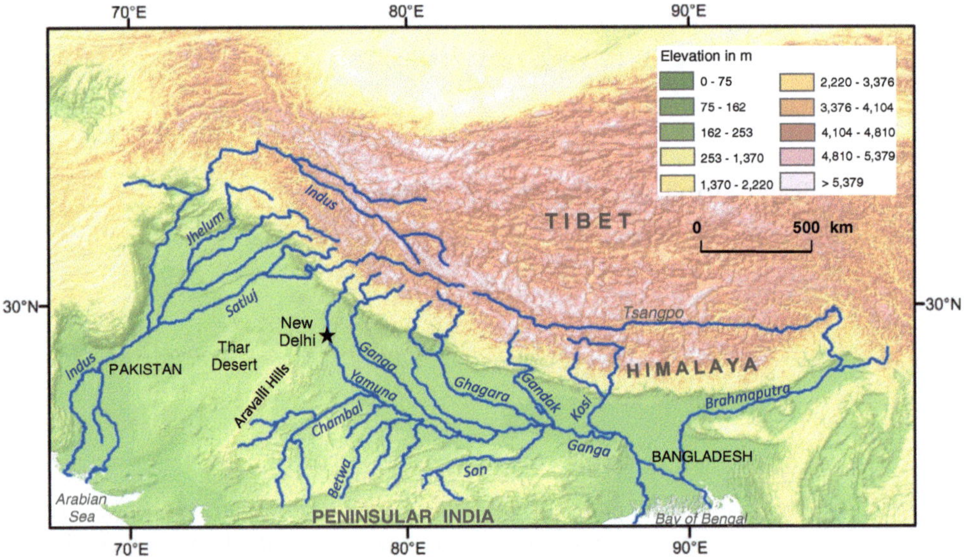

Fig. 1 Major drainage of the Indus, Ganga and Brahmaputra basins

Table 1 Drainage basin characteristics and hydrology of Indus, Ganga and Brahmaputra Rivers

Parameter	Indus	Ganga	Brahmaputra
Catchment area (10^3 km^2)	960	980	580
Total length (km)	3180	2150	2880
Average annual discharge (m^3/s)	7610	11,600	19,300
Annual sediment load at river mouth (million tonnes/year)	291	599	580–650
Unit discharge (10^3 m^3 s^{-1} km^{-2})	8	11.8	33.27
Sediment yield (million tonnes/km^2/year)	0.30	0.61	0.85–1.12
Contribution of snowmelt (%)	>50	22	<25
Major tributaries	Shyok, Shigar, Hunza and Gilgit, Jhelum, Chenab, Ravi, Beas and Satluj	*Himalayan*—Yamuna, Ramganga, Ghaghra, Gandak and Kosi *Cratonic*—Chambal, Sindh, Ken, Betwa, Son and Punpun	*Himalayan*—Dibang, Lohit, Jiadhal, Ranganadi, Puthimari, Pagladiy *North Bank*—Subansiri, Jia Bharali and Manas *South Bank*—Burhi Dihing, Dhansiri, Dikhow, Kopili, Kulsi and Krishnai

Source Goswami 1985; Hovius 1998

northern margin and the Indian Peninsula occurring to the south. These plains are richly endowed with surface and groundwater resources and fertile soils and form one of the important monsoon rainfed agricultural regions of the world supporting a population of >500 million, almost a third of the population of the subcontinent.

2 Geological Setting

The IGB Plains predominantly consist of alluvial and deltaic surfaces that are an integral part of the Himalayan foreland. The India-Eurasia collision led to the formation of NW-SE trending Himalayan Arc as well as the development of

Fig. 2 Geology of the Indus Plains

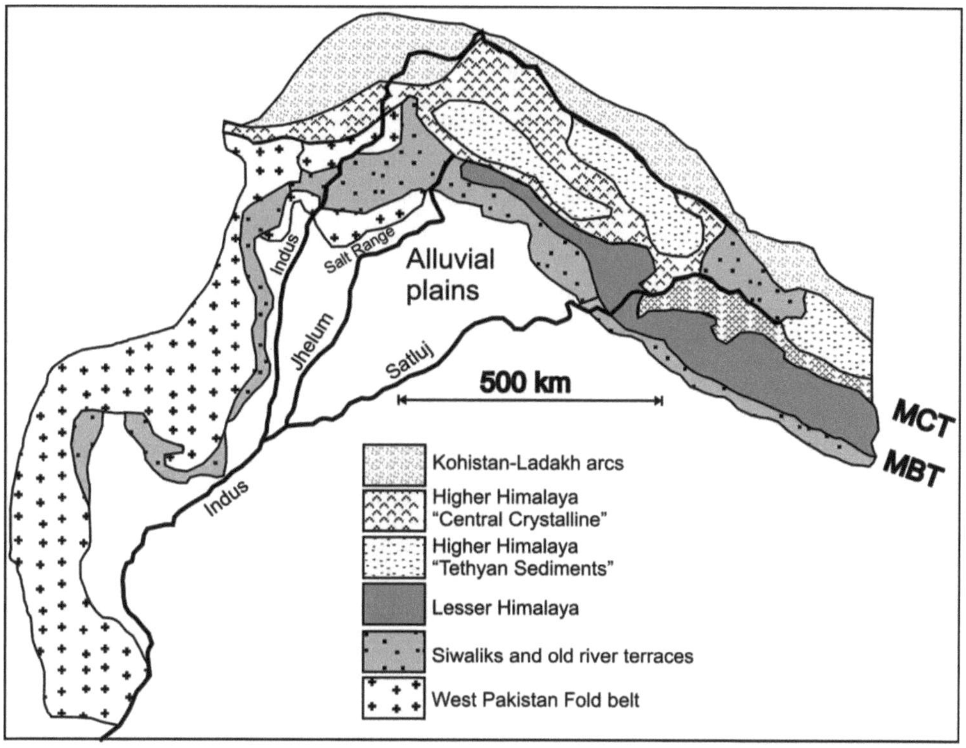

syntaxial bends in the western and eastern extremities in western Pakistan and in Arunachal Pradesh (Gansser 1964). Drainage development in the western part led to the formation of the NE-SW trending Indus Basin, and the ENE-WSW trending Brahmaputra Valley in the eastern part—both basins being structural depressions whose evolution was largely governed by the western and eastern Himalayan syntaxis, respectively. In the central part, the Ganga Basin was formed as a WNW-ESE oriented elongated depression that is bounded by the outer Himalayan thrust sheets in the north and over significant areas to the south by the Bundelkhand Craton and its constituents.

The main geologic units that occur in the region of the Indus alluvial plains include the Himalayan thrust sheets, the Indus Suture Zone and the Kohistan Arc (Fig. 2). In its lower segment, the Indus flows parallel to the NE-SW trending west Pakistan fold belt. In the Ganga Plains, the depth to the basement is variable from a few kilometers in the northern and central parts to a few tens of meters towards the southern margin of the basin where the Pre-Cambrian outcrops are observed in the Indian Shield. West of Allahabad, the Archean rocks comprise the Banded Gneissic Complex and the Bundelkhand Complex (Fig. 3). The Proterozoic rocks include the Vindhyan Supergroup of rocks south and east of the Great Boundary Fault (Fig. 3 in Chap. 1)—a major tectonic feature which separates the Meso- to Neo-proterozoic Vindhyan sedimentary rocks from the Late Archean (2500 Ma) Berach Granite and the low-grade metamorphic Palaeoproterozoic volcano-sedimentary rocks of the Hindoli Group. Deccan Traps cover considerable tracts in the headwaters of the Chambal, Betwa and Ken Rivers. In the north, the Ganga and Yamuna catchments include the meta-sedimentary rocks of the Lesser Himalaya, and large granitic intrusive bodies. Southwards in the foothills, the foreland fold-thrust belt includes both the Paleogene Dharamshala Group and the Neogene Siwalik Group (Fig. 3).

The major regional geological features of the relatively narrow Brahmaputra Basin include the NE-SW trending Himalayan thrust sheets, the NW-SE trending Mishimi Thrust Belt, as well as the NE-SW trending Naga-Haflong-Dibang Thrust Belt in the south (Fig. 4). Further downstream, the Mikir Hills are followed by the Meghalaya or Shillong Plateau. The Brahmaputra River follows the northern margin of the Meghalaya Plateau and then bends to take its southerly course along the western margin of the plateau. The Brahmaputra Basin is divided into six zones from source to mouth into: (1) high plateau of Tibet (2) the Eastern Syntaxis (3) Mishimi Hills (4) the Himalayan Mountains (5) the Indo-Burman and Naga Patkai Ranges and (6) the plains of Assam and Bangladesh (Singh 2007a).

Fig. 3 Geology of the Ganga Plains

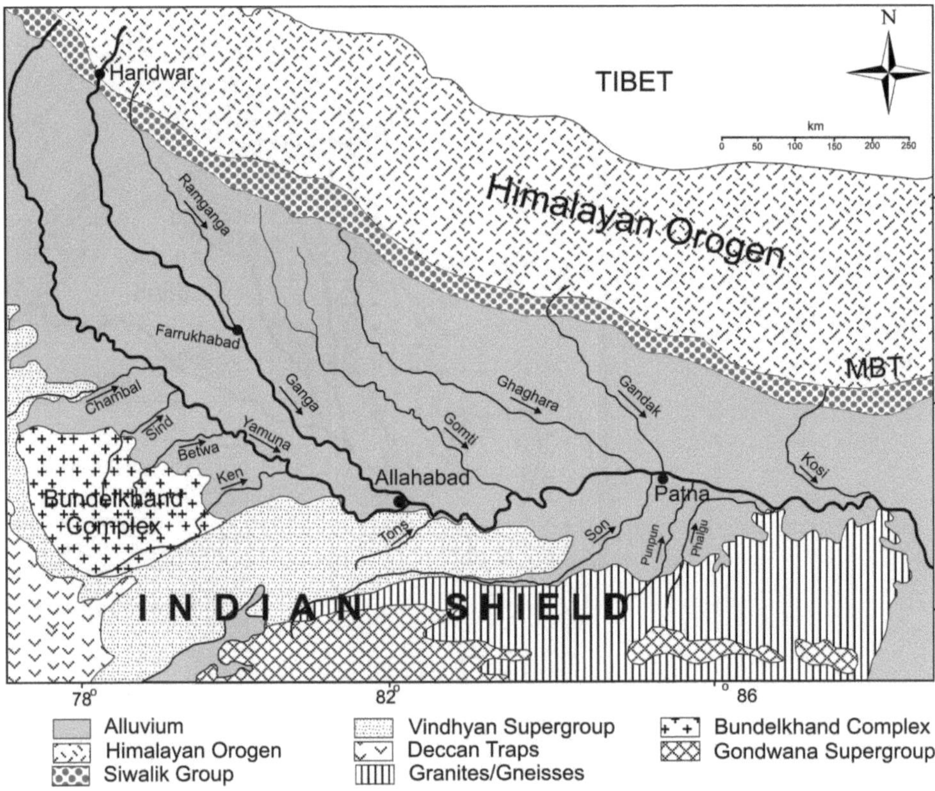

3 Basin Characteristics and Drainage

3.1 Indus

The Indus River (Fig. 5a), famous for supporting the Harappan Civilization (4.8–3.5 ka), is one of the largest rivers in the world in terms of its length, drainage area and average annual discharge (Table 1). Out of the total drainage area, ~506,753 km^2 of area lies in the semi-arid region of Pakistan and the rest in mountains and foothills. The Indus originates at an altitude of 5486 m a.s.l. from the Mount Kailas Range in the Tibetan Plateau on the northern side of Himalaya (Inam et al. 2007). The Indus falls into the Arabian Sea and forms an extensive alluvial plain and delta as well as the world's second largest submarine fan (Milliman and Meade 1983). The average annual suspended sediment load of 291 million tonnes/year ranks the Indus as one of the highest sediment load carrying rivers in the world. The Indus has a much lower water discharge compared to other rivers in this region, averaging ~3,000 m^3/s but can reach up to ~30,000 m^3/s during summer monsoons.

3.2 Ganga

The Ganga Plains have been built by sediments derived from two distinct hinterlands—the Himalaya in the north and the cratons to its south—which have had varying relative influence through geological time (Sinha et al. 2009). Mountain-fed tributaries of the Ganga such as the Yamuna, Ramganga, Ghaghra, Gandak and the Kosi are generally multi-channel, braided systems, characterized by discharge and sediment loads that are many times higher than those of the single-channel, sinuous foothills-fed and plains-fed river systems (Table 1). They also transfer a large quantity of sediments from their high relief catchments to the plains and consequently form large depositional areas (megafans). Some reaches of the Ganga e.g. between Allahabad and Varanasi show a meandering pattern possibly due to the presence of a shallow basement in the vicinity of the southern cratonic margin and decrease in sediment to water discharge ratio. Downstream of its confluence with the Chambal, the Yamuna also shows meandering pattern due to the relatively higher hydrological and sediment inputs from the cratonic hinterland. The foothills-fed (e.g. Baghmati, Rapti) and plains-fed (e.g. Burhi Gandak, Gomti) tributaries derive their sediments from the foothills and from within the plains, and a large proportion of this material is re-deposited in the plains after local reworking.

The largest of the cratonic tributaries of the Yamuna/Ganga (Table 1) is the Chambal which is ~1,000 km long and has a catchment area of 140,000 km^2, larger than any of the Himalayan tributaries of the Ganga. Flow volumes from the main six cratonic tributaries constitute ~20 % of total

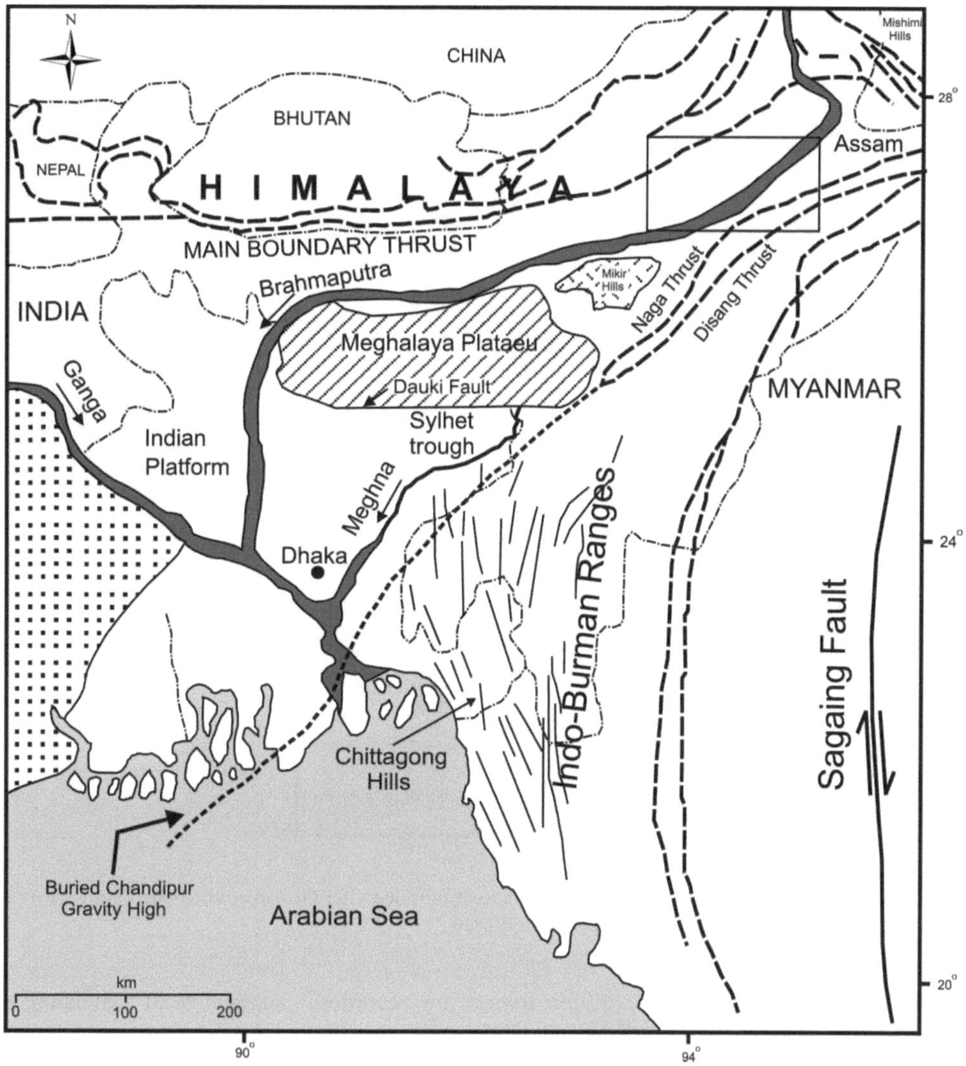

Fig. 4 Geology of the Brahmaputra Plains. *Box* shows the area covered in Fig. 7a

Ganga flow but can be very significant in certain reaches. The average annual sediment load of the Ganga at its mouth in Bangladesh is 599 million tonnes/year and together with the Brahmaputra it transports more than one billion tonnes of sediments into the Bay of Bengal every year.

3.3 Brahmaputra

The Brahmaputra Valley is the narrowest amongst the three river valleys being confined by the Himalayan thrust sheets to its northwest and the Naga-Patkai Hills and the Mikir Hills to the northeast as a result of convergence of multiple tectonic regimes. It can be subdivided into the North Brahmaputra Valley adjacent to the Himalayan foothills, and the South Brahmaputra Valley adjacent to the Naga foothills. The South Brahmaputra Valley extends southwestwards across the Dhansiri Valley after a break in the region of Barail Ranges into Cachar and Tripura Ranges.

The Brahmaputra, with a drainage area of 580,000 km^2 (50.5 % in China, 33.6 % in India, 8.1 % in Bangladesh and 7.8 % in Bhutan), is one of the world's largest rivers originating from the Chema-Yung-Dung Glacier in the Kailas Range of southern Tibet at an elevation of 5300 m a.s.l. (Goswami 1985). Two major tributaries of the Brahmaputra, the Dibang and the Lohit, join the upper course of the Brahmaputra, a little south of Pasighat and the combined drainage flows westward through Assam for about 640 km until near Dhubri where it abruptly turns south and enters Bangladesh. The Brahmaputra is then joined by several smaller tributaries both along the north as well as the south bank (Table 1).

The Brahmaputra ranks fifth among the world's largest rivers in terms of its annual mean discharge of 19,300 m^3/s measured at Bahadurabad, Bangladesh (Hovius 1998) and first in terms of its sediment yield (0.85–1.12 $t/yr/km^2$) (Table 1). Estimates of sediment load of the Brahmaputra at its mouth vary from 581 to 650 million tonnes/year and

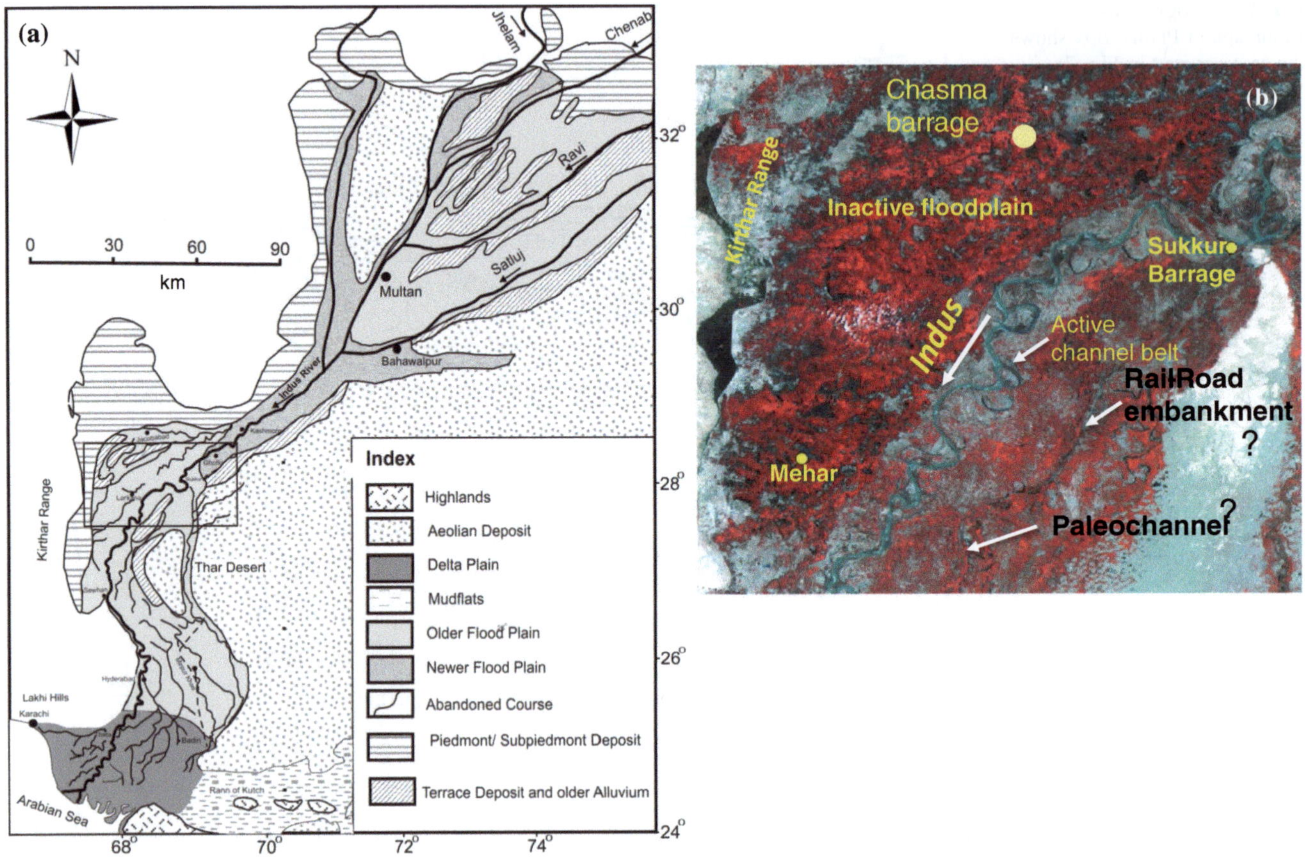

Fig. 5 **a** Geomorphological map of the Indus plains (Compiled after Giosan et al. 2006; Valdiya 2010). **b** Landsat image of a part of the Indus basin (marked as a *box* in **a**) showing major landforms

transport rates as high as 26 million tonnes are recorded during peak flows at Pandu (Goswami 1985). Amongst the largest rivers of the world located in monsoon-controlled regions, e.g. the Indus, Ganga, Brahmaputra, Irrawaddy, Mekong, Yangtze and Huanghe, the Ganga-Brahmaputra system presently contributes about 11 % of the total sediment flux to the world oceans (Goodbred and Kuehl 1999). About 30 % of the annual sediment load is accommodated within the floodplains and delta plains, ~40 % remains in the sub-aqueous delta and the remaining 30 % is transported to deep sea Bengal fans (Goodbred and Kuehl 1999).

4 Geomorphology and Landforms

The Indus has built a broad alluvial valley in the plains (~150 km wide on average) in the lower reaches and is bound by the Kirthar Hills to its west and Thar Desert to its east (Fig. 5a). Abandoned channels and frequent meanders of the Indus (Fig. 5b) have been mapped on both sides of the present course (Giosan et al. 2006; Gaurav et al. 2011) although most of these have now been obliterated due to large-scale agricultural activities. These palaeochannels are suggestive of a dynamic regime of the river in the past and most of these avulsions took place well upstream of the delta region between Kashmore and Sehwan (Jorgensen et al. 1993). A major westward avulsion of the Indus around Hyderabad (Pakistan) occurred in 1758–1759 that resulted in the establishment of the present-day course within the delta. The present course of the Indus is confined within the embankments and the modern floodplain widths are less than 5 km in most reaches.

The alluvial reaches of the Indus have undergone large-scale human interventions in terms of barrages and dams that have reduced the sediment discharge into the delta to only ~250 million tonnes instead of 300–675 million tonnes under natural conditions (Milliman et al. 1984). This has impacted the development of the Indus Delta significantly and has pushed the apex of the delta from its Holocene position between Hyderabad (Pakistan) and Sehwan to a much downstream position. The delta extends to the east into the Rann of Kachchh and forms an extensive mudflat that is often inundated during monsoons.

The rivers draining the Ganga Plains display significant geomorphic diversity in east-west as well as north-south transects (Fig. 6a) manifested as variability of fluvial

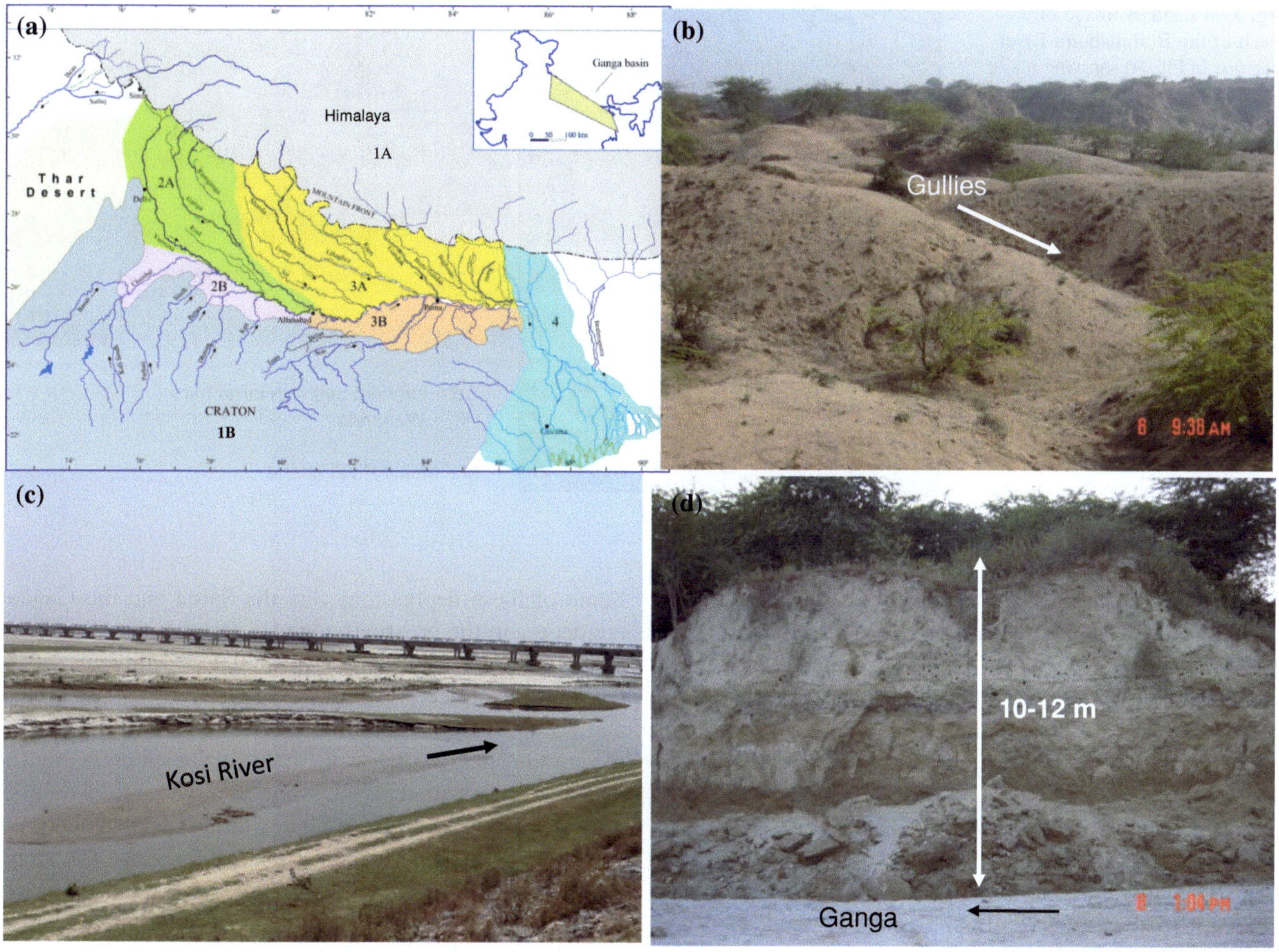

Fig. 6 **a** Genetic classification of the Ganga Plains (after Tandon et al. 2008) showing major geomorphic entities (*1A*, *1B*—Himalayan and cratonic hinterlands; *2A*, *2B* Western Ganga Plains divisible into Himalayan sources and cratonic sourced, *3A*, *3B*—Eastern Ganga Plains divisible into Himalayan-sourced and cratonic-sourced, *4*—Lower Ganga Plains and deltaic plains. **b** Dissected topography (badlands) in the Yamuna floodplain at Kalpi in WGP (*2B*). **c** Flat aggradational channel belt of the Kosi at Nirmali in EGP (*3A*). **d** Incised valley of the Ganga at Bithur (Kanpur) in the WGP (*2A*)

processes, spatial distribution of different geomorphic units and frequency of geomorphic elements (Sinha et al. 2005; Singh 2007b; Tandon et al. 2008). The northern and southern plains are influenced by two distinct catchments, the high elevation ranges of the Himalaya in the north and low-elevation cratons in the south respectively, the former resulting into piedmont plains and intermontane valleys and the latter in flat alluvial plains which are highly dissected (Fig. 6b). In the eastern Ganga Plains such as north Bihar, low stream power combined with higher sediment supply has resulted in less prominent valleys, frequent avulsion of rivers, and the inundation of large areas during monsoon floods. Many of these rivers such as the Kosi and the Gandak have formed large fans separated by interfan areas and flat aggradational channel belts (Fig. 6c). On the other hand, high stream power and lower sediment supply in the western Ganga Plains have resulted in incised valley systems (Fig. 6d) manifested as narrow valleys and wide interfluves. Such geomorphic diversity is in turn related to higher uplift rates and higher precipitation regimes in the hinterland of the eastern plains that not only result in higher sediment production but also in greater mobilization of sediments into the plains.

The Brahmaputra has a highly braided-anabranching channel in the plains of Assam and Bangladesh, marked by the presence of numerous mid-channel and lateral bars and islands, locally known as *chars* (Fig. 7a). Most of these bars are submerged during high flows and the river attains exceptional widths of the order of 18–20 km. The north-eastern region, especially the floodplains of the Brahmaputra, is dotted with a large number of wetlands or *beels*, which have tremendous ecological significance as unique habitats for an exquisite variety of flora and fauna. The *beels* function as floodwater retention basins and traditional fisheries. Over 3500 such wetlands have been identified in Assam, of which 177 are more than 100 ha in size. One of the most distinctive

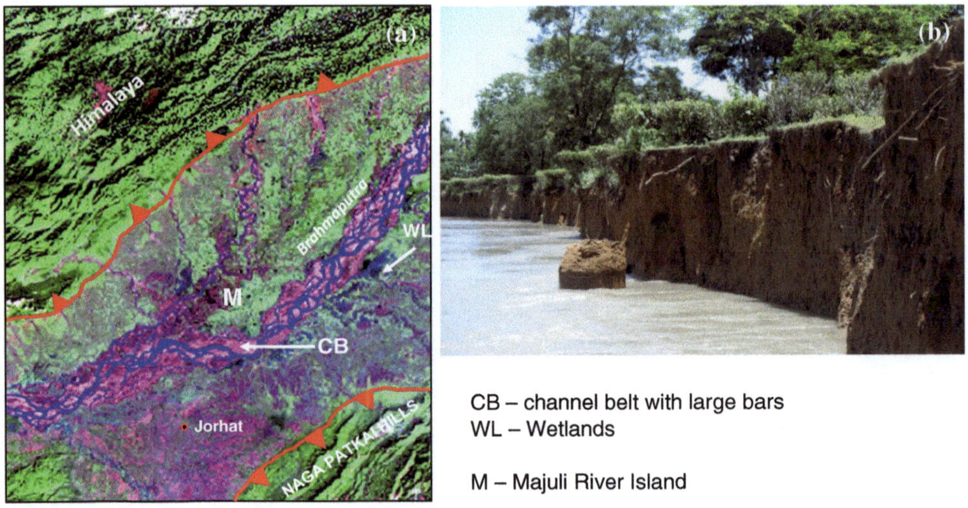

Fig. 7 **a** Landsat image of a reach of the Brahmaputra River (see *box* in Fig. 4) showing major geomorphic features. **b** Severe bank erosion along Brahmaputra River at Rohomaria, ~20 km NE of Dibrugarh

geomorphic elements of the Brahmaputra system are large riverine islands (Fig. 7a), some of which are more than 100 years old. Locally called '*Majuli*' (meaning land locked large island), these islands differ from the usual sand bars in terms of their size and evolutionary history. Most reaches of the Brahmaputra are prone to large-scale channel dynamics and bank erosion (Fig. 7b).

The Ganga-Brahmaputra (G-B) Delta (Fig. 8) is considered as the largest active delta in the world formed by frequent avulsion of both Ganga and Brahmaputra over a time scale of hundreds of years. The channel courses of the Ganga and the Brahmaputra Rivers were separate until about 200 years ago; the Brahmaputra first occupied its present position as late as 1830 and has switched between the present course and an eastern course (Meghna River) more than once in the historical time period (Goswami 1985). Systematic geomorphic mapping of the G-B Delta has distinguished three major units (a) lower floodplain and delta margin (b) upper delta plain and (c) lower delta plain. These geomorphic units show significant differences in terms of sedimentary facies and stratigraphic development (Goodbred and Kuehl 2000). One of the interesting features is the development of thick estuarine deposits and the persistence of intertidal facies in the deltaic successions which indicates that sediment supply to the delta system has been sufficient to infill accommodation created by the rapid sea-level rise (Sinha and Sarkar 2009).

5 Basin Fills and Alluvial History

Geophysical surveys together with drilling have revealed that the IGB basement is uneven and cut up by faults (Sastri et al. 1971). This has resulted in the formation of individual depressions separated by ridge like features or 'highs'. Some of these depressions like the Sarda and the Gandak are deeper in the proximal zone i.e. towards the Himalayan Mountain Front and have accumulated sediments up to 6–7 km thick. Deep drilling in the Ganga Basin by the Oil and Natural Gas Corporation (ONGC) of India, allowed the recognition of Siwalik rocks at many locations below the Ganga Plains sediments. Despite the large thickness of alluvial sediments on these plains, little is known about the sub-surface Quaternary sediments of the plains. The stratigraphic units include the Older Alluvium (Banda, Varanasi, and Bhangar) and the Newer Alluvium (Bhur and Khadar) of earlier classifications (Table 2). The Bengal Plains constitute the eastern extreme of the Ganga Plains. In its eastern side, the Lalmati Terrace is ~30 m a.s.l. The Barind Tract lies in the north and has three topographic levels. The Madhupur Terrace is made up of reddish brown sediments with abundant ferruginous/calcareous concretions that are similar to the Bhangar of the Ganga Plains.

In the last few years, the shallow subsurface in the southern part of the Ganga Plains (Ganga-Yamuna-Betwa) has been studied using a combination of resistivity surveys and shallow sediment coring (Sinha et al. 2007, 2009). Based on these studies, it has been established that: (1) the Ganga has been near to its present location since at least 30 ka (2) the Ganga Valley and the interfluve to the south between the Ganga and Yamuna have existed in their spatial domains for at least tens of thousands of years (3) thick wedges of red feldspathic sand and gravel underlie much of the southern foreland basin at shallow depth where the uppermost red feldspathic strata in the Kalpi (on Yamuna) section have yielded an age of ~119 ka (Gibling et al. 2005) (4) widespread fluvial activity took place in Marine Isotope Stages (MIS) 5 and 3 in several parts of the basin (Singh 2007b; Srivastava et al. 2003), and (5) the fluvial aggradational events are bounded by discontinuities marked

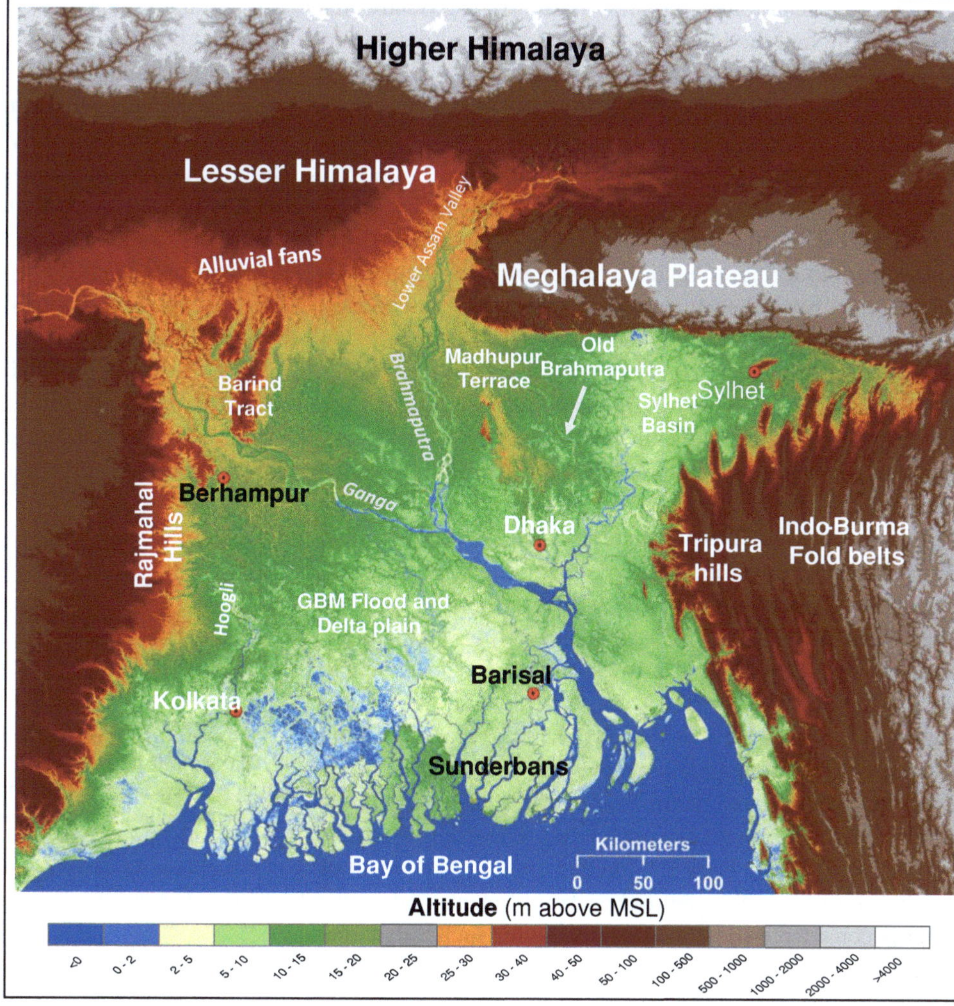

Fig. 8 Digital elevation map of the lower Brahmaputra and Ganga Basins showing major landforms

Table 2 Comparative morphological characteristics of the three major drainage systems

Basinal setting	Indus	Ganga		Brahmaputra
		Western Ganga Plain	Eastern Ganga Plain	
Proximal	Piedmont sub-piedmont	Piedmont (Bhabhar[a])	Piedmont (Bhabhar[a])	Piedmonts bordering various thrust sheets (Himalayan, Mishimi, Naga-Haflong)
Medial	Meandering-Braided	Sandy-Braided, locally sinuous reaches	Sandy-Meandering-Braided	Sandy-Braided-anabranching channels Southern tributaries highly sinuous
	Alluvial plain	Valley-interfluve (*Bhangar-Khadar-Bhur*[b–d]) Incised channels; Badland development in interfluves of cratonic rivers	Dominant fan-Interfan setting Megafans, Avulsive channels common e.g. Kosi, Baghmati	Wide channel belts (~20 km), wetlands (*beels*), and large bars and alluvial islands (*majulis*), no large alluvial fans possibly due to limited space
Distal	Deltaic plain		G-B deltaic plain	G-B deltaic plain, relict terraces (Barind tract, Madhupur surface, Lalmai surface)

[a] *Bhabhar*: Small fans and cones of gravelly deposits that form an apron along the frontal margin of the foothills and the zone of the mountain front

[b] *Khadar* (Newer Alluvium): Grey micaceous fine to medium-grained sand with clay intercalations and terraced alluvium—grey to light brown, quartz feldspathic fine grained sands

[c] *Bhur sands*: 3–10 m thick fine-grained aeolian sand accumulations

[d] *Bhangar* (Older Alluvium): G-B Ganga Brahmaputra

by degradational relief surfaces, palaeosols, groundwater cementation, and the local development of lakes, ponds and aeolian (*bhur*) deposits (Gibling et al. 2005).

6 Major Concerns and Issues

All three basins, the Indus, Ganga and Brahmaputra, are known for rapid migration and avulsion of their channels as manifested by several palaeochannels and abandoned meanders on older floodplain surfaces. The sediment-laden water of the Indus has created many water resource management problems, mainly in the upper Indus basin, and the construction of embankments on both sides of the river, dams and barrages have worsened the situation. The Indus flood of 2010 was triggered by unusual rainfall but the influence of engineering structures such as barrages and embankments compounded the problem due to accelerated aggradation of the river bed and reduced the carrying capacity of the channel (Gaurav et al. 2011). The eastern parts of the Ganga Plains are well-known for the dynamic rivers draining the region. Apart from the Ganga, several tributaries such as the Kosi, Gandak and Baghmati draining the north Bihar Plains have created havoc in the region due to their hyperavulsive nature and extensive flooding (Gole and Chitale 1966; Jain and Sinha 2004). The Brahmaputra is notorious for severe bank erosion due to rapid movement of its thalweg and once again excessive sediment load of the river is considered to be the main cause for such a dynamic regime (Goswami 1985).

Continuous interventions and excessive extraction of water for human use have impacted all three river basins. Large tracts of the Ganga Basin suffer from water quality problems due to industrial and municipal effluents being discharged into the river. At the same time, the upper reaches of the Ganga River have undergone loss in biodiversity due to dams and barrages. Deforestation and land cover changes are common problems in all three basins and recent results in the upper Ganga Basin (Yu et al. 2007) indicate that a large part of forest and barren land has been converted to agriculture areas during the last two decades.

7 Summary and Conclusions

The Indus-Ganga-Brahmaputra Plains are a part of the Himalayan foreland basin that is drained by three mountain-fed large river systems of the world. The plains are marked by high geomorphic diversity as they are influenced variably by different geologic, tectonic and climatic regimes over wide tracts that extend along the length of the Himalayan Arc and the Himalayan syntaxial bends. The Indus is characterized by a narrow alluvial plain and a dynamic regime; the Ganga Plains in the western part are marked by the presence of valley-interfluves formed over several tens of thousands of years and by megafans in the eastern part; the Brahmaputra system being the wettest part of the plains is marked by very wide channel belt with the development of large bars, exceptionally large alluvial islands, and large number of wetlands. Several parts of the plains are impacted severely by avulsive shifts of the drainages as well as flooding. Extended use of the plains over thousands of years by large populations has impacted the plains through many forms of anthropogenic disturbances including large-scale land use/land cover changes, pollution linked to agricultural practices, urbanization and river engineering.

References

Gansser A (1964) Geology of the Himalayas. London, Interscience Publishers, p 289

Gaurav K, Sinha R, Panda PK (2011) The Indus flood of 2010 in Pakistan: a perspective analysis using remote sensing data. Nat Hazards 59:1815–1826

Gibling MR, Tandon SK, Sinha R, Jain M (2005) Discontinuity-bounded alluvial sequences of the southern Gangetic plains, India: aggradation and degradation in response to monsoonal strength. J Sediment Res 75:373–389

Giosan L, Constantinescu S, Clift PD, Tabrez AR, Danish M, Inam A (2006) Recent morphodynamics of the Indus delta shore and shelf. Cont Shelf Res 26:1668–1684

Gole VC, Chitale VS (1966) Inland delta building activity of Kosi river. J Hydraul Div ASCE 92:111–126

Goodbred SL, Kuehl SA (1999) Holocene and modern sediment budgets for the Ganges-Brahmaputra river system: evidence for high stand dispersal to flood-plain, shelf and deep-sea depocentres. Geology 27(6):559–562

Goodbred SL, Kuehl SA (2000) Enormous Ganges-Brahmaputra sediment discharge during strengthened early Holocene monsoon. Geology 28(12):1083–1086

Goswami DC (1985) Brahmaputra river Assam, India: physiography, basin denudation and channel aggradation. Water Resour Res 21:959–978

Hovius N (1998) Controls on sediment supply by larger rivers. In: Shanley KW, McCabe PJ (eds) Relative role of eustasy, climate, and tectonism in continental rocks, vol 59. Special Publication, Society for Sedimentary Geology (SEPM) pp 3–16

Inam A, Clift PD, Giosan L, Tabrz AR, Tahir M, Rabbani MM, Danish M (2007) The geographic, geological and oceanographic setting of the Indus River. In: Gupta A (ed) Large rivers: geomorphology and management. Wiley, New York, pp 333–346

Jain V, Sinha R (2004) Fluvial dynamics of an anabranching river system in Himalayan foreland basin Baghmati river, north Bihar plains, India. Geomorphology 60:147–170

Jorgensen DW, Harvey MD, Schumm SA, Flam L (1993) Morphology and dynamics of the Indus river: implications for the Mohen Jo Daro site. In: Shroder JF Jr (ed) Himalaya to the sea. Routledge, New York, pp 288–326

Milliman JD, Meade RH (1983) World-wide delivery of river sediments to the oceans. J Geol 91:1–22

Milliman JD, Quraishee GS, Beg MAA (1984) Sediment discharge from the Indus river to the ocean: past, present and future. In: Haq BU, Milliman JD (eds) Marine geology and oceanography of Arabian Sea and Coastal Pakistan. Van Nostrand Reinhold, New York, pp 65–70

Sastri VV, Bhandari LL, Raju ATR, Dutta AK (1971) Tectonic framework and subsurface stratigraphy of the Ganga Basin. J Geol Soc India 12:222–233

Singh SK (2007a) Erosion and weathering in the Brahmaputra system. In: Gupta A (ed) Large rivers: geomorphology and management. Wiley, Chicheste, pp 373–393

Singh IB (2007b) The Ganga river. In: Gupta A (ed) Large rivers: geomorphology and management. Wiley, New York, pp 347–372

Sinha R, Sarkar S (2009) Climate-induced variability in the Late Pleistocene-Holocene fluvial and fluvio-deltaic successions in the Ganga plains, India. Geomorphology 113:173–188

Sinha R, Jain V, Prasad Babu G, Ghosh S (2005) Geomorphic characterization and diversity of the fluvial systems of the Gangetic plains. Geomorphology 70:207–225

Sinha R, Bhattacharjee P, Sangode SJ, Gibling MR, Tandon SK, Jain M, Godfrey D (2007) Valley and interfluve sediments in the southern Ganga plains, India: exploring facies and magnetic signatures. Sed Geol 201:386–411

Sinha R, Gibling MR, Kettanah Y, Tandon SK, Bhattacharjee PS, Dasgupta AS, Ghazanfari P (2009) Craton-derived alluvium as a major sediment source in the Himalayan foreland basin of India. GSA Bulletin 121(11/12):1596–1610

Srivastava P, Singh IB, Sharma M, Singhvi AK (2003) Luminescence chronometry and late Quaternary geomorphic history of the Ganga plain India. Palaeogeogr Palaeoclimatol Palaeoecol 197:15–41

Tandon SK, Sinha R, Gibling MR, Dasgupta AS, Ghazanfari P (2008) Late Quaternary evolution of the Ganga Plains: myths and misconceptions, recent developments and future directions. Memoir Geol Soc India 66:259–299

Valdiya KS (2010) The making of India: geodynamic evolution. Bloomsbury Publication, New Delhi

Yu H, Joshi PK, Das KK, Chauniyal DD, Melick D, Yang X, Xu J (2007) Landuse/cover change and environmental vulnerability analysis in Birhi Ganga sub-watershed of the Garhwal Himalaya, India. Tropical Ecology 48:241–250

The Indian Peninsula: Geomorphic Landscapes

Vishwas S. Kale and R. Vaidyanadhan

Abstract

The Indian Peninsula is a collage of many terranes, where ancient rocks, denudational surfaces and rivers predominate. The peninsula displays amazing diversity of landscape features, from tiny koppies to the ~1,500-km long Great Escarpment of India (the Western Ghat/*Sahyadri*). By and large, bedrock landforms and partially to deeply weathered rocks dominate the landscape. Following the Gondwanaland breakup, the landscape of the Indian peninsula has been chiefly fashioned by divergent weathering and differential erosion. Two types of terrains are conspicuous—the granite-gneissic landscape characterized by undulating plains dotted with koppies, boulder inselbergs and bornhardts in the south, and the Deccan Trappean landscape dominated by flat-topped, stepped hills separated by wide, open valleys in the north and west. Laterite-capped plateaux and mesas are prominent features in some parts. The peninsular rivers are rainfed and are active only during the monsoon season. Only the east-flowing larger rivers have developed deltas and deltaic plains. Although the human history of the Peninsula covers a span of nearly 4,000 years, human presence within it is known for a greater part of the Quaternary.

Keywords

Peninsula • Cratons • Granite-gneissic landscape • Traps landscape • Duricrust • Plateaux • Western Ghat • Gorges • Deltas

1 Introduction

The Indian Peninsula (Fig. 1), also called the Indian Shield, is one of the ancient and stable parts of the Earth's crust, where ancient rocks, denudational surfaces and rivers predominate. The palimpsest landscape of the triangular-shaped peninsula is the result of its long, chequered geological history. The cratonic landscape has been exposed to weathering and fluvial erosion for many millions of years under long periods of relative tectonic stability compared to those of the Himalayan terrain. Major geological and tectonic events such as separation from Gondwanaland and migration of the Indian landmass through different latitudinal (climatic) zones, rifting along the eastern and western margins, Deccan volcanism, and superimposed effects of climatic and tectonic vicissitudes have all left an indelible imprint on this fragment of the Gondwana landscape. Post-Neogene differential uplift and reactivation of faults linked to Himalayan Orogeny and the onset of monsoon conditions over south Asia (~8–10 Ma) have also played a key role in shaping the scenery of this ancient landmass.

V. S. Kale (✉)
Department of Geography, University of Pune, Pune 411007
India
e-mail: vskale.unipune@gmail.com

R. Vaidyanadhan
B 1203 Neel Sidhi Atlantis, Plot 49/1 Sector 19A, Nerul (E), Navi Mumbai 400706, India
e-mail: rvaidya31@gmail.com

V. S. Kale (ed.), *Landscapes and Landforms of India*, World Geomorphological Landscapes,
DOI: 10.1007/978-94-017-8029-2_6, © Springer Science+Business Media Dordrecht 2014

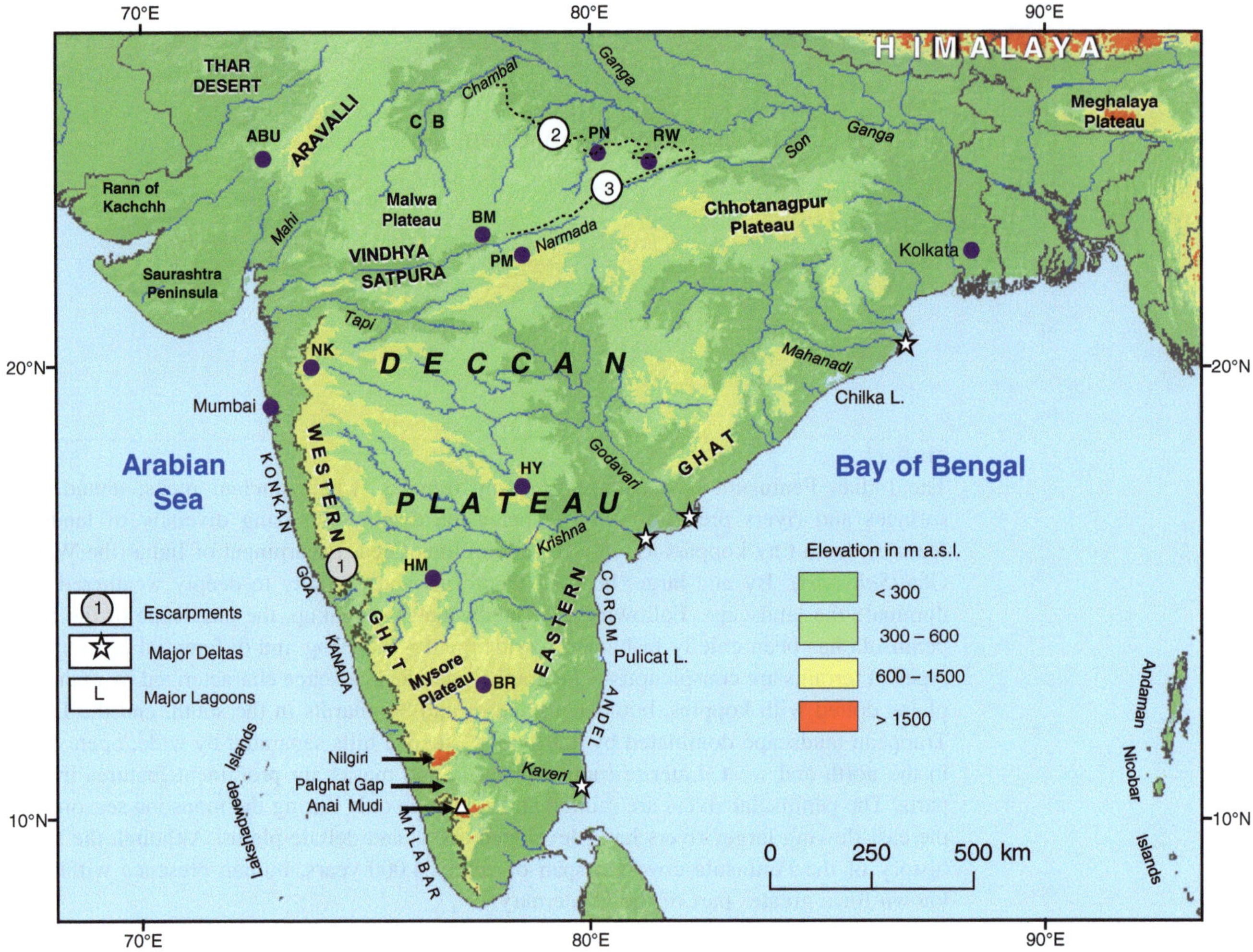

Fig. 1 The Indian Peninsula and its major geomorphic features. *1* Western Ghat Escarpment, *2* north-facing Kaimur Scarp (*dashed line*) overlooking the Yamuna–Ganga Plains, 3 Scarp (*dashed line*) overlooking the Narmada-Son trough. Some places mentioned in the text are also shown. *ABU* Mt. Abu, *PN* Panna, *RW* Rewa, *BM* Bhimbetka, *PM* Pachmarhi, *NK* Nasik, *HY* Hyderabad, *HM* Hampi, *BR* Bengaluru, *CB* Chambal Badlands

The Indian Peninsula is a collage of many terranes (such as greenstone-gneiss-granite, volcanic and sedimentary) and displays amazing diversity of landscape features owing to spatial differences in lithology, geological history, topography, and hydrology. The terrain presents a landscape of wide, open valleys, bedrock gorges, series of plateaux, pediments or rocky surfaces, residual hills, inselbergs and hill ranges. Nearly 1.2 million km^2 area of the peninsula lies over 300 m a.s.l. (Fig. 1). South of the Palghat Gap, the Western Ghat (*Sahyadri*) contains Anai Mudi (2,695 m a.s.l.), the highest point in the Peninsula. By and large, bedrock landforms and partially to deeply weathered rocks dominate the scenery. Divergent weathering, mass wasting and differential erosion during its long geological history have primarily fashioned the landscape. The mean Mesozoic denudation rates across the southern Dharwar Craton have been in the range of 14–20 m/million years, and the modern and long-term denudation rates are not very dissimilar (Gunnell 1998a).

2 Geology and Geological History

The Indian Peninsula is made up of a number of Precambrian cratonic blocks bordered by rifts and Proterozoic fold belts. Following lithologies characterize the Peninsula (Fig. 2)—(1) the Archean (primarily gneisses, granites and volcanics) representing some of the oldest rocks on the Earth (~2.5–4.0 Ga), including metamorphosed sedimentary rocks of the Dharwar Supergroup (2) the Cuddapah and the Vindhyan Supergroups (Proterozoic) containing quartzites, sandstones, slates, shales, limestones, volcanics and a few

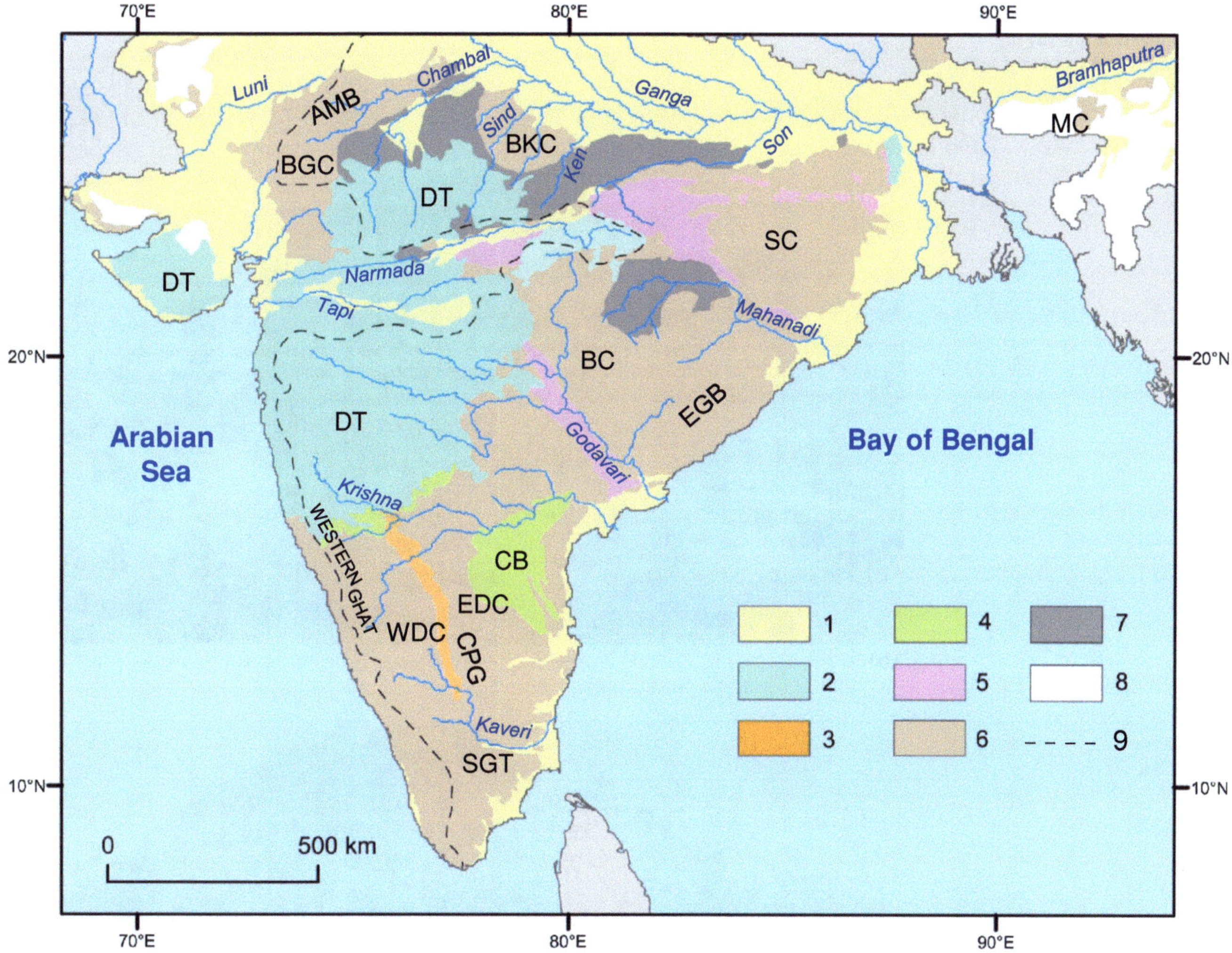

Fig. 2 Generalized geological map of the Indian Peninsula. *1* Quaternary deposits, *2* Deccan and Rajmahal Traps, *3* Closepet Granite (CPG), *4* Cuddapah and Kaladgi Supergroups, *5* Gondwana Rocks, *6* Peninsular gneisses (undifferentiated), *7* Vindhyan and Chattisgarh Supergroups, *8* Others rock types, *9* major water divide. *BC* Bastar Craton, *BKC* Bundelkhand Craton, *CB* Cuddapah Basin, *DT* Deccan Traps Basalts, *EDC* Eastern Dharwar Craton, *SC* Singbhum Craton, *SGT* Southern Granulite Terrain, *WDC* Western Dharwar Craton, *MC* Meghalaya Craton, *BGC* Banded Gneissic Complex, *AMB* Aravalli-Delhi Mobile Belt, *EGB* Eastern Ghat Mobile Belt. Map generalized from GSI's geological map of India

metamorphosed patches (3) the Gondwana Supergroup of rocks largely confined to rift grabens of the Godavari, Mahanadi and Damodar Valleys, and (4) the Deccan Volcanic Group of rocks, in the western and central India.

As an evolved landmass, no part of the Peninsular India is considered to be older than mid-Jurassic (~ 170 Ma). In the Mesozoic, the Peninsula formed a part of Gondwanaland and was sandwiched between Africa and Antarctica, before it broke away from the Gondwanaland and drifted northwards to ultimately collide with the Eurasian landmass in early Cenozoic (Gunnell and Radhakrishna 2001).

Both the eastern and western continental margins of India are passive margins, where volcanic, tectonic and orogenic activity is relatively subdued. It is believed that the eastern margin of the peninsula was created by the breakup of India from Antarctica–Australia around early Cretaceous (~ 130 Ma). In comparison, the western continental margin is relatively younger and was formed in two stages—i) breakup of Greater India–Madagascar (Malagasy) at ~ 80–90 Ma, and (ii) separation of India–Seychelles ~ 65 Ma. During the northward movement, the Indian Plate passed over the Réunion hotspot, which caused widespread extrusion of thick piles of lava flows. The emplacement of Deccan flood basalts at ~ 65 Ma was pene-contemporaneous with the rifting and separation of the Seychelles micro-continent (Widdowson 1997).

The Peninsula is presently made of relict mountains, block mountains, plateaux, cuestas, escarpments and valleys. Whereas the Aravalli (Rajasthan) and Mahendragiri Hills (Odisha) are examples of relict and residual mountains,

Fig. 3 A bornhardt (**a**) and a boulder mantled hill (**b**) in charnockites in Krishnagiri area, Tamil Nadu

the Satpura, the Nilgiri, the Shevaroy and the Biligirirangan Hill Ranges are examples of horst or block mountains. The Saurashtra Horst is also a large fault block. Evidence of vertical movements is also present in the form of rift valleys of Narmada and Tapi, the Godavari, Mahanadi and Damodar River Grabens as well as Kachchh and Khambhat (Cambay) Grabens. The Deccan Plateau is the largest plateau in the Peninsula. Apart from Western Ghat Escarpment, the two other prominent escarpments within the Peninsula include the north-facing Kaimur Scarp of the Vindhyan sandstones and quartzites overlooking the Yamuna–Ganga Plains, and the south-facing scarp flanking the Narmada-Son fault-trough (Fig. 1).

3 Principal Geomorphic Terrains

3.1 Landforms of the Granite and Gneissic Terrain

Granitic outcrops and boulder-strewn hills (called nubbins) rising from the undulating plains create a dramatic landscape in the cratonic parts of the Peninsula. The inselbergs, bornhardts (Fig. 3), koppies (kopjes), surrounded by mantled or bare pediments, whether formed by backwearing of hills or exhumation of the weathering fronts, dominate many areas underlain by granite and gneisses (i.e. Peninsular gneisses) as

Fig. 4 Block- and boulder-strewn granite nubbins at Hampi (Photo courtesy of H. R. Mungekar)

well as charnockites (chapter "Granite Landforms of the Indian Cratons" this monograph). Charnockites are most resistant to weathering and erosion and hence form prominent highlands in south India. All the Indian cratons namely, Dharwar, Bastar, Singhbhum, Bundelkhand, and Meghalaya as well as the Banded Gneissic Complex, Chhotanagpur Gneiss Complex and the Southern Granulite Terrain (Fig. 2) have sizable areas under granite and granitoid rocks favourable for the formation of such distinctive landforms. Sravanabelagola, Madhugiri, Savandurga, Shivagange, Nandi, Ramnagara and Kabbala Durg Hills near Bengaluru (Ranganathan and Jayaram 2006), Moula Ali and Bhongir Hills near Hyderabad, and Narlai in the heart of the Aravalli Hills are some examples of domed inselbergs or bornhardts. In Rajasthan, inselbergs of Jalore, Siwana and Sirohi are well known. Elsewhere, the hills display characteristic rounded boulders (core stones) and knob-shaped outcrops formed as a result of spheroidal weathering and erosion of fractured granites. Boulder-strewn hills (Fig. 3) and castle koppies are also common, such as near Hampi (Fig. 4) and Hyderabad. Another example is of the Barabar and Nagarjuni Hills in Bihar. Gigantic blocks forming bizarre shapes are also present over the Mt. Abu (a granitic inselberg) and interesting sculpted forms in granite are observed at Sendra near Pali in Rajasthan. Minor forms include tafoni, split rocks, flakes and spalls, etc.

3.2 Landforms of the Basaltic Terrain

Presently exposed over an area of about half a million square km, the Deccan Basalt Province (DBP) or the Deccan Traps is the second largest basalt province in the world (next to Paraná Province in Brazil). The DBP is thought to have been originally almost three times as extensive as at present. The extrusion of Deccan Traps lavas $\sim$65 Ma, not only partly buried the pre-Cenozoic low-relief landscape of the Dharwars, the central Indian cratons and the Vindhyan rocks (Sheth 2007), but also created a completely new land surface over a vast area that was gradually sculpted by tropical weathering and differential fluvial erosion as India passed through different latitudinal (climatic) zones. Along the present east–northeastern margin of the Deccan Traps, Vindhyan rocks have been exhumed in a few places owing to the erosion and removal of the overlying Deccan Traps flows (Choubey 1971).

The keynote of Trappean landscape (Fig. 5) is its flat-topped hills and ridges, separated by v-shaped or box-shaped valleys (Kale and Rajaguru 1988). Hillslopes display steps with rocky benches and cliffs, reflecting control by resistant lava flows (Figs. 5 and 6). This landscape form is generally referred as "treppen" (step-like). The thickness of the resistant lava flows determines the height of rocky cliffs. Spectacular high cliffs in compound lava flows can be seen around Nasik (Fig. 6a). Erosion along fractures generally gives rise to jagged peaks. Once the harder top flows are stripped, the flat-topped hills become rounded and/or conical-shaped (Fig. 6c). In the rainshadow of the *Sahyadri*, the differences in lava flows are enhanced and pediments occur at the foot of the hill ranges, at times covered by late Quaternary colluvium. In the more humid parts, the landscape is more dissected, with narrow valley floors. Thick weathering profiles and/or ferricrete duricrusts occur at a number of places in the Western Ghat zone. The soils developed from the Deccan Traps basalts are vertisols, commonly known as

Fig. 5 The box-shaped valley of the upper Krishna River in the Deccan Traps Region. *M* Mahabaleshwar, *L* laterites, *K* source of the Krishna River on the edge of Western Ghat, *D* backwaters of Dhom Dam on Krishna (Photo courtesy of H. R. Mungekar)

Fig. 6 **a** Spectacular high cliffs in resistant compound basalt flows at Vani near Nasik **b** The Western Ghat Escarpment as seen at Mahabaleshwar **c** View of the flat-topped hill, conical shaped hill and wide valley of Vel River near Rajgurunagar, north of Pune

Fig. 7 A view of the Ajanta Caves excavated in the gorge wall of the Waghur/Waghora River. The caves, listed as a World Heritage Site, are excavated in compound lava flows. The Buddhist cave monuments date from ~2nd century BCE to 6th century CE (Photo courtesy of Francesca Lugeri)

black soils. The black soils are widespread in the drier parts and generally occur in low-lying areas.

Within the Deccan Traps, most of the renowned early historical Buddhist, Jain and Hindu rock-cut caves—such as Ajanta, Ellora and Karla—are excavated in the high cliffs of the compound lava flows (Figs. 7 and 8).

3.3 Landforms of the Sandstone and Limestone Terrains

Landforms developed in quartzites, sandstones, shales and limestones (Figs. 9, 10, 11 and 12) are conspicuous in areas underlain by rocks of Vindhyan and Cuddapah Supergroups (Fig. 2) as well as in Kachchh (Jurassic-Cretaceous). Resistant sandstones cap hills and form plateaux, mesas and high scarps (for example, the Kaimur Scarp; Fig. 1). Some of most impressive landscapes are observed in the Pachmarhi (Venkatakrishnan 1984) and Rewa areas in Madhya Pradesh. The Gondwana rocks within the half-grabens usually present subdued ridge (cuesta) and valley topography. Erosion by sub-aerial processes at times produces strange, weird shapes, mushroom-like structures, alcoves and rock shelters. Some of these can be seen around Bhimbetka near Bhopal, which is well known for its pre-historic rock paintings.

Typical karst topography is not known from any part of the Peninsula, although modest karstic features and caves in limestones and dolomites have been reported from the Cuddapah Basin, the Rewa-Maihar-Satna area in Madhya Pradesh and in Jaintia Hills in Meghalaya. Several caves with active and inactive speleothems are known from these areas. Some of the karst caves, which are popular tourist destinations, are Belum, Borra and Kurnool Caves in Andhra Pradesh, Gupteswar and Dandak Caves in Odisha, and Mawsmai Caves near Cherrapunji. The Kurnool Caves have yielded rich assemblages of fossils and Palaeolithic stone tools.

3.4 Duricrusted Landforms

Iron-rich crusts or laterites and aluminous (bauxitic) duricrusts occur in many parts of Peninsular India and form distinctive elevated flat-topped landforms such as plateaux, tablelands and mesas (chapter "The Laterite-Capped Panchgani Tableland, Deccan Traps" this monograph). Laterite-capped mesas are usually the product of circum-denudation (erosion from all sides).

Laterite cappings are conspicuously present along the east and west coasts, and also over the Western and Eastern Ghats as well as inland (Widdowson and Gunnell 1999). Laterites/bauxites have developed over different lithologies such as the Deccan Traps basalts (Amarkantak, Mahabaleshwar), the Rajmahal Traps (Santhal Parganas), the Vindhyan rocks (Semaria), granite and gneiss (Chiro, Richiguda), schists (Kharagpur), khondalites (Anantagiri, Panchpatmali, Gandhamardan), and charnockites (Shevaroy, Gopalswami Betta, Palni Hills). Although the duricrusts can be found from close to sea level (Goa) to >2,400 m a.s.l. (Nilgiri), the high-level laterites/bauxites over the Deccan Plateau are usually, but not exclusively, confined to areas above 900 m a.s.l.

Fig. 8 A rock-cut cave temple at Ellora excavated in compound basalt flows. The caves and monuments at Ellora date from 6–7th century to 11–12th century CE. Ellora is listed as a World Heritage Site (Photo courtesy of Francesca Lugeri)

Typically, the laterite cap (1–20 m thick) is underlain by deeply weathered kaolinized saprolite of variable thickness. Detrital laterites show lack of such saprolitic horizon, and occur at lower levels. Whereas the laterites developed over the Deccan Traps (Fig. 5) are certainly of the Cenozoic period, there is uncertainty about the age of laterites developed over the Precambrian rocks. $^{40}Ar/^{39}Ar$ ages suggest intensive lateritic weathering over Mysore Plateau between ~26 and 36 Ma (Nicolas et al. 2014). Occurrence of laterites over Vindhyan rocks, below the Deccan Traps near Sagar (Choubey 1971) indicates that some of the laterites/bauxites in the peninsula may be much older. Where thick laterite duricrusts are found outside the present humid areas, they denote inheritance from a former humid climate as in Bidar (Babechuk et al. 2014), Udgir and Jath along the eastern margin of the Deccan Traps.

4 The Western Ghat: The Great Escarpment of India

A distinctive landform associated with passive margins is present on the western continental fringe of India. This well-defined, coast-parallel, and nearly 1,500-km long great escarpment is known as the *Sahyadri* (Sanskrit for "benevolent mountain") or the Western Ghat (Fig. 1). The escarpment forms the western edge of the Deccan Plateau and cuts across Precambrian gneisses, granites, and charnockites in the south (roughly south of 16°N latitude) and Cretaceous-Eocene Deccan basalts in the north (Fig. 6b). Despite significant lithological and structural variations there are some marginal deviations in the configuration and morphology of this stupendous escarpment. The Western Ghat crest coincides with the major waterdivide between the Arabian Sea and the Bay of Bengal drainage, except where the plateau drainage has been diverted westward by river capture (e.g. Sharavathi River near Gersoppa). The Palghat Gap (Fig. 1) is the only prominent break in the escarpment, which coincides with the Palghat–Kaveri Shear Zone. The Western Ghat is one of the world's major biodiversity hotspots. Due to high altitude, cooler climate, and breathtaking views, the Western Ghat zone is dotted with popular tourist resorts, medieval forts, and ancient temples (Kale 2010).

The precipitous western edge (~1-km high) is an assemblage of valley heads and finger-like ridges. Other associated landforms are plateau outliers, deep gorges, beheaded plateau valleys, and laterite plateaux or tablelands (Kale 2010). Three major east-flowing rivers and their tributaries, namely Godavari, Krishna and Kaveri, and numerous short, steep, coastal rivers have their origins here. The stupendous escarpment was formed by rifting at the time of the separation of the Seychelles micro-continent in the late Cretaceous (Gunnell and Radhakrishna 2001). According to Valdiya (2001) the scarp recession is mainly

Fig. 9 **a** Channel of the Khari River near Bhuj, Kachchh, incised into late Cretaceous sandstones belonging to the Bhuj Formation. **b** Limestone terrain drained by Gosthani River near Borra Cave, Araku Valley

aided by strike slip faults in an en echelon disposition. Although it is generally agreed that the rifted margin was driven inland by erosion due to the higher elevation of the interior (Radhakrishna 1993; Widdowson 1997), there are differences of opinion about the mode and rate of scarp recession (Kale 2010).

5 Interior Plateaux and High-level Surfaces

The geomorphology of much of the Indian Peninsula is dominated by the presence of number of plateaux and high-level surfaces at different elevations (100 to >1,000 m a.s.l.),

Fig. 10 The Marble Canyon on the Narmada River downstream of Dhuandhar Falls at Bedhaghat near Jabalpur. The gorge is cut in nearly vertically dipping dolomite rocks

sometimes occupied by ferruginous or aluminous duricrusts. The Deccan (meaning south) Plateau (Fig. 1), bounded by the Western and Eastern Ghats, constitutes the largest geomorphic sub-unit of the Indian Peninsula. This Plateau is made up of several plateaux, apart from valleys and hill ranges. Whereas some plateaux extend over large areas, others are highly dissected and exist as fragments or outliers with accordant heights. Some of the outstanding plateaux are (Fig. 1)—Malwa (500–600 m a.s.l), Chhotanagpur (600–1,100 m a.s.l.), Mysore (900–1,200 m a.s.l.), and Meghalaya (900–1,900 m a.s.1.). Some plateaux show multiple surfaces at different levels (Vaidyanadhan 1977, 2002). For example, the Chhotanagpur Plateau displays a series of plateaux at 1,100 m (Pat, with laterite-capped mesas), 600 m (Ranchi and Hazaribagh), and 300 m a.s.l. (Kodarma). In addition, there are a number of modest-sized plateaux such as Rewa, Bijawar-Panna, Rohtas, Pachmarhi, Basatar, Kalahandi, Telangana, Coimbatore, etc. Laterite-capped plateaux constitute major coastal landforms in Konkan, Goa, Kanada (Kanara) and Malabar. The Kas Plateau (Bamnoli Range) in Maharashtra and the Manmanhara Plateau in Karnataka are the two largest known laterite plateaux in the Western Ghat zone and on the west coast, respectively. The highest plateau in the Peninsula is the Nilgiri Plateau (1,800–2,400 m a.s.l.), which is developed within charnockites and is capped by laterites in places. Other well-known high plateaux are Amarkantak (~1,000 m a.s.l.), Mt. Abu (~1,200 m a.s.l.) and Mahabaleshwar (~1,400 m a.s.l., Fig. 5). Rivers descending from the plateau surfaces usually flow across waterfalls and via deep gorges, such as in the Rewa-Panna, Pachmarhi and Chhotanagpur regions.

The widely held view is that most of these plateaux or high-level surfaces are uplifted erosional or denudational surfaces. These surfaces have been identified as peneplains, pediplains, etchplains or planation surfaces by different workers (Vaidyanadhan 1977; Gunnell 1998b), due to the remarkable flat appearance of the plateau-tops and occurrence of isolated hills or laterites/bauxites over the surfaces. The age of these plateaux and surfaces is uncertain but generally the highest surfaces are the oldest, such as the remnant of surfaces in the Nilgiri and Palni Hills (~2,200 m a.s.l.) which are believed to be late Cretaceous-Paleocene in age (Gunnell 1998b).

6 Fluvial Landforms

The control of the tectonic fabric (NW-SE Dharwarian trend) and the eastward tilt of the Indian Plate is evident on the asymmetric drainage network of the Peninsula, particularly to the south of the Narmada-Son-Tapi Rift. Even the Gondwana grabens roughly have an eastward trend. As a result, all the major rivers flow from west to east (Fig. 1). Another view is that the easterly drainage was a consequence of regional domal uplift caused by the Deccan plume head (Cox 1989). There is evidence to suggest that some drainage re-organization has taken place after the establishment of easterly drainage due to river capture, diversion and beheading (Radhakrishna 1993).

The Peninsular rivers are characterized by broad and shallow valleys with low gradients. Only the coastal rivers, heading in the Western or Eastern Ghats are short, steep and swift-flowing. All the rivers draining the Indian Shield are monsoon-fed and, therefore, are active only during the monsoon season. Many of them are known for extraordinary floods (peak discharges in the range of 70,000–100,000 m^3/s) associated with Bay of Bengal depressions and cyclones. Bedrock reaches and erosional features are common. Both

Fig. 11 The Punasa Gorge on Narmada River downstream of Dardi Falls, near Khandwa. Quartzites of the Vindhyan Supergroup constitute the bedrock. A large boulder berm is seen in the foreground. Some of the boulders are >5 m in diameter

Fig. 12 The Krishna Gorge at Srisailam through Nallamala Hills of the Eastern Ghat. The rocks exposed on gorge walls are sub-horizontally-bedded quartzites belonging to the Cuddapah Supergroup

gravel-bed and sand-bed rivers occur. Only the east-flowing larger rivers have developed deltas and deltaic plains (Fig. 1) (Vaidyanadhan 2002).

Almost all the rivers are in a phase of incision. Rivers such as Chambal, Narmada, Tapi, Mahi, Sabarmati, Godavari, Krishna and Kaveri are incised in bedrock or late Quaternary alluvium. Incision and formation of bank-gullies in alluvial reaches have given rise to badlands, such as along the Chambal, Narmada, Tapi and Mahi.

Rivers incised in rock have developed bedrock gorges in certain reaches, oftentimes with waterfalls at the gorge head. Examples of spectacular gorges are: the Marble Canyon (Fig. 10), downstream of the Dhuandhar Falls on Narmada, Punasa Gorge on Narmada, downstream of Dardi Falls

Fig. 13 Papikonda Gorge developed through Eastern Ghat by the Godavari River, upstream of Polavaram. The rocks exposed on gorge walls are khondalites

(Fig. 11), the Jaldurg and the Srisailam (Fig. 12) Gorges on Krishna, Papikonda Gorge on Godavari (Fig. 13), the Kaveri Canyon between Shivasamudram and Hogenakal Falls, the Satkosia Gorge on Mahanadi, the Chambal Gorge (between Chourasigarh and Kota), etc. Most of these gorges appear to be pre-Holocene in age. Some magnificent gorges and waterfalls are also seen in the Pachmarhi area and along the Bhander, Panna and Binji Scarps in central India. Some of the gorges are incised across structural and topographic highs, such as across the Eastern Ghat (by Godavari, Krishna and Pennar), Khainjua Ridges (by Son) and Biligirirangan Hills (by Kaveri) suggesting that either the rivers are antecedent or superimposed.

Late Quaternary alluvial deposits, forming river terraces, are confined to a narrow belt along river channels. The areal extent and thickness of the Quaternary alluvial fill is remarkably high in the Narmada and Tapi Basins.

7 Coastal Landforms

Some of the youngest landforms, sculpted by waves, currents and tides, can be seen along the ~5,500 km coastline of the Indian Peninsula. This is due to the fact that the sea-level reached the present position around mid-Holocene (5–6 ka). At the time of the Last Glacial Maximum (~21 ka) the global sea level was 100–120 m below that of the present day. This means that many of the older coastal features could be submerged now.

The pattern of geological outcrops, the tidal range and the supply of sediments to the coastal zone are among the three fundamental controls on the character of the peninsular coastline (chapter "The Indian Coastline: Processes and Landforms" this monograph). The west coast (Konkan to Malabar) is dominantly featured by rocky headlands, narrow bays with bay-head beaches, spits and tombolos, drowned river estuaries and extensive tidal flats. The Rann of Kachchh (Fig. 1), with an area of ~16,000 km^2, is the largest saline marshland in India. The backwaters (*kayals*) are the characteristic feature of the Malabar Coast (Fig. 1). In comparison, the east coast is characterized by extensive deltas of the Mahanadi, the Godavari, the Krishna and the Kaveri (Fig. 1), as well as straight, long beaches, beach ridges, dune ridges, spits, sand bars at the river mouths, lagoons (Chilka and Pulicat are the largest lagoons; Fig. 1) and coastal dunes. The Gulf of Mannar is a biosphere reserve best known for corals and mangroves. Wherever hills approach the coastline, sea cliffs and wave-cut platforms are also present (Fig. 14).

Due to Bay of Bengal cyclones and depressions the eastern coast is frequently impacted by storm surges and sea water flooding. The December 2004 tsunami, one of the most extreme events in recent geological history, has left an indelible impact on the coastal features along the east coast as well as on the Andaman and Nicobar Islands.

Evidence of sea level fluctuations due to eustatic changes in the sea level or local tectonics, such as abandoned sea cliffs, marine terraces and raised beaches, oyster beds and

Fig. 14 Wave-cut platform in khondalites 10 km southwest of harbor channel in Visakhapatnam on the east coast. Note the discordance between the platform and the foliation of the gneisses

coral reefs, etc. have been reported particularly from Saurashtra, Konkan, Goa and Malabar coasts, Kanyakumari, Rameswaram, Visakhapatnam, Andaman and Nicobar Islands, etc.

8 Human Impact

In spite of the fact that recent dating of Acheulian artefacts indicates the presence of human ancestors (hominins) in southern India during a greater part of the Quaternary (Pappu et al. 2011), evidence of Neolithic sedentary settlements, crop cultivation and animal husbandry is present only for the last 3–4 ka. As a result, the last four millennia have witnessed remarkable changes in the land use and the cultural landscape of this ancient landmass. Peninsular India is also rich in natural resources and is the main store-house of mineral and fuel resources of India. Many of the Peninsular landforms are vulnerable to human alterations. Human activities and land use changes have added greatly to the process of land erosion (Reddy 2003). Landslides, soil erosion, expansion of gullies, alterations in the sediment load and discharge (due to multiple weirs and dams), etc. are significantly affecting the landforms, particularly the meso-scale features. Most rivers are now altered (Hema Malini and Nageswara Rao 2004; Panda et al. 2011). Over-exploitation of the natural resources, destruction of natural habitats and increased frequency of natural hazards are some of the key environmental challenges facing the population of this ancient and diverse tropical landscape.

9 Conclusions

The Indian Peninsula constitutes the oldest and the largest geomorphic province of India. Apart from wide, open valleys, residual hills, inselbergs, and hill ranges, the geomorphology of much of the Peninsula is dominated by the presence of number of plateaux and high-level surfaces at different elevations, sometimes occupied by ferruginous or aluminous duricrusts. Although landforms developed within sedimentary rocks are present in certain areas, the peninsular landscape is overwhelmingly dominated by landforms of the granite-gneissic terrain and basaltic terrain. Major geological, tectonic and climatic events since mid-Mesozoic have left an indelible imprint on this fragment of the Gondwana landscape. Post-Neogene differential uplift and reactivation of faults linked to Himalayan Orogeny and the onset of monsoon conditions over south Asia as well as Quaternary climatic and sea level changes have played a key role in shaping the scenery of this ancient landmass.

References

Babechuk MG, Widdowson M, Kamber BS (2014) Quantifying chemical weathering intensity and trace element release from two contrasting basalt profiles, Deccan Traps, India. Chem Geol 363:56–75

Choubey VD (1971) Pre-Deccan Trap topography in central India and crustal warping in relation to Narmada rift structure and volcanic activity. Bull Volc 35:660–685

Cox KG (1989) The role of mantle plumes in the development of continental drainage patterns. Nature 342:873–877

Gunnell Y (1998a) Present, past and potential denudation rates: is there a link? Tentative evidence from fission-track data, river sediment loads and terrain analysis in the South Indian shield. Geomorphology 25:135–153

Gunnell Y (1998b) The interaction between geological structure and global tectonics in multistoreyed landscape development: a denudation chronology of the South Indian shield. Basin Res 10:281–310

Gunnell Y, Radhakrishna BP (eds) (2001) Sahyadri, The Great Escarpment of the Indian subcontinent. Geol Soc India, Mem 47, Bangalore, India

Hema Malini B, Nageswara Rao K (2004) Coastal erosion and habitat loss along the Godavari delta front—a fallout of dam construction (?) Curr Sci 87:1232–1236

Kale VS (2010) The Western Ghat: The Great Escarpment of India. In: Migon P (ed) Geomorphological landscapes of the world. Springer, New York, pp 257–264

Kale VS, Rajaguru SN (1988) Morphology and denudation chronology of the coastal and upland river basins of western Deccan Trappean landscape India: a collation. Z Geomorph 32:311–327

Nicolas J, Bonnet NJ, Beauvais A, Arnaud N, Chardon D, Jayananda M (2014) First $^{40}Ar/^{39}Ar$ dating of intense Late Palaeogene lateritic weathering in Peninsular India. Earth Planet Sci Lett 386:126–137

Panda DK, Kumar A, Mohanty S (2011) Recent trends in sediment load of the tropical (Peninsular) river basins of India. Global Planet Change 75:108–118

Pappu S, Gunnell Y, Kumar A, Braucher R, Taieb M, Demory F, Thouveny N (2011) Early Pleistocene presence of Acheulian hominins in South India. Science 331:1596–1599

Radhakrishna BP (1993) Neogene uplift and geomorphic rejuvenation of the Indian Peninsula. Curr Sci 64:787–793

Ranganathan N, Jayaram S (2006) Geomorphology of Karnataka. Karnataka Geologists' Association, Department of Mines and Geology, Bangalore, 350 p

Reddy VR (2003) Land degradation in India: extent, costs and determinants. Econ Polit Weekly 38:4700–4713

Sheth HC (2007) Plume-related regional pre-volcanic uplift in the Deccan Traps: absence of evidence, evidence of absence. Geol Soc Am Spec Pap 430:785–813

Vaidyanadhan R (1977) Recent advances in geomorphic studies of Peninsular India: a review. Ind J Earth Sci S Ray volume:13–35

Vaidyanadhan R (2002) Geomorphology of the Indian sub-continent. Monograph published by the Indian Society of Remote Sensing, Dehra Dun, 29 p

Venkatakrishnan R (1984) Parallel scarp retreat and drainage evolution, Pachmarhi area, Madhya Pradesh, central India. Jour Geol Soc India 25:401–413

Valdiya KS (2001) River response to continuing movements and the scarp development in central Sahyadri and adjoining coastal belt. Jour Geol Soc India 57:13–30

Widdowson M (1997) Tertiary palaeosurfaces of the SW Deccan, Western India: implications for passive margin uplift. In: Widdowson M (ed) Palaeosurfaces: recognition, reconstruction and palaeoenvironment interpretation. Geol Soc Spec Publ Lond 120:221–248

Widdowson M, Gunnell Y (1999) Lateritization, geomorphology and geodynamics of a passive continental margin: the Konkan and Kanara lowlands of western Peninsular India. Spec Publ Int Assoc Sedimentol 27:245–272

The Thar or the Great Indian Sand Desert

Amal Kar

Abstract

Located along the eastern margin of the vast Indo-Saharan hot arid belt the monsoon-driven Thar Desert has a dominantly sandy landscape, where a range of other landforms sculpted by the fluvial and aeolian processes are also found. Several fluctuations in climate between drier and wetter phases during the Quaternary period and periodic earth movements have decided about the type and intensity of geomorphic processes. Most of the landform assemblages are, therefore, polygenetic in nature and have interesting evolutionary history. Parabolic dunes cover the largest area in the Thar, while numerous ranns (playas) act as some major receptors of water and sediments. Presently the human activities are becoming a major driver of geomorphic process acceleration, resulting in significant modifications in the landforms and consequent implications for environment and livelihood both within and outside the desert.

Keywords

Landforms • Fluvial • Aeolian • Lacustrine • Quaternary • Climate change • Sand dunes • Tectonism • Human pressure

1 Introduction

The Thar Desert, or the Great Indian Sand Desert ($\sim$285,000 km^2), is situated in the arid western part of Rajasthan state in India and includes the adjoining sandy terrain of Pakistan till the Indus River. It forms a distinctive, but integral part of the arid lands of western India that runs through the states of Punjab, Haryana, Rajasthan and Gujarat (Fig. 1). The eastern limit of Thar Desert is defined by the moisture availability index (Im, defined by India Meteorological Department as: ((P-PET)*100)/PET, where P is precipitation and PET is potential evapo-transpiration) of −66.6, which divides the arid from the semi-arid tract. This boundary roughly passes through the foothills of the degraded Aravalli Hill Ranges through Rajasthan, which has the distinction of being one of the world's oldest mountains ($\sim$2500 Ma). In the west, the desert extends up to the fertile alluvial plains of the Indus River in Pakistan, beyond which the arid lands continue across West Asia and North Africa up to the Atlantic coast. In the south, the Thar boundary lies along the Great Rann of Kachchh and the sandy plain and fossil dune system of north Gujarat. The fossil dunes and other old dissected aeolian bedforms in the semi-arid tracts beyond the Thar mark its previous extension (i.e., Megathar; Fig. 2).

2 Landforms

Broadly, the basement for Quaternary sedimentation and landform development in Thar Desert was a vast pediplaned surface, composed largely of the Precambrian meta-sediments and igneous rocks in the east and gradually thickening

A. Kar (✉)
A-1/6, Baitalik Housing Society, KMDA Housing Complex, Baghajatin, Kolkata 700094, India
e-mail: akarcaz50@gmail.com

V. S. Kale (ed.), *Landscapes and Landforms of India*, World Geomorphological Landscapes,
DOI: 10.1007/978-94-017-8029-2_7, © Springer Science+Business Media Dordrecht 2014

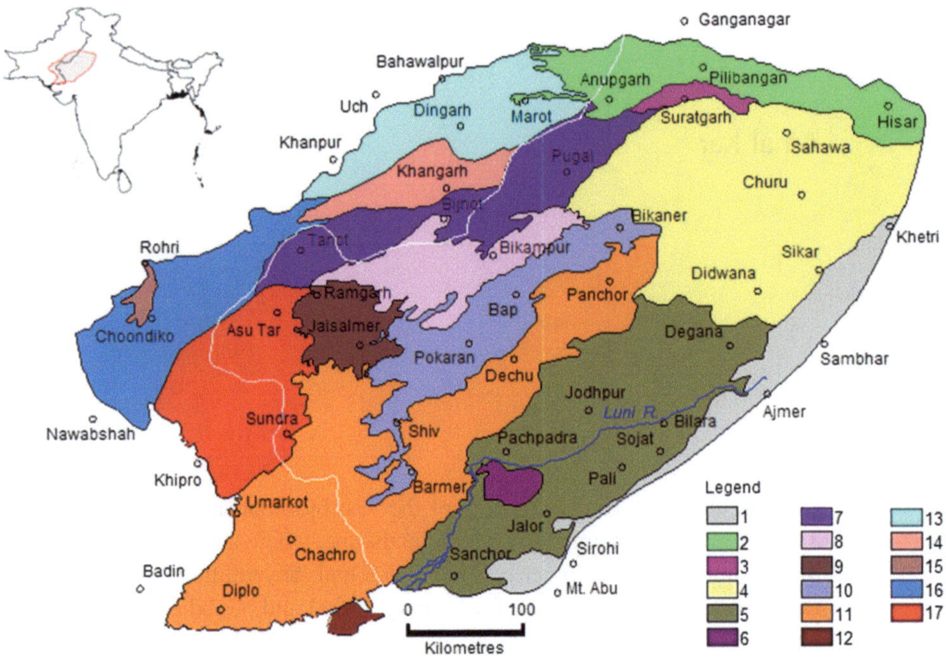

Fig. 1 Major geomorphic provinces in Thar Desert, with an emphasis on dune landscape. *1* Aravalli Hills with/without obstacle dunes, *2* Alluvial plain of N, *3* Star dune field of N, *4* Transitional parabolic dune field of NE, *5* Luni alluvial plain in S with isolated hills and colluvial plains, *6* Siwana Hills with obstacle dunes, *7* Transverse dune field of NW, *8* Parabolic dune field of NW, *9* Hamada landscape of Jaisalmer with sand streaks and zibars, *10* Gravel pavements with isolated sand dunes and sand streaks, *11* Parabolic dune field of S, *12* Nagarparkar upland, *13* Saline alluvial plain of NW, *14* Transitional parabolic dune field of NW, *15* Rohri upland, *16* Network dune field of W, *17* Linear dune field with megabarchan fields in W. The *white line* across the map shows the boundary between India and Pakistan

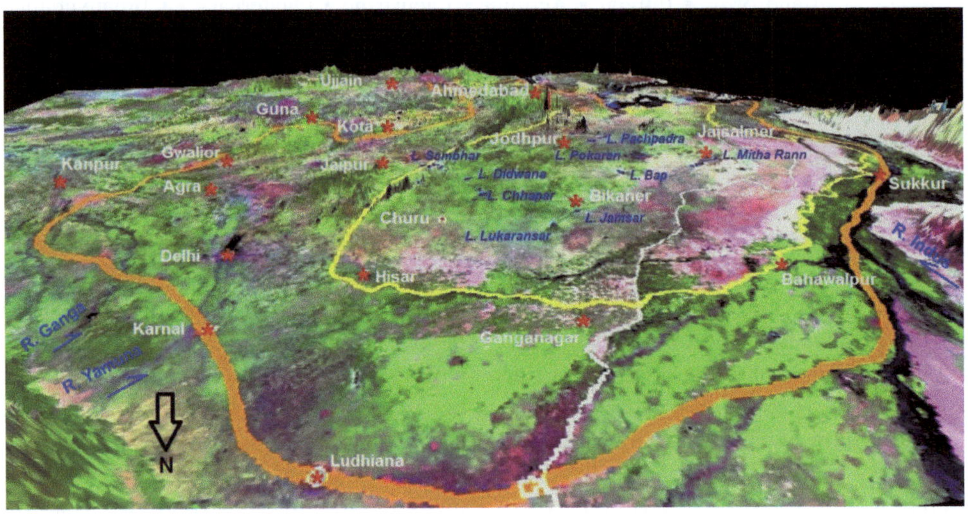

Fig. 2 Looking south into the Thar—Megathar Region. *Thin yellow line* demarcates Thar Desert. *Thick saffron line* demarcates Megathar. L (*in blue*) stands for major ranns (saline lakes). This *3D image* is generated from Landsat TM and ETM + FCC mosaic of the region (freely available from Global Land Cover Facility, University of Maryland, USA), draped over the SRTM DEMs (*sourced* from Global Land Cover Facility)

deposits of Mesozoic and Tertiary periods westward. Although most landform units have experienced interplay of different processes over time, it is possible to classify them according to the processes that dominated for a fairly long period as fluvial, lacustrine and aeolian. The major geomorphic provinces are shown in Fig. 1, whereas the broad physiography of the region and location of some major saline lakes is provided in Fig. 2.

Fig. 3 A sandstone mesa with gravel-strewn pediment near Osian, north of Jodhpur

Fig. 4 Exfoliation dome of granite near Mt. Abu

2.1 Hills and Pediments

The major landform sequence is hills and uplands—rocky/gravelly pediments (and pavements)—buried pediments (colluvial plains)—flat older alluvial plains—younger alluvial plains—river beds. The hillslopes usually have a general paucity of debris. Concave and straight segments dominate over the convex segment. Lithological and structural variations are faithfully replicated in the slope configurations, as noticed in the cuestas and mesas (Fig. 3). The limestone hills (e.g. between Bilara and Sojat to the south of Jodhpur) have a much subdued topography with convexo-concave slope. Most granite hills (mostly in the southeast) have a convex outline due to exfoliation (Fig. 4). The rhyolite hills (mainly in the central part), on the other hand, have a craggy outline. Cavernous weathering that produces tafoni and alveolar relief along joints is dominant along many granite hills and some ferruginous sandstone hills. The Siwana Hills in south-central part, with circular arrangement of an outer rim of rhyolite peaks and inner bosses of granites and porphyries, are the remnants of a major Proterozoic volcanic cone (735–505 Ma) that collapsed due to cauldron subsidence after the ejection of magma, leaving a saucer-shaped caldera interspersed with granitic bosses (Kar 1995).

The pediments at the base of hills usually have a slope of <4°, where the debris character is highly influenced by local lithology. Usually the pediment slope is concave, but joint-controlled weathering and erosion along the granite pediments produce a multi-convex profile. The piedmont angle is more pronounced on sandstone and granite, and the least on rhyolite. Some pediments in western part are formed over very old lithic calcretes (e.g., in Shiv-Barmer-Chohtan area). The long history of pediplanation in south-eastern Thar has left the remnants of many pediments on granite-gneiss as broadly convex and dissected gravely surfaces without any trace of an adjacent hill (Kar 1995). A bevelled edge along many pediments with fringing gully networks suggests a phase of pediment regradation upon stream rejuvenation.

2.2 Desert Pavements and Hamadas

The desert pavements with a broadly convex outline occur mainly between Bikaner, Bap, Pokaran and Fatehgarh, where the surface is characterised by closely packed gravels and pebbles ($\sim$2–30 cm size) in a matrix of fine sand and silt-sized particles. Large boulders are few. At many places the gravels have a gradational contact with a conglomerate bed of Lathi Sandstone (Jurassic), from which the loose

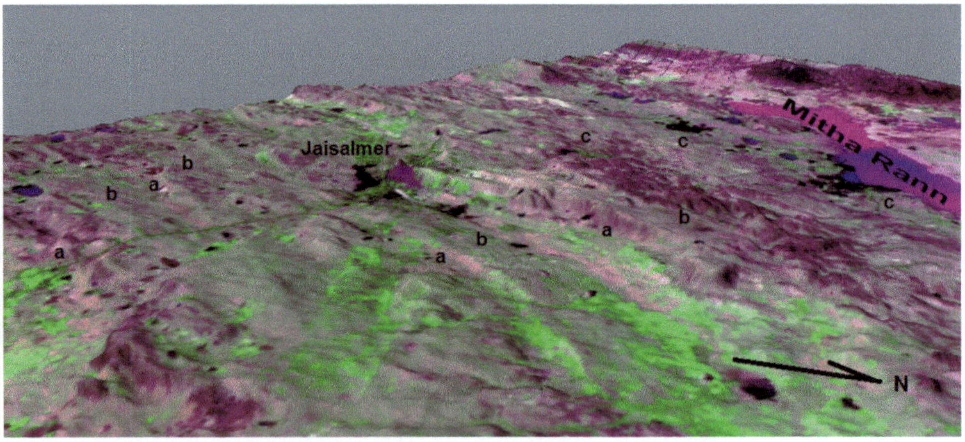

Fig. 5 Landsat ETM + FCC dated 19 September 2006, draped over SRTM data, to show the Hamada landscape around Jaisalmer. The major distinctive features are a series of steep SE-facing escarpments (*a*), each followed northwards by a gentle slope that faithfully follows the dip of the rock, and ends at the foot of another escarpment (*b*), thus forming a cuesta landform. A number of shallow ephemeral channels originating along the dip slope translocate rainwater and sediments towards the foot of the next escarpment. Traditional practice of the farmers is to regulate the flow along valleys through a series of anicut-like structures, and also to prevent it from meeting the saline Rann (*c*). After the rains, as water percolates through soil, the conserved moisture is used for winter cropping of wheat, mustard, etc. This is called 'Khadin' cultivation. *Greenery* in the image is largely due to rejuvenation of range vegetation (*lighter tone*) and invasion of the aggressive mesquites, *Prosopis juliflora* (*dark tone*), but occasionally due to crops (*pearl millet*) also. *Black parallel strips* in far right are due to the skipping of pixels by ETM + sensor

gravels are derived, but at others (e.g. near Bap, Jayal and Kathoti, yielding a rich hoard of Palaeolithic artefacts) no such association is noticed.

In the non-sandy areas of the very dry western part the Tertiary and pre-Tertiary beds of sandstone, limestone and shale in Jaisalmer area with typical near-horizontal or gently north-westward-dipping disposition, have been sculpted into high level structural plains (or hamadas) and cuestas, bounded by steep southeast-facing escarpments. Usually, behind the escarpment a cuesta or a hogback is followed by a narrow gravel plain where ephemeral channels from the cuesta converge. Then a broadly convex and north-westward dipping rocky surface follows, the continuation of which is broken by another escarpment (Fig. 5). The hamadas on the Eocene Khuiyala and Bandah Limestone beds have some palaeo-karstic features like shallow dolines and sink holes (e.g., near Khinya; Kar 1995). A narrow arcuate belt of dissected ferricrete in the area (with subsequent calcretization at places) possibly demarcates the boundary of a Miocene coast under a warm and humid tropical climate, followed by post-Tertiary drier climates. The pavements and hamadas are replete with desert varnish, case hardening and box weathering features, as well as cleaved gravels and ventifacts that are often etched and grooved by aeolian sand blasting on limestone (Fig. 6). The shifting of coarse sand by wind across the rocky surface often creates numerous waves of closely-spaced 1–2 m high sandy bedforms that are called zibars.

Fig. 6 Etched and grooved limestone formed due to aeolian sand blasting on a hamada surface near Jaisalmer

2.3 Colluvial and Alluvial Plains

Down the slope from pediments and pavements, the buried pediments/colluvial plains are composed of heterogeneous sediments. The thickness is more than a few meters near the Aravalli Hills, but decreases gradually westward where it varies from 30 to 100 cm. The flat older alluvial plains occur downslope of the colluvial plains, and are usually characterized by zones of illuviated soft nodular kankar, or gypsum at 30–300 cm depth within the sand and alluvium. Often these are followed down the profile by alternate sequences of aeolian and fluvial sediments, and sometimes lacustrine beds, which are followed further down by hard and compact carbonates/fluvial gravels resting on basement rocks. These plains are widely distributed in the south-east, but also form the base of dune landscape in most parts of the desert. The alluvial stratigraphy provides clues to interesting history of the early drainage systems from the Himalaya and the Aravalli Hills, as well as the changing depositional environment with time (Kar and Ghose 1984). The younger alluvial plains occur downslope of the older alluvial plains, especially as narrow stripes along the major ephemeral channels, including the Luni and its tributaries in the southern Thar and the Ghaggar in northern Thar. The sandy alluvium of the unit gets periodically replenished by floods, making them the most fertile areas for cultivation.

2.4 Ranns (Playas)

A number of inland ranns (ephemeral playa lakes), with a flat salt-encrusted surface that get inundated during monsoon rains, are found within the desert. The largest among these is the Sambhar Lake ($\sim$293 km^2; 32 km E–W; 17 km N–S) along the eastern fringe, followed by Bap ($\sim$51 km^2) in the western part, whereas other notable ones with approx. sizes of 21–30 km^2 are at Pachpadra, Thob, Chhapar, Sanwarla, Mitha Rann, Kanodwala Rann, and those below 20 km^2 size at Didwana, Sujangarh, Parihara, Lunkaransar, Jamsar, Rakhi, Bassi, Kandara, Pokaran, Lawan, Redana, etc. Some of these are associated with an ephemeral stream network (e.g., Pachpadra, Thob), but several have been formed in the lee of high hills through deflation and capturing of an ephemeral stream into the resultant depressions by the hill-margin dunes (e.g., at Didwana and Chhapar in E; Redana in W), or through a long period of etching and grooving of the softer rocks (e.g., Mitha Rann and Kanodwala Rann in Jaisalmer area). Some of the ranns have a distinct neotectonic connection (e.g., Sambhar Lake; Rakhi, Bassi and Kandara along an E–W lineament in S) (Kar 1990, 1995). Most of the ranns are used for commercial salts or gypsum production.

2.5 Aeolian Bedforms

The dominant landforms in the Thar Desert are the aeolian bedforms, especially sand dunes and interdune plains, and the low sandy hummocks and sand sheets on the vast plains. Presently the strong pre-monsoon summer winds from the southwest and the west are mostly responsible for dune-building. Over the millennia enormous quantity of medium to fine sand and silt has been brought into the region and its western fringes through the Indus and the Satluj River systems, with the Himalayan provenances, and enough has also been produced locally through sub-aerial processes and contributions from the Aravalli provenances, which get recycled several times to construct the aeolian landforms.

The sand dunes could be broadly classified as 'old' and the 'new' dunes, where the old ones with an average height of 15–30 m are mostly the partly stabilized and vegetated parabolic, linear, transverse, star, network and major obstacle dunes. Most of these old dunes occur in coalescing chains. Parabolic dunes cover the largest area, and their shape reflects the changing wind strength from southwest to northeast. In the southwest, where the wind strength is high, the dune chains usually occur in hair-pin and chevron patterns with 3–5 km long arms that join together at sharp angles. As the wind strength gradually declines in the central and the northern parts, the arms become shorter ($\sim$0.5–1.0 km in the northeast), and the sharp junction is replaced by a curved nose (Fig. 7). Further downwind, as the wind strength becomes feeble along the fringe of the desert the parabolics transit into two major kinds of network dunes: (a) compound hooked dunes with linear and transverse forms to the right-hand side of the main dune arm (when looking downwind; Fig. 8), and (b) a wide, long-arm parabolic form with small network pattern within it (Fig. 9; Kar 1993). The linear dunes occur mainly in the western part of the Thar, especially in Jaisalmer-Tharparkar region, and are broadly oriented SW–NE, i.e., in the direction of wind (Fig. 10). The dunes were earlier thought to have originated from parabolic dunes, but studies since then traced their origins from streams of barchans in the high wind energy zone in the west (Fig. 11), and from lee vortices behind major obstructions, or along major stream valleys through funnelling effect (Kar 1987). The dunes are characterised by a broadly convex summit. The length varies from more than 10 km in the extreme west of the field to 1–2 km in the east. The high transverse dunes occur mainly in the NW part and were formed astride the path of sand-laden wind (Fig. 12). Linked star dunes of 15–35 m height occur in a narrow strip along the northern margin of the Thar (Fig. 13). Simple and compound obstacle dunes have been formed on the windward and leeward sides of the hills. The dunes are highly dissected by rills and gullies, especially along the eastern margin of the desert (Fig. 14).

Apart from the old dunes, numerous fields of new dunes also exist within the desert, especially in the high wind energy regime of the west, where fields of 1–8 m high barchans and 20–40 m high compound megabarchans form under natural settings (Fig. 15). Elsewhere within the desert high human pressure on the semi-stabilized old dunes and sandy plains are accelerating the aeolian process, leading to

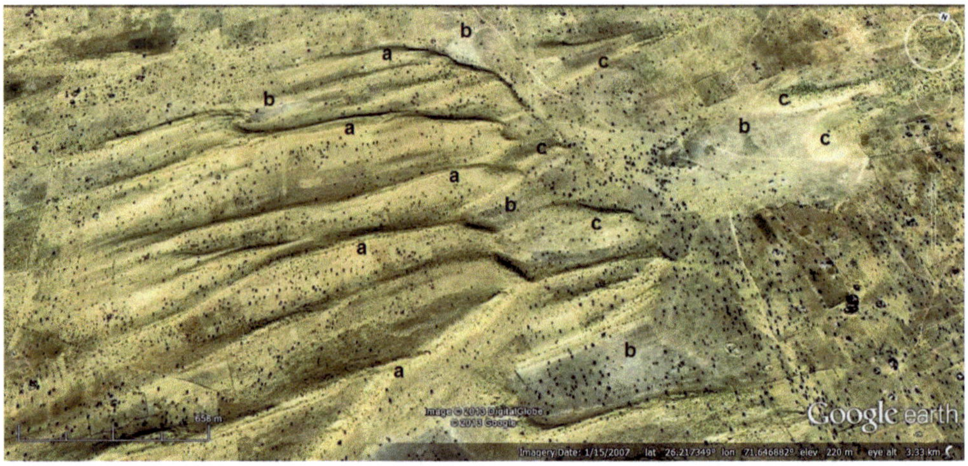

Fig. 7 Google Earth image of a typical parabolic dune cluster near Bafrau in the central part of Thar Desert, showing *a* semi-stabilized/reactivated arms of old dunes; *b* deflation hollows in the wake of the dunes; and *c* formation of new aeolian bedforms downwind

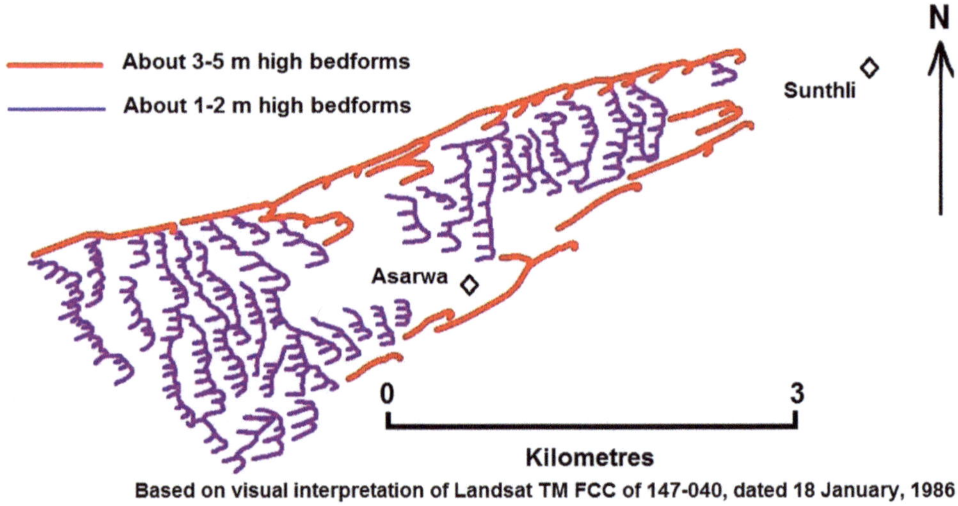

Fig. 8 Outline of compound hooked dunes with low transverse bedforms to the right-hand side near Didwana. Dominant wind is from SW during summer months

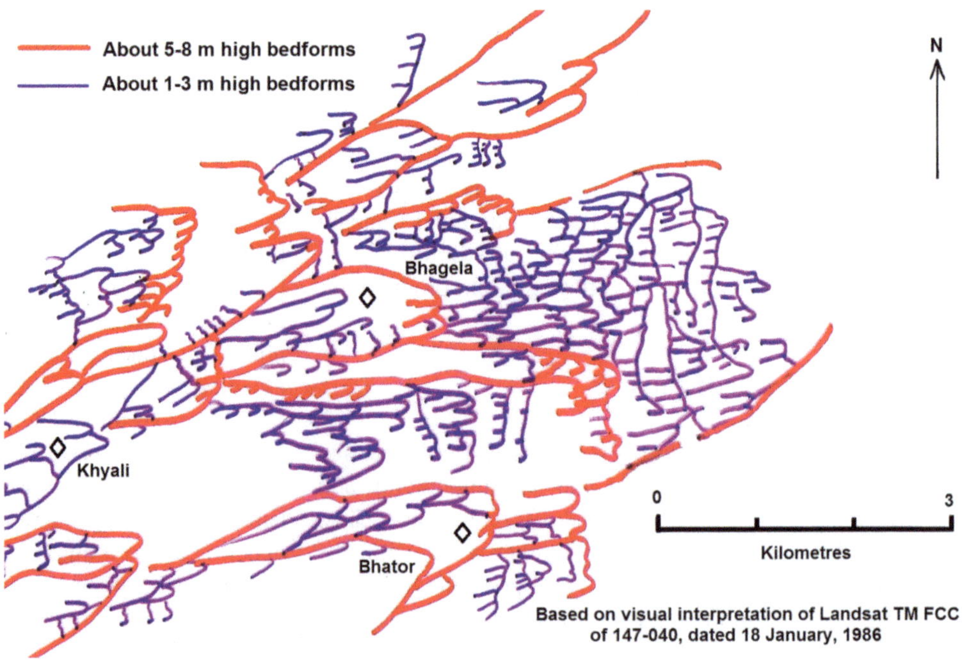

Fig. 9 Outline of the wide, long-arm parabolic bedforms with small network dunes within them near Churu

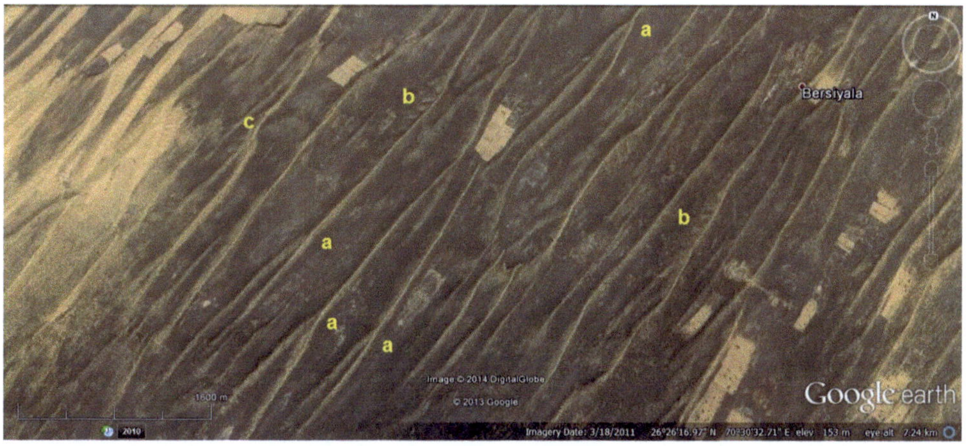

Fig. 10 Google Earth image of 6–10 m high linear dunes near Bersiyala (SW of Jaisalmer), with *a* Y-junctions, and *b* small feathers on left flank. When reactivated, the dunes crests are infested by a chain of crescentic bedforms (*c*). Dominant wind is from SSW during summer months

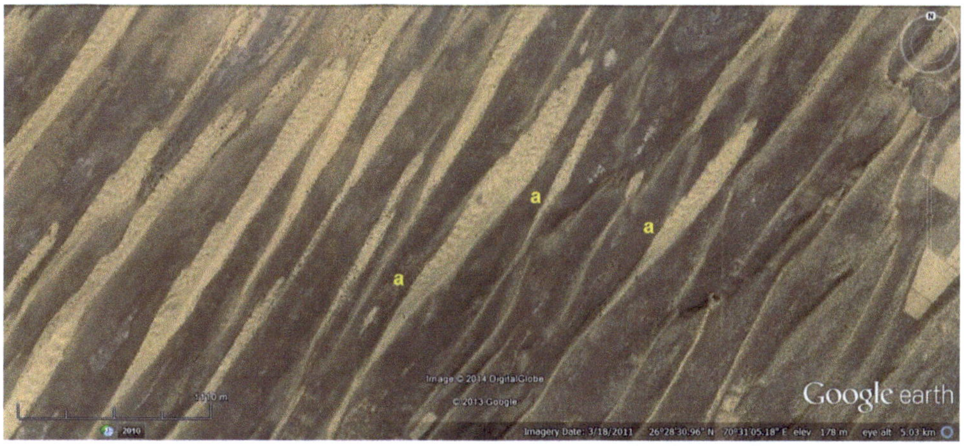

Fig. 11 Google Earth image of 1–2 m high new linearly-arranged bedforms (*a*), being formed to the NW of Bersiyala through chains of barchans downwind of the older, semi-stabilized linear dunes

Fig. 12 Google Earth image of 18–25 m high transverse dune chains to the west of Bikaner, with narrow interdune plains where strong deflation has exposed the subsurface gypsum layer (*a*). Some of the interdunes are now irrigated (*b*) by the Indira Gandhi Canal network

either advancement of the arms of old dunes or formation of new bedforms as barchans, sand streaks or nebkhas (Fig. 16), which ultimately evolve into the dominant bedform type that triggers their origin and growth. Examples of such sand reactivation abound near the settlements and in areas of deep ploughing.

3 Evolutionary History

Antiquity of Thar Desert is enigmatic, but stratigraphic records bear testimony of fluctuating arid to humid climate from as early as late Jurassic (as carbonaceous, near-shore

Fig. 13 Google Earth image of 12–20 m high linked star dunes to the SE of Suratgarh. The peaks occur at major nodes of the dune arms (*a*)

Fig. 14 Stabilized old obstacle dunes (*background*) and thick sandy plains (*foreground*) dissected by gullies along the margin of the Thar Desert near Jaipur. The sand profile often shows a faint hint of palaeosol formation nearer the surface, and yields medium to large calcified plant roots below, suggesting antiquity of the edifices

Fig. 15 A field of 20–30 m high megabarchans with tiers of smaller barchans to the SW of Asu Tar during winter when the sharp NE-facing slip face is replaced by a much-rounded crest due to wind blowing from NE direction

aeolian sand beds of Bhadasar Sandstone, impregnated with case-hardened and ferricreted nodules), to the early Quaternary (as ferruginous sandstone of Shumar formation, duricrusted calcrete pediments of Barmer area, widespread gypsum/halite beds, etc.) as the Indian landmass drifted northwards and monsoon climate evolved. Discovery of the wide, dry bed of the Ghaggar River in the northern fringe of the Thar as that of the legendary Saraswati River from the Himalaya, and establishment of the contribution of a major Himalayan river, the Satluj, in its survival (Oldham 1893), provided the initial clues to the role of climate change and tectonism in the desert landscape evolution. Studies since then have established the role of some large, now-extinct stream networks in construction of the thick sandy alluvial plains. Notable among these was the Saraswati-Drishadvati River system, which used to be nourished by the snowfed Satluj and the Yamuna Rivers at different periods of time between 10 and 60 ka BP or beyond, especially during the vigorous monsoon phases (Ghose et al. 1979; Kar and Ghose 1984; Clift et al. 2012). The ephemeral Luni River system

Fig. 16 A typical shrub-coppice dune, or Nebkha, associated with a clump of 3 m high *Capparis decidua* plants near Pokaran. Usually these dunes attain a height to length ratio of 1:7 to 1:10 during the period of strong summer wind from SW. The ratio gradually decreases after the monsoon (July–September) as the SW wind is gradually replaced by the weaker NE wind, and by winter the Nebkhas are reduced to small humps around the bushes

from the Aravalli Hills is now the only surviving system in the south-east where several of its tributaries are becoming defunct and partially buried under aeolian sand (e.g., the Lik, the Jojri), but periodic floods sometimes revive them (e.g., floods during 1979, 1991 and 2006).

Observation of several Quaternary sequences suggests widespread occurrence of alternate fluvial and aeolian activities across the desert, the lacustrine and fluvio-lacustrine deposits being noticed in some favoured locations only. Radiocarbon and luminescence dating of some sequences provided a chronology of events within the limits of those methods (typically ~0.5–150 ka) for late Quaternary period (Singh et al. 1974; Singhvi and Kar 2004). Electron Spin Resonance (ESR) dating of calcretes provided dates of ~100 ka, but the process of calcretization homogenises a range of previous depositional environments, often making it difficult to filter the original environments out of the mass. Since the dated sequences are few and landscape variability large, the history of dominantly fluvial, aeolian and lacustrine events summarized below is provided with a caution that the dates and the deduced environments may be considered as a broad first-level framework, and not as definitive region-wide phenomena.

3.1 Fluvial Sequences

Stratigraphic records suggest that the beginning of the Quaternary period in the desert might have been marked by a wetter climate with high-intensity rainfall that was responsible for basal gravels and conglomerates. Pre-Quaternary aeolian deposits, if any, were either washed away by repeated floods and/or were admixed with fluvial deposits and became calcretized. A somewhat consistent chronology of fluvial deposition could be framed for the lower part of the Luni River only, between Tilwara and Khudala, where the major pluvial periods were: ~80–90 ka, ~55 ka, ~30 ka and 7–6 ka (Kar et al. 2001; Bajpai 2004; Jain et al. 2005), possibly yielding more frequent monsoon season flows along the ephemeral sand-bed stream networks. The pluvial peaks were followed along their rising and falling arms by significant aeolian activities as wind strength responsible for the vigorous monsoon continued beyond the peak phases. The Last Glacial Maximum (LGM; 24–18 ka) perhaps witnessed the driest phase during the last 100 ka as monsoon circulation receded. A tectonic event intervened at ~10 ka when a post-LGM, semi-viscous, fluvio-lacustrine deposit of sand-silt at Khudala (Fig. 17) was shaken vigorously to form oblong mud balls and tilted. Older fluvial records from the region are beyond the conventional dating range, but suggest a high-energy braided stream system in a subsiding basin under a strong monsoon regime during Pliocene (Jain et al. 2005). Analysis of several slack-water flood deposits along a tributary junction with the Luni in its gorge section near Tilwara revealed many monsoon-driven high magnitude floods during 0.5–1.0 ka period (i.e., Medieval Warming), which became less frequent during 0.4–0.2 ka, the Little Ice Age (Kale et al. 2000).

3.2 Aeolian Sequences

Dating of sediments from a number of dune sections across the Thar and its eastern margin suggests that the aeolian history is at least 160 ka old (Singhvi and Kar 2004; Dhir et al. 2010). The major periods of enhanced sand mobilization were 100–115 ka, ~75 ka, ~50 ka, 30–25 ka and 14–7 ka, when aeolian deposition took place even beyond the present Thar boundary (i.e., into the Megathar). Each of these episodes was followed by a period of relative landscape stability. The stability after the ~75 ka flux was perhaps associated with a period of sustained monsoon rainfall and strong seasonality, as attested by reddening of sediments (Kar et al. 2001; Singhvi and Kar 1992; 2004). The main driving force

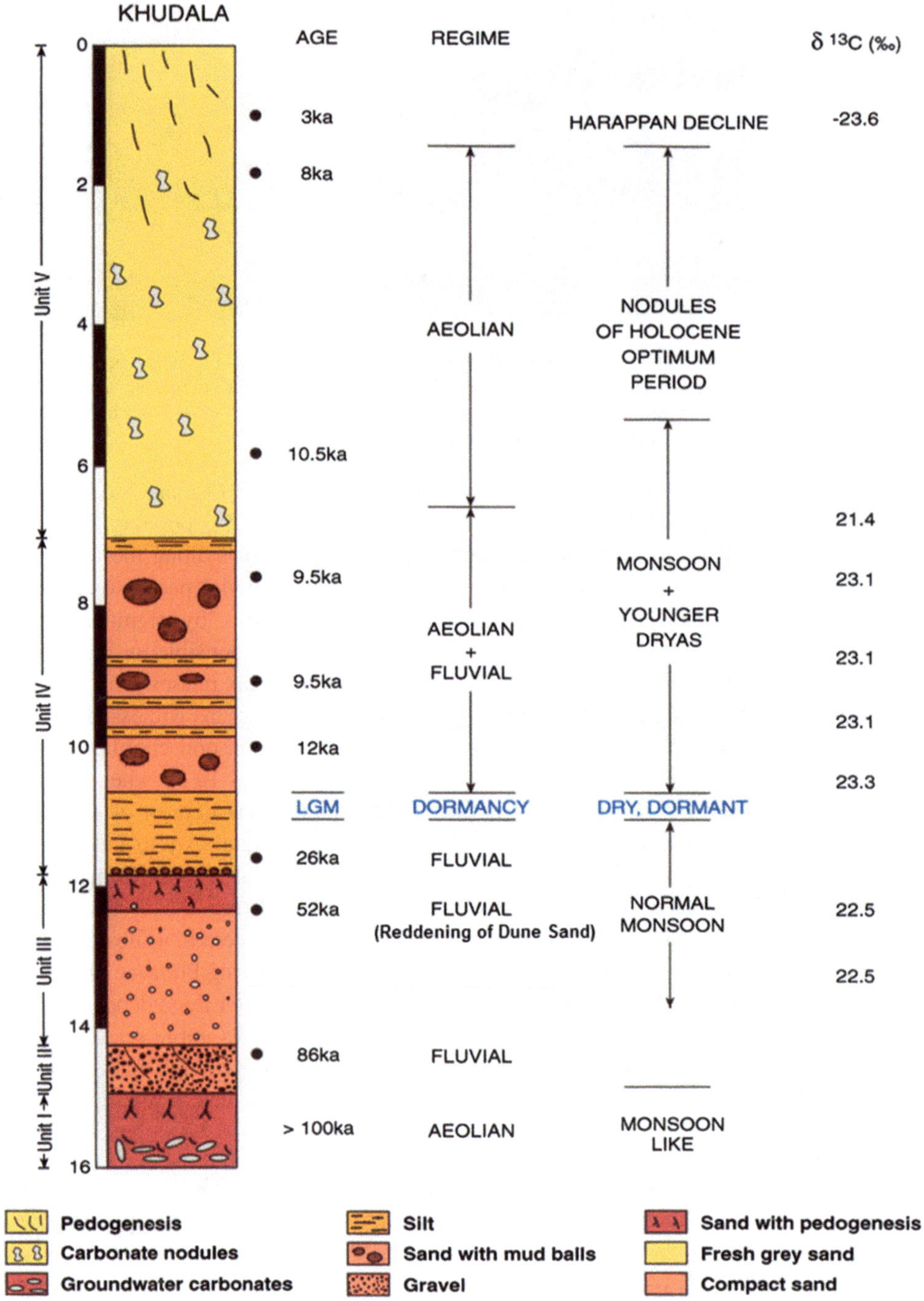

Fig. 17 Generalized late Quaternary sedimentary sequence at Khudala (25° 21′ 55″ N, 71° 48′ 38″ E), Luni River, with approximate age and interpreted environment (modified after Kar et al. 2001)

for aeolian activities all through was the SW monsoon. Contrary to popular view, the LGM was not a period of high aeolian activities because the SW monsoon was weak. Also, no significant aeolian activities were noticed in the Megathar region after the post-LGM aeolian peak of 14–7 ka. After this, significant aeolian activities were recorded only from within the desert at 5.0–3.5 ka, and 2.0–0.6 ka. The latest phase of sand mobility in the desert started around 0.3 ka, when the rates of dune mobility and sand accretion began to surpass the geological rates due to increasing human pressure (Kar et al. 1998).

3.3 Lacustrine Sequences

Analysis of sediments from several ranns within the desert suggested a hyper-saline condition in most of them during the LGM, and till ~13 ka, when the summer monsoon got re-established and the freshwater conditions started dominating over the dry and hyper-saline conditions, especially in the ranns in the east and the north–west. The Climatic Optimum helped most lakes to receive a sustained and higher fresh-water inflow by 7.5–6.0 ka, but a decreasing trend was noticed from about 5 ka (Singh et al. 1974). The ranns to the west of the present 200 mm isohyet appeared to have been much less affected by such shifts in the monsoon behaviour, as the rainfall pattern did not change appreciably. The present ephemeral condition of the ranns under a shifting climate between wet and dry phases was established from 3 ka in the east, and from 6 ka in the west (Singhvi and Kale 2009).

4 Human Impact

Thar Desert is one of the world's most thickly populated arid regions with a density of human population of $\sim 110\ \text{km}^{-2}$, and livestock population of $\sim 120\ \text{km}^{-2}$. While the livestock population has doubled over the last five decades, human population has registered a four-fold increase. To cater to the needs of these increasing populations, especially for cropping, wood collection and grazing, significant land use decisions have been taken by the stakeholders with environmental consequences. This includes deep tractor-ploughing of the sandy plains across the desert, as well as of the sand dunes and sandy hummocks that earlier used to function as either rangelands or as long fallows, replacement of the traditional agro-forestry-cum-land fallow system that used to provide food and fodder especially during droughts, with cropping alone with/without irrigation, and un-regulated use of the pastures for fuelwood collection and grazing. Since the landforms are in a very fragile state due to climatic and terrain constraints, such activities are increasing the erosion risk, sediment mobility and degradation of land quality. Many of the semi-stabilized old dunes have become highly reactivated and are advancing to newer areas, while new barchans are forming in parts of the sandy plains where none existed earlier. Gully erosion is increasing in the groundwater-irrigated old sandy plains and obstacle dune slopes along the eastern margin of the desert, while water-logging and salinization is spreading in the palaeochannel-dominated canal-command areas. Falling groundwater level and higher input cost of irrigated agriculture along the eastern margin of the Thar is compelling many farmers to switch back to rainfed farming, but the semi-stabilized sandy landscape that was levelled previously for irrigated farming and was denuded of natural vegetation, is becoming more prone to sand reactivation. The net result is a gradual increase in the dust load in the atmosphere, which when viewed in the context of global warming, may get enhanced further and significantly modify the monsoon behaviour over Indian subcontinent (Bollasina et al. 2011). Indeed, the scale at which the changes are taking place, the landscape of the region is perhaps transiting gradually from the Holocene to Anthropocene! It is, therefore, necessary that land use decisions and land management in the region are based on proper understanding of the land forming processes and their vulnerability to induced pressures.

5 Conclusions

Landforms in the Thar Desert have evolved through a fairly long period of interplay between fluvial and aeolian processes, especially during the Quaternary period, but periodic tectonic activities have also played a crucial role. Broadly, the base mosaic of hill/upland—pediment—colluvial plain—alluvial plain—river sequence has been acted upon by the contemporary processes under a dominantly dry but fluctuating climate such that the aeolian processes and a variety of aeolian bedforms are now most dominant, modifying or burying many fluvial landforms, but also aiding the formation of some playas. The stability of most landforms in the current era of Anthropocene depends largely on their resource use potentials and robustness in the face of ill-managed human exploitation. As human pressure is accelerating the geomorphic processes, the fertile and otherwise usable agricultural lands, both within the desert and along its wetter margins, are becoming more threatened. Unless appropriate policy interventions are made early towards the use of land the consequences may prove disastrous under a warmer climate.

References

Bajpai VN (2004) Hydrogeological evolution of the Luni River basin, Rajasthan, western India: a review. Proc Ind Acad Sci (Earth Planet Sci) 113:427–451

Bollasina M, Ming Y, Ramaswamy V (2011) Anthropogenic aerosols and the weakening of the South Asian summer monsoon. Science 334:502–505

Clift PD, Carter A, Giosan L, Durcan J, Duller GAT, Macklin MG, Alizai A, Tabrez AR, Danish M, VanLaningham S, Fuller DQ (2012) U-Pb zircon dating evidence for a Pleistocene Sarasvati River and capture of the Yamuna River. Geology 40:211–214

Dhir RP, Singhvi AK, Andrews JE, Kar A, Sareen BK, Tandon SK, Kailath A, Thomas JV (2010) Multiple episodes of aggradation and calcrete formation in Late Quaternary aeolian sands, central Thar Desert, Rajasthan, India. J Asian Earth Sci 37:10–16

Ghose B, Kar A, Husain Z (1979) The lost courses of the Saraswati River in the Great Indian Desert—New evidence from Landsat imagery. Geog J 145:446–451

Jain M, Tandon SK, Singhvi AK, Mishra S, Bhatt SC (2005) Quaternary alluvial stratigraphical development in a desert setting: a case study from the Luni River basin, Thar Desert of western India. Int Assoc Sedimentol Spec Publ 35:349–371

Kale VS, Singhvi AK, Mishra PK, Banerjee D (2000) Sedimentary records and luminescence chronology of Late Holocene palaeofloods in the Luni River, Thar Desert, northwest India. Catena 40:337–358

Kar A (1987) Origin and transformation of longitudinal sand dunes in the Indian desert. Z Geomorph 31:311–337

Kar A (1990) A stream trap hypothesis for the evolution of some saline lakes in the Indian desert. Z Geomorph 34:37–47

Kar A (1993) Aeolian processes and bedforms in the Thar Desert. J Arid Environ 25:83–96

Kar A (1995) Geomorphology of arid western India. Mem Geol Soc Ind 32:168–190

Kar A, Ghose B (1984) Drishadvati River system in India: an assessment and new findings. Geog J 150:221–229

Kar A, Felix C, Rajaguru SN, Singhvi AK (1998) Late Holocene growth and mobility of a transverse dune in the Thar Desert. J Arid Environ 38:175–185

Kar A, Singhvi AK, Rajaguru SN, Juyal N, Thomas JV, Banerjee D, Dhir RP (2001) Reconstruction of the late Quaternary environment of the lower Luni plains, Thar Desert, India. J Quat Sci 16:61–68

Oldham CF (1893) The Saraswati and the lost river of the Indian desert. J Roy Asiatic Soc 34:49–76

Singh G, Joshi RD, Chopra SK, Singh AB (1974) Late Quaternary history of vegetation and climate of the Rajasthan desert, India. Phil Trans Roy Soc 267B:467–501

Singhvi AK, Kale VS (2009) Palaeoclimate studies in India: last Ice Age to the present. Indian National Science Academy, New Delhi

Singhvi AK, Kar A (eds) (1992) Thar Desert in Rajasthan. Geological Society of India, Bangalore

Singhvi AK, Kar A (2004) The aeolian sedimentation record of the Thar Desert. Proc Ind Acad Sci (Earth Planet Sci) 113:371–401

The Indian Coastline: Processes and Landforms

Ranadhir Mukhopadhyay and S. M. Karisiddaiah

Abstract

The nearly 7,500 km long coastline of India is remarkably varied and dynamic and displays diverse rock-based, sediment-based and coral-based landforms. Coastal processes along the Indian coasts are controlled largely by monsoons. The Arabian Sea coast differs from the Bay of Bengal coast in several respects. The east coast is wider, with several large deltas, large lagoons, one of the world's largest mangrove wetland (Sundarbans) and long stretches of sandy beaches backed by dunes or ridges. In comparison, the west coast is more indented with rocky headlands, intervening sandy bays and multiple estuaries. Cliffs and associated features are relatively more common. A large saline marshland and lagoon-barrier complexes (kayals) are some of the noteworthy features along the west coast.

Keywords

Indian coast • Monsoon • Tidal range • Waves • Landforms • Beaches • Arabian Sea • Bay of Bengal

1 Introduction

The varied tectonic history, underlying lithology, monsoon climate, sea level fluctuations in the Quaternary and modern littoral processes have created a great variety of the coastal landforms along the ~7,500 km long coastline of the Indian Peninsula. The coastline displays diverse rock-based (headlands, sea cliffs and shore platforms), sediment-based (beaches, dunes, sand bars, spits, tidal flats) and coral-based (reefs and atolls) landforms, which vary significantly in their spatial scale and form.

Coastal erosion is a widespread problem along much of the Indian coastline. About 23 % (~1,248 km) of shoreline along the Indian mainland is affected by erosion to various degrees (ICMAM 2009). The problem is relatively more severe on the west coast than on the east coast of India (Table 1). Apart from natural factors, coastal structures (groynes, breakwaters, seawalls, etc.) constructed for coastal protection works and seaport operations are the common causes of coastal erosion in many areas (ICMAM 2009).

2 Coastal Morphology

The Indian coasts comprise nearly 43 % of sandy beaches, 11 % of rocky features with headlands, platforms and cliffs, and 46 % of mud flats and marshy wetlands (ICMAM 2009). Characteristic differences between the east and the west coasts of India in terms of coastal topography, landforms and processes are apparent from Figs. 1 and 2, and

R. Mukhopadhyay (✉) · S. M. Karisiddaiah
National Institute of Oceanography, Dona Paula, 403004 Goa, India
e-mail: ranadhir@nio.org

S. M. Karisiddaiah
e-mail: kari@nio.org

V. S. Kale (ed.), *Landscapes and Landforms of India*, World Geomorphological Landscapes,
DOI: 10.1007/978-94-017-8029-2_8, © Springer Science+Business Media Dordrecht 2014

Table 1 Nature of coast and physical processes along the Indian coastline

States /Union Territory	Total length (km)	Marshy (%)	Muddy (%)	Rocky (%)	Sandy (%)	Stretch affected by erosion in %	Wind speed (m/s)	Wave height (m*)	Current speed (m/s)	Mean sea level (m)	Sediment transport 10^5 (m^3/year)	
											Southerly	Northerly
Arabian Sea / West Coast												
Gujarat	1214.7	22	29	21	28	12.78	47	0.1–2.9	0.729	3.64	10.88	9.11
Daman & Diu	9.5	–	–	–	–	–	44	0.1–6.0	–	–	–	–
Maharashtra	652.6	–	46	37	17	40.30	39	0.2–5.1	0.01–0.5	2.43	6.93	7.74
Goa	151.0	–	35	21	44	12.70	39	0.3–5.9	0.337	1.20	8.20	5.30
Karnataka	280.0	–	14	11	75	89.14	39	*4.1–6.1*	0.475	0.90	11.51	3.56
Kerala	569.7	–	15	05	80	83.93	39	0.3–1.9	0.475	0.50	12.66	5.99
Lakshadweep	132.0	–	–	–	–	100.00	39	0.2–2.5	–	0.96	–	–
Bay of Bengal / East Coast												
Tamil Nadu	906.9	–	38	05	57	16.74	50	0.2–2.1	0.227	0.45	5.03	10.59
Pondicherry	30.6	–	–	–	–	31.03	50	0.3–2.1	–	0.70	6.92	9.39
Andhra Pradesh	973.7	07	52	03	38	6.75	50	1.81	–	0.75	4.45	14.88
Odisha	476.4	10	33	–	57	22.58	50	1.97	0.3937	1.50	2.72	8.93
West Bengal	157.5	49	51	–	–	79.36	50	*2.1*	–	3.13	–	–
Andaman and Nicobar	1962.0	–	–	–	–	–	44	–	–	1.10	–	–
Mainland	5422.6	10	36	11	43	29.95	–	–	–	–	–	–
Islands	2094.0	–	–	–	–	–	–	–	–	–	–	–
TOTAL	**7516.6**	–	–	–	–	**23.37**	–	–	–	–	–	–

Modified after Kumar et al. (2006) and ICMAM (2009)
* Number in italics is 100-year return period
– Data unavailable
Southerly covers November to February, northerly spans from April to September

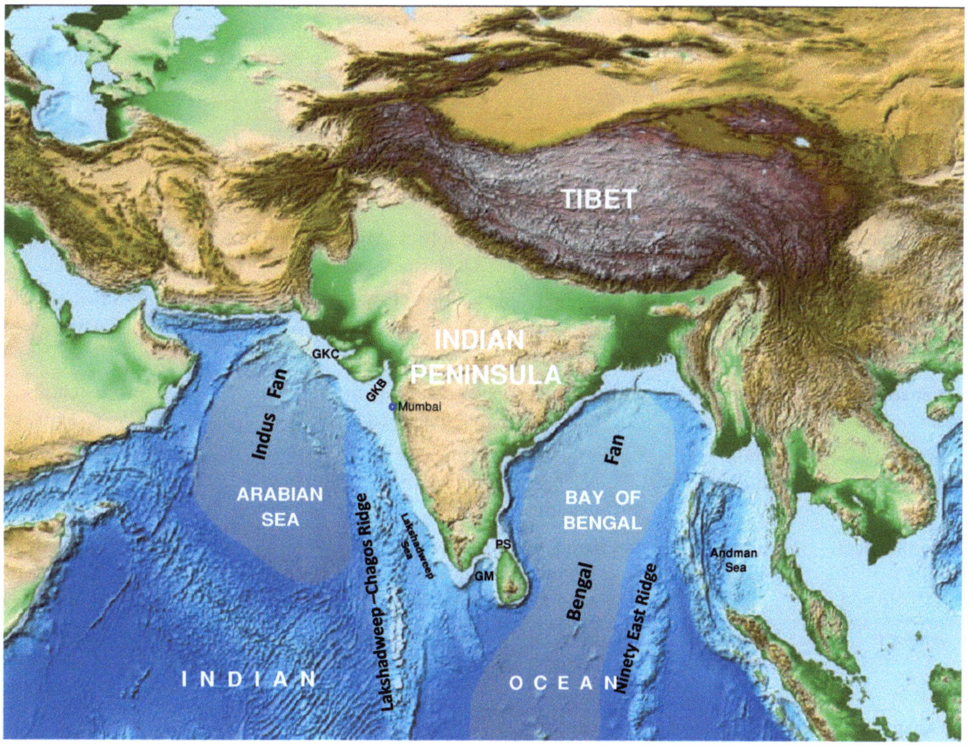

Fig. 1 Coastal and offshore morphology along the Indian subcontinent. *GKC* Gulf of Kachchh, *GKB* Gulf of Khambhat, *GM* Gulf of Mannar, *PS* Palk Strait. The boundary of the Indus and Bengal Fans is approximate. The Bengal Fan, with an area of ~ 3 million km^2, is the largest deep-sea fan in the world. The area of the Indus Fan is ~ 1.1 million km^2. *Source of the base map* Amante and Eakins (2009). http://www.ngdc.noaa.gov/mgg/image/color_etopo1_ice_low.jpg

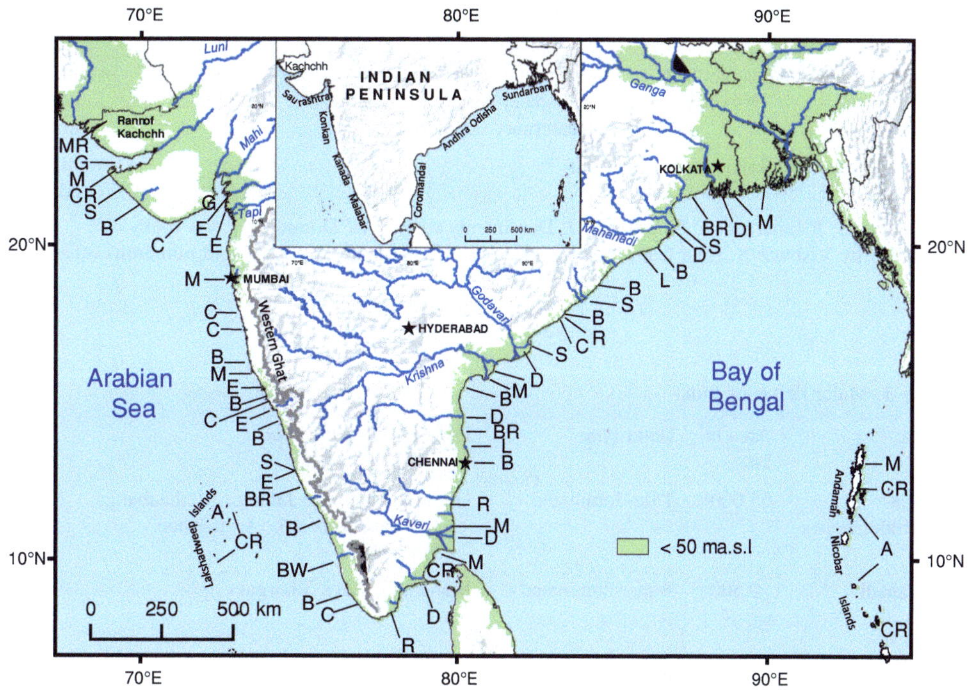

Fig. 2 Map showing the locations of the noteworthy coastal landforms along the Indian coastline. The inset map shows the major geomorphic coastal divisions discussed in the text. *A* Archipelago, *B* Beach, *BR* Beach/Dune ridges, *BW* Back water, *C* Cliff, *CR* Coral reef, *D* Delta, *DI* Deltaic islands, *E* Estuary, *G* Gulf, *L* Lagoon, *M* Mangrove, *MR* Marshland, *R* Red Sand, *S* Spit (Map courtesy of Vishwas S. Kale)

Tables 1 and 2. In general, the eastern seaboard comprises long stretches of sandy beaches backed by coastal dunes (active or stabilized) and interrupted by large prograding deltas, whereas the west coast is more indented with rocky headlands, pocket beaches and multiple estuaries. There are also two archipelagos—one in the Bay of Bengal and another in the Arabian Sea (Fig. 2).

The eastern coastal plain of India is a 100–130 km wide stretch of land lying between the Eastern Ghat Ranges and the Bay of Bengal. The rivers on the east coast are long with wide and extensive delta formations (Vaidyanadhan 1991) (Table 3). The Ganga-Brahmaputra, Mahanadi, Godavari, Krishna and Kaveri Rivers drain this eastern coastal plain, giving rise respectively to four major offshore sedimentary

Table 2 Coastal geomorphic divisions and associated landforms

Geomorphic divisions	Name of states/union territory	Dominant lithology	Coastline type	Major landforms and features
Kachchh	Gujarat	Miocene sandstones and limestone, Tertiary formations and Quaternary deposits	S, NR	Large saline marshland, estuaries, cliffs and beaches, coastal sand dunes, mangroves
Saurashtra	Gujarat	Deccan Traps, Miliolites, Mio-Pliocene formations, Quaternary sediments	S, R	Estuaries, cliffs, beaches, spits, coastal sand dunes, mudflats, raised marine benches
Konkan (and south Gujarat)	Maharashtra-Gujarat	Deccan Traps, laterites, Quaternary sediments	S, R	Estuaries, cliffs, wave-cut platforms, pocket beaches, coastal sand dunes, spits, tombolo
Kanada-Goa	Karnataka-Goa	Peninsular Gneiss, Metasediment, laterites, Quaternary sediments	S, R	Estuaries, cliffs, wave-cut platforms and beaches, spits, coastal sand dunes, tombolo
Malabar	Kerala	Charnockites, Peninsular gneiss, Neogene and Quaternary sediments, laterites	S, NR, R	Estuaries, lagoon-barrier complexes, spits, beaches, coastal sand dunes, cliffs
Coromandal	Tamil Nadu-Puducherry	Charnockites, gneisses, granite, Cuddalore sandstone, Mio-Pliocene and Quaternary sediments	E, P, NR	Large deltas, long beaches, spits, mangroves, coral reefs, dunes, chenniers, beach and dune ridges, red sediments (teri)
Andhra	Andhra Pradesh	Charnockites, khondalite, Quaternary deltaic and coastal deposits	E, P, R, NR	Large deltas, long beaches, spits, dunes, beach and dune ridges, mangroves, red sediments, sea cliffs, shore platforms
Odisha	Odisha-West Bengal	Quaternary deltaic and coastal/marine deposits, laterite	E, P, NR	Large deltas, long beaches, spits, dunes, beach and dune ridges, chenniers, mangroves
Sundarbans	West Bengal	Quaternary deltaic and coastal deposits	E, P, NR	Large delta, thick mangrove forests, tidal channels, deltaic islands, beaches, dunes
Andaman and Nicobar Islands	Andaman and Nicobar	Sandstones, conglomerates, limestone and shale, basalt, and coral reefs, Quaternary deposits	–	Archipelago, islands and islets, cliffs, caves, beaches, sand dunes, fringing coral reefs, mangroves, lagoons
Lakshadweep Islands	Lakshadweep	Corals	–	Archipelago, coral reefs and atolls, submerged coral banks, beaches, lagoons

S Submerged; *E* Emerged; *P* Prograding, *R* Dominantly rocky; *NR* Dominantly non-rocky
Compiled by Vishwas S. Kale from Vaidyanadhan (1987, 1991), SAC (2012) and numerous other sources

Table 3 Major deltas of India

Delta	Area in km^2	Delta type	Principal distributaries	Major geomorphic features
Ganga-Brahmaputra	65,000	Tide-dominated	Gorai-Madhumati, Jalangi, Methaabanga, Bhairab, Kobadak, Arial-Khan, Hoogly	Tidal channels, tidal islands, mudflats, large mangrove swamps, palaeo-channels
Mahanadi	9,500	Wave-dominated	Mahanadi, Devi, Bhargavi	Beaches, palaeo-channels, mangrove swamps, beach ridges, strandlines
Godavari	6,300	Wave-dominated	Vasishta, Gautami, Gostanadi	Palaeo-channels, beach ridges, mangrove swamps, Kakinada spit, delta lobes, strandlines
Krishna	5,500	River-dominated	Gulumuttapaya, Medimurru, Krishna	Palaeochannel, tidal islands, mangrove swamps, levees, five delta lobes, strandlines
Pennar	1,470	Wave-dominated	–	Palaeo-channels, spits, lagoons, levees, beach ridges, tidal flats,
Kaveri	8,770	Wave-dominated	Coleroon, Kaveri, Arasalar, Vettar	Palaeo-channels, mangrove swamps

Modified after Vaidyanadhan (1991) and articles therein and other sources basins—Bengal, Mahanadi, Krishna-Godavari, and Kaveri, with sediment thickness reaching at places to more than 5 km (Murthy et al. 2012).

The eastern continental margin is a mature margin, and came into existence between 140 and 120 Ma, when India broke away from East Antarctica at the 'bight' (knee-fold along the east coastline) where present-day Krishna-Godavari Basin is located. The break probably occurred in two stages giving rise to NE-SW oriented Krishna-Godavari rift segment and N–S oriented Kaveri shear segment (Murthy et al. 2012). The shelf along this margin is narrow with the deep ocean floor encountered within 50–60 km distance from the coastline (Fig. 1). In the deeper areas, the Bengal Fan (Fig. 1) sediments are underlain by two hotspot traces, viz., the 85 °E Ridge, which runs through the central part of the fan and the Ninety-East Ridge farther east, close to the Andaman Subduction Zone.

The western coastal lowland on the other hand, is a relatively narrow strip of land (50–100 km) sandwiched between the Western Ghat and the Arabian Sea. In contrast to the east, the west coast is fed by short and swift flowing rivers with hardly any delta formations. The estuaries of these rivers (Narmada, Tapi, Mandovi, Zuari, Sharavathi, Periyar, etc.) and backwaters characterize the west coast and form five offshore sedimentary basins—Kachchh, Saurashtra, Mumbai (Bombay) Offshore, Konkan and Kerala.

The west coast has a wider continental shelf compared to the east coast, particularly off Mumbai (Fig. 1). Compared to the east coast, the sedimentation is moderate here, usually not exceeding 3 km in thickness. The western continental margin was developed through two major continental separations—Madagascar breaking away from India at ~90–85 Ma and Seychelles parting away at ~64–62 Ma. The western continental margin has several ridge-like features, which run parallel to the coast and also have significantly controlled the sedimentation pattern along the margin. The architecture of this margin seems to have been affected by three events—two episodes of continental separation associated with episodic eruptions from two plumes—the older Marion Plume in the south and the younger Reunion in the north, and the Deccan volcanism (Mukhopadhyay et al. 2012).

3 Coastal Processes

Modern coastal processes along the Indian coasts are controlled largely by three meteorological seasons—fair weather (February to May), southwest monsoon (SWM, June to September) and northeast monsoon (NEM, October to January). Among the processes that shape the Indian coasts, affect the dynamics in the near-shore region and contribute substantially to alter the coastal environment; the monsoonal rainfall, winds, waves, tides, currents, and extreme events. Sediment input and sub-surface sedimentary structures also play a role in moulding the coastal tracts.

The east coast experiences a subtropical monsoonal climate with an annual rainfall of 1,600–1,800 mm and severe cyclonic storms (Gopal and Chauhan 2006). The SWM splits into two branches, the Bay of Bengal branch and the Arabian Sea branch. The Bay branch moves northwards in early June, while the Arabian Sea branch remains active all along the west coast as the moisture laden winds hit the Western Ghat Escarpment. The NEM, on the other hand, largely affects the east coast and is characterized by a high pressure over the landmass and a persistent northeasterly wind. While the west coast is visited by SWM alone, the east coast is drenched both by SWM and NEM.

Apart from the monsoons, the Indian coasts are constantly impacted by cyclones and storm surges. When such surges occur during the spring tide they cause devastation on low-lying coastlines and river mouths. The west coast experiences comparatively homogenous wave regime throughout the year (except during SWM), and the current pattern is usually influenced by the shifting of the Inter-Tropical Convergence Zone (ITCZ). Compared to this, east coast has a varying climatic regime. Every year, low-pressure systems form in the Bay of Bengal during October–January and move towards the land to strike the coast. The occurrence of cyclones in northern Bay of Bengal has increased by ~26 % during the last 120 years (Kotal et al. 2009).

The average wind speed during the NEM and SWM varies considerably (Table 1). For example, the average wind speed attains ~20 km/h (kmph) during the NEM, while during the SWM the speed attains ~35 kmph, frequently rising up to 45–55 kmph (Kumar et al. 2006). Both the east and west coasts show seasonal reversal in ocean currents, particularly with the change in monsoon pattern. For example, along the west coast, the water masses circulate clockwise for about 8–10 months of the year, and reversal occurs at the end of the SWM and continues for the next 2–4 months (Shankar 2000). The circulation along the east coast is characterized by the clockwise flow during most months but strong counter clockwise currents occur with the onset of the SWM. In the west coast waters, evaporation exceeds precipitation and runoff, thereby adding to the salinity of the Arabian Sea while low salinity prevails in the east coast waters due to higher runoff and precipitation from two monsoons. Currents near the river mouths and in the Gulf of Kachchh and Gulf of Khambhat are highly influenced by tides. The measured current speed is found to vary from ~1.4 m/s in the open ocean to about 3.2 m/s in the Gulf of Khambhat (Table 1).

Fig. 3 Sandy-gravelly beach along the Kachchh coast near Khuada. The shore platform is developed over Tertiary conglomerates (*red in colour*). The gravels derived from this unit constitutes shingle beach. The *light coloured* backshore dunes are also seen (Photo courtesy of Nilesh P. Bhatt)

The west coast of India experiences high wave activity during the SWM, whereas on the east coast, wave activity is significant both during SWM and NEM (Kumar et al. 2006). Along the west coast, waves approach from west and west-southwest during SWM, west and west-northwest during NEM and southwest during fair weather period. In contrast, along the east coast, waves approach from southeast during SWM and fair weather periods and from northeast during the NEM.

Tidal range along the Indian coastal region varies from 8.5 m at Bhavnagar (Gulf of Khambhat) to 0.5 m along the southern tip of the Indian Peninsula (Kanyakumari). The tidal range increases to over 6.0 m in Hoogly River and Haldia. Large-scale currents along the outer shelf and beyond reverse seasonally with the monsoon winds, and these along with tides, winds and river runoff play important roles in influencing currents in the inner shelf (Gopal and Chauhan 2006; Kumar et al. 2006).

The gross longshore sediment transport rate has been one million cubic meter per year along south Kerala and Odisha coasts. Net transport of sediments along east coast is towards north whereas that along west coast is south (Table 1). The Ganga-Brahmaputra system alone contributes sediment to the tune of one billion tonnes/year to the Bay of Bengal. The suspended sediment discharge into the Bay of Bengal is estimated to be about 1.4 trillion tonnes (Milliman and Meade 1983). Annual gross sediment transport rate is high (15–20 $\times$ 10^5 m^3) along the coasts of south Odisha, north Tamil Nadu, south Kerala, north Karnataka and south Gujarat, whereas it is comparatively less (5 $\times$ 10^5 m^3) along the southern Tamil Nadu and Maharashtra coasts. The annual net transport is northerly on the east coast and southerly on the west coast, except along south Gujarat coast (Chandramohan et al. 2001).

Fig. 4 Cliffs in miliolite limestone along with raised tidal notch at Diu, Saurashtra (Photo courtesy of Nilesh P. Bhatt)

4 Coastal Landforms

The landforms along the Arabian Sea and Bay of Bengal coasts display characteristic variations (SAC 2012), which are briefly described division-wise (Fig. 2) below and summarized in Table 2.

4.1 The Arabian Sea Coast (Fig. 2)

(a) *Kachchh*: A large saline desert, known as the Rann of Kachchh, is the most prominent feature in this sector. Coral reefs occur in the Gulf of Kachchh and mangroves are present around Kori Creek. Whereas sandy beaches and dunes are present along the northern coastline of the Gulf, the southern coastline is dominated by mudflats. Along the northern coast, shore platforms developed over Tertiary conglomerates (Fig. 3) and longitudinal sand dunes are some of the salient features.

(b) *Saurashtra*: Bioclastic carbonate deposits (miliolite limestone) of late Quaternary age characterize land–sea interaction in the form of cliffs (Fig. 4), marine terraces, wave-cut platforms and marine notches along the Saurashtra coast. These features have been used to ascertain the magnitude of sea level changes during late Quaternary. Sandy beaches occur as, more or less, continuous strip between Dwarka and Diu. Cliffs and shore platforms are prominently seen between Varval, Diu and beyond Jafrabad.

(c) *Konkan*: The Konkan coastline is dominantly rocky, with rock promontories and intervening sandy, pocket beaches and intertidal mudflats. Sediments are predominantly derived from Deccan Traps basalts. This stretch includes estuaries, headlands, sea cliffs, shore platforms, pocket beaches, tidal flats, mangroves, and

Fig. 5 Wave-cut platform in laterite at Dona Paula, Goa

spits. Well-developed wave-cut platforms are present at Harihareshwar and Murud Janjira, and sea caves can be seen around Ratnagiri. Large patches of mangroves occur especially along the Thane Creek and Malvan coast. The Malvan Marine Sanctuary is rich in corals, seaweeds and mangroves. Beach rocks (*karal*) and aeolianites have been reported from several places along the Konkan and Goa coasts.

(d) *Goa-Kanada*: Goa's coastline is a scenic combination of bays and headlands broken by large estuaries of the Mandovi and Zuari rivers. Of the bays, the Baga, Calangute, and Colva are long curved stretches of white beach sands and palm fringes. The Uttar (north) Kanada coast is more rocky than that of Dakshin (south) Kanada. Netravati, Kalinadi, Sharavati, etc. rivers have prominent estuaries and Kundapur Estuary is the largest estuarine complex. Sand bars are present at mouths of most of the estuaries. Wave-cut platform in laterite at Dona Paula (Fig. 5), cliffed coastline around Someshwar, columnar basalt at St. Mary Island near Udupi and crescent-shaped beaches near Gokarna, are some of the noteworthy features in this stretch.

(e) *Malabar*: The Malabar Coast in Kerala is distinguished by a chain of brackish lagoons and lakes (*kayals*) lying parallel to the coast. The backwaters were formed by the action of waves and shore currents creating low barrier islands across the mouths of the many rivers flowing down from the Western Ghat. Cliffed shorelines developed in Neogene sediments, laterites and crystalline rocks are observed at several places such as near Kovalam, Varkala and especially between Vettor and Tangasseri. Wave cut-platforms and stacks have been reported from Kannur, Kollam, Quilandy and Thalassery.

(f) *Lakshadweep Islands:* This group of over 27 coral islands is associated with the north–south aligned submarine Lakshadweep Chagos Ridge in the Arabian Sea (Fig. 1). Apart from coral reefs and atolls, lagoons and beaches are other distinctive features that attract large number of tourists. The archipelago experienced severe coral mortality and bleaching in 1998, and the live coral cover in the reef lagoons was reduced to <10 %. Today the live coral cover is ~20 %. The severely damaged Kadmat and Agatti atolls are showing good recovery.

Fig. 6 Sea cliffs and shore platform in khondalites near Visakhapatnam on the east coast of India (Photo courtesy of Vishwas S. Kale)

4.2 The Bay of Bengal Coast (Fig. 2)

(a) *Coromandal*: The Gulf of Mannar is endowed with three marine ecosystems—the corals, sea grass beds and mangroves with swampy regions that are surrounded by highly productive fringing and patchy coral reefs (Table 2). Rich coral reefs also occur in the Palk Bay. Long Marina Beach in Chennai, coastal dunes along Mahabalipuram sector, unusually straight Kaveri Delta and the Pulicat Lagoon are some of the noteworthy features along this coastline. Beach rocks have been reported from Rameswaram and Mandapam, and sea caves, wave-cut platforms and marine terraces are present at some places, for example at Mandapam and Tiruchendur. The red (teri) sand in Tamil Nadu consists of economically viable placer minerals including ilmenite, zircon and sillimanite. The entire Coromandal coast was impacted severely by the December 2004 tsunami.

(b) *Andhra*: The ~300 km long Andhra coast is dominated by the Godavari-Krishna Deltas complex and long, extensive beaches. A prominent sand spit (~17 km long) enclosing the Kakinada Bay is one of the interesting features along the Godavari Delta coast. More than four strandlines have been identified in the delta area. Rocky sea cliffs, shore platforms and sea caves are present along the northern Andhra coast, particularly between Revu Polavaram and Vishakhapatnam (Fig. 6).

(c) *Odisha*: Much of the ~480 km coastline in this sector consists of long sandy beaches and muddy shorelines. The coastal plain of Odisha is called the "hexadeltaic region" formed by six rivers (Subarnarekha, Budhabalanga, Baitarani, Brahmani, Mahanadi, and Rushikulya). The Bhitarkanika Sanctuary is one of the largest mangrove forests outside the Sundarbans, and the Chilka Lake is the largest brackish water coastal lagoon in the subcontinent. Several strandlines have been identified along the Mahanadi Delta. Beaches are extensive (e.g. Puri) and beach-dune complexes are common. The Odisha coast is frequently affected by coastal flooding and high storm surges associated with Bay of Bengal depressions and cyclones.

(d) *Sundarbans*: With an area of ~10,000 km^2, the Sundarbans tidal delta forms a part of the world's largest

Fig. 7 Marshy, muddy and mangrove rich Sundarbans (Photo courtesy of R. L. Chavan)

delta fed by Ganga-Brahmaputra, with several tidal inlets, creeks and deltaic islands. The tidal delta is very swampy, with extensive mangroves (Fig. 7). Major features include mud flats, mangrove swamps, dune complexes, estuaries, creeks, beaches and islands. The sinuous tidal channels separating the numerous islands are fed by the backwaters of Bay of Bengal.

(e) *Andaman and Nicobar Islands*: This archipelago, consisting of over 570 islands and islets, is characterized by ophiolite, epiclastic/volcaniclastic sandstones, tuffs, bioclastic limestones and patchy reefs of late Cretaceous to Pleistocene age. Narcondam and the Barren Island are two small volcanic islands in the Andaman Sea. Sandy beaches, mangrove-lined creeks, tidal inlets and bays, fringing reefs and rocky shores are some of the salient features of the islands. The sediment-based coastal features (beaches, tidal flats), mangroves and coral reefs experienced intense damage during the December 2004 tsunami.

5 Late Quaternary Sea Level Changes

The coastal zone has been alternately flooded and exposed during the sea-level fluctuations during the late Quaternary period as well as earlier. In general, lower sea levels were associated with the glacial periods while warmer interglacial periods were characterized by higher sea levels and transgression. During the Last Glacial Maximum (LGM) the sea level along the Indian coastline was lower by more than 120 m. Available evidence suggests that the sea level rose rapidly during the early Holocene flooding low lying areas and penetrated inland via rivers creating estuaries (Bruckner 1988; Hashimi et al. 1995; Banerjee 2000).

Evidence of higher sea levels in the past in the form of raised marine terraces, beach rocks, oyster beds, corals, tidal deposits, etc. have been reported from some parts of Saurashtra, Konkan, Malabar, Coromandal and Andhra coasts. Marine shells, beach rocks and coral reefs occurring ~1–2 m above the high tide level have been reported from Rameswaram and Kanyakumari (Bruckner 1988; Banerjee 2000). Beach rocks at Manori (Mumbai) and Mirya Bhatti (Ratnagiri) are present ~3–6 m above present sea level (Agrawal and Guzder 1972).

Instrumental records provide evidence for onset of sea level rise during the 19th century. Estimates for the 20th century show that in response to global-warming the mean sea level is rising by ~1.30 mm/yr along the Indian coasts and future global projections suggest a rise of about 480 mm by the turn of the 21st century (Mehra et al. 2013).

6 Concluding Remarks

Occurrence of mineral resources along the coasts of India is of great importance to its economy. For example, the Quaternary sediments of two important delta-basins along the east coast- Krishna-Godavari (KG) Basin and Kaveri Basin, and the Tertiary (Paleogene-Neogene) sediments off Mumbai along the west coast, hold rich potential of gas and petroleum reserves. Additionally, several placer minerals—magnetite, ilmenite, hematite, monazite, rutile, garnet, sillimanite, thorium, zircon, and chromite—are found along the Indian coasts. Eventually, mining of several of these valuable minerals as well as beach sand could pose threat to the coastal environment.

Apart from coastal erosion (Table 1), pollution is another major threat to the coastline. Most pollution arises from man-made sources—industrial, domestic wastes, agricultural runoff, shipping, ship-building, and bio-invasion through release of ballast water. Oil spills are also a huge risk. Yet, the pollution data acquired over the few decades indicate that the Indian coasts have well circulated oxygenated healthy waters, and that pollution-hotspots remain contained within reasonable limits (Gopal and Chauhan 2006).

References

Agrawal DP, Guzder SJ (1972) Quaternary studies on the western coast of India: preliminary observations. The Palaeobotanist 21: 216–222

Amante C, Eakins BW (2009) ETOPO1 1 Arc-minute global relief model: procedures, data sources and analysis. NOAA Technical Memorandum NESDIS NGDC-24, 19 pp, March 2009

Banerjee PK (2000) Holocene and late Pleistocene relative sea level fluctuations along the east coast of India. Mar Geol 167:243–260

Bruckner H (1988) Indicators for formerly higher sea-levels along the east coast of India and on the Andaman Islands. Hamburger Geogr stud 44:47–72

Chandramohan P, Jena BK, Kumar SV (2001) Littoral drift sources and sinks along the Indian coasts. Curr Sci 81:292–297

Gopal B, Chauhan M (2006) Biodiversity and its conservation in the Sundarban mangrove ecosystem. Aquat Sci 68:338–354

Hashimi NH, Nigam R, Nair RR, Rajagopalan G (1995) Holocene sea-level fluctuations on western Indian continental margin: an update. J Geol Soc India 46:157–162

ICMAM (2009). Report on use of satellite data for detection of violation of land use along the Coastal Regulation Zone and impact of port structures on shoreline changes—ICMAM Project Directorate, Chennai and Indian National Centre for Ocean Information Services (INCOIS), Hyderabad

Kotal SD, Kundu PK, Roy-Bhowmik SK (2009) An analysis of sea surface temperature and maximum potential intensity of tropical cyclones over the Bay of Bengal between 1981 and 2000. Meteorol Appl 16:169–177

Kumar SV, Pathak KC, Pednekar P, Raju NSN, Gowthaman R (2006) Coastal processes along the Indian coastline. Curr Sci 91:530–536

Mehra P, Desai RGP, Joseph A, Vijay Kumar K, Agarvadekar Y, Luis R, Nadaf L (2013) Comparison of sea-level measurements between microwave radar and subsurface pressure gauge deployed at select locations along the coast of India. J Appl Remote Sens 7(1):16. doi: 10.1117/1.JRS.7.073569

Milliman JD, Meade R (1983) Worldwide delivery of river sediments to oceans. J Geol 91:1–21

Mukhopadhyay R, Karisiddaiah SM, Ghosh AK (2012) Geodynamics of Amirante Ridge and Trench complex, western Indian Ocean. Int Geol Rev 54:81–92

Murthy KSR, Subrahmanyam AS, Subrahmanyam V (2012) Tectonics of the eastern continental margin of India. The Energy & Resources Institute New Delhi, 180 p, ISBN 9788179934081

SAC (2012) Coastal zones of India. Space Applications Centre, Ahmedabad, p 597

Shankar D (2000) Seasonal cycles of sea level and currents along the coasts of India. Curr Sci 78:279–288

Vaidyanadhan R (1987) (ed) Coastal geomorphology of India. Journal Geological Society of India, special volume 29, Bangalore

Vaidyanadhan R (1991) (ed) Quaternary deltas of India. Memoir 22, Geological Society of India, Bangalore, 291 p

Part III

Landforms and Landscapes

The Siachen Glacier: The Second Longest Glacier Outside the Polar Regions

Mahendra R. Bhutiyani

Abstract

The Nubra Valley in Karakoram Himalaya is a highly glacierized valley with about 33 valley glaciers of different lengths and sizes. The glaciers occupy nearly 2/3rd of the total basin area. The most prominent and longest amongst them is the Siachen Glacier. This well-known glacier is a fine example of a compound and piedmont glacier. Glacier erosion has resulted in the development of a myriad of erosional and depositional features. These include numerous tributary glaciers with cirques, arêtes, horns, bergschrunds, moraines, supra-glacial streams and glacier lakes. Being a temperate glacier, the Siachen produces copious amount of runoff during ablation season in its proglacial streams. Heavy sediment load carried by the Nubra River has given rise to many glacio-fluvial depositional landforms and features such as outwash plains, braided rivers, alluvial fans, varved clays, etc. Evidence indicates three periods of glaciation in the area during the last ~145 ka. The role of tectonics in the late Quaternary period is also evident by way of at least three episodes of uplift, resulting in the formation of river terraces. There is no indication of significant recession of the Siachen Glacier snout during the last millennium.

Keywords

Siachen glacier • Karakoram Himalaya • Piedmont glacier • Valley glaciers • Moraines • Quaternary glaciations • Snout recession

1 Introduction

The Himalaya, one of the youngest folded mountain systems, is one of the most dynamic regions (seismically and geologically) on the surface of the Earth. With about 17 % of the total area occupied by approximately 15,000 glaciers, it has a large influence on the meteorological and hydrological conditions over the Indian subcontinent, particularly on the Indo-Ganga Plains where majority of the rivers derive significant portion of their runoff from either seasonal snowmelt and/or glaciermelt.

The Karakoram Range (Fig. 1) in the northern-most part of the Himalaya is highly glacierized with about 37 % of its area covered by a number of valley and piedmont glaciers of varying lengths, ranging in altitude from 2,800 to 7,600 m a.s.l. The Shyok River, with the Nubra River as a tributary, is a major river system in the area (Fig. 1). The Shyok is a tributary of the Indus River.

Closeted in a lap of an archetypal U-shaped glacier valley with the Saltoro Hills to the west and Karakoram Range to the east in the Ladakh District of the state of Jammu and Kashmir, the Siachen Glacier is the longest (76 km) glacier in Himalaya and the second longest glacier in the world outside the Polar Regions. A north-west-southeast trending glacier, it originates from a cirque

M. R. Bhutiyani (✉)
Defence Terrain Research Laboratory, Metcalfe House, New Delhi 110054, India
e-mail: mahendra_bhutiyani@yahoo.co.in

V. S. Kale (ed.), *Landscapes and Landforms of India*, World Geomorphological Landscapes,
DOI: 10.1007/978-94-017-8029-2_9, © Springer Science+Business Media Dordrecht 2014

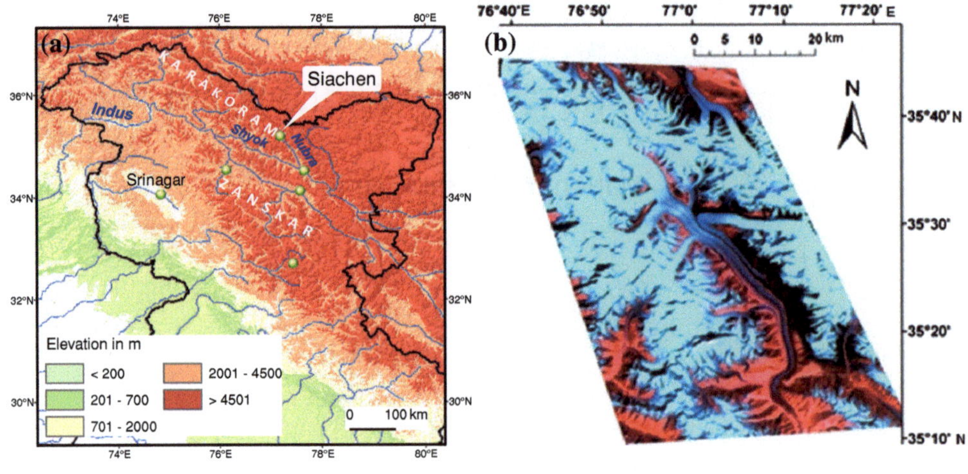

Fig. 1 **a** Location of the Siachen Glacier and **b** LISS-III satellite image of the WNW-ESE trending Siachen Glacier showing its snout and tributary glaciers

at the base of the ridge called "Indira Col" at an altitude of 6,115 m a.s.l, follows a well defined WNW-ESW trending valley and culminates in a snout located at an altitude of 3,570 m a.s.l (Fig. 1). Two proglacial melt-water streams emerge out of two ice caves in the snout region and merge into a single stream one km down-stream. The Nubra River thus formed, traverses its own outwash plain for a distance of about 90 km in a braided channel pattern up to its confluence with the Shyok River near Deshkit.

High relief, sub-zero temperatures, scanty precipitation in the lower reaches and moderate to heavy snowfall in the upper reaches and presence of numerous valley glaciers have made the terrain extremely rugged and difficult to traverse. The lower altitudes are marked by arid climate and high climatic gradient, resulting in the formation of typical dry desertic landscape features such as sand dunes and sand bars within the river-bed.

The Siachen Glacier has fascinated many explorers in the past two centuries who had organized a number of expeditions to this region and generated significant amount of information in the form of their travelogues and technical reports. First-ever recorded report on the Siachen Glacier was authored by Henry Strachey in October 1848. In 1909, Dr. Tom Longstaff, Dr. Arthur Neve and Lt. A M Slingby crossed over Bila Fond La Pass and later over to the Siachen Glacier snout in a pioneering effort to establish the length and exact locations of various passes. Dr Tom Longstaff was the explorer who named this glacier as Siachen. "*Sia*" in the Balti language refers to the rose family plant widely dispersed in the region. "*Chun*" refers to any object found in abundance. Thus, the name Siachen refers to a land with an abundance of roses.

2 Geology and Climate

Geologically, the Nubra Valley can be sub-divided into three sub-divisions: (1) The Northern Sedimentary Zone (2) The Central Crystalline Zone, and (3) The Southern Volcanic Schist Zone.

The northern portion of the valley is almost entirely covered by ice and has very few rock outcrops. However, the moraines originating from this part of the valley have abundance of sedimentary rocks like dolomitic limestones, carbonaceous shales and slates. These moraines, which form the middle three rows on the glacier body have a whitish grey tone on the satellite imagery. This northern sedimentary zone has a thrusted contact with the central crystalline zone to its south (Ganser 1964) which has pink granites with phenocrysts of pink feldspars and quartz at its centre and granodiorites and granite-gneisses towards the periphery. The granodiorites and granite-gneisses can be demarcated from granites by their dark greyish tone (Fig. 2). This zone is traversed by number of pegmatitic and quartz veins and lit-par-lit injections of the calcitic material. It has a thrusted contact with volcanic schist zone in the southern part of the valley consisting mainly of volcanic ultrabasics with serpentinite lenses, ferruginous shales and quartzites.

A major portion of the precipitation in this part of the north-western Himalaya occurs under the influence of extra-tropical low pressure systems called 'western disturbances' (WDs). The lower elevations in the Karakoram Himalaya receive an annual snowfall of about 75–150 cm (snow depth) whereas higher elevations receive more than 1000 cm (snow depth). Higher and lower portions of the area remain in sub-zero temperatures for 63–90 % and

Fig. 2 An aerial view of the Siachen Glacier and the Nubra Valley. The central crystalline zone with *pink* granites and granite-gneisses (*light grayish* tone) at centre and granodiorites towards the *top* portion (*dark greyish* tone)

35–40 % of the days respectively in a year. Minimum temperatures in the range of −20 to −40 °C have been recorded in the months from December to February. The area receives comparatively very low precipitation during the monsoon months from July to September.

3 Landforms

Majority of the glaciers are aligned in WNW-ESE direction followed by a few in NNW-SSE and ENE-WSW direction. Three major sets of fractures appear to have shaped the morphology of the valleys hosting these glaciers. WNW-ESE trending fractures, which are generally parallel to NW-SW strike/trend of the Himalaya, appear to have given rise to strike valleys which now house some of the longest glaciers in the world (e.g. Siachen, Baltoro, Batura, etc.). The morphology of transverse glaciers, such as Lagongma and Layongma indicates that WNW-ESE and ENE-WSW trending fractures have given rise to these valleys. The glaciers occupying strike valleys tend to have longer lengths than the glaciers in transverse valleys because the lineaments are also longer in WNW-ESE direction as compared to other two directions.

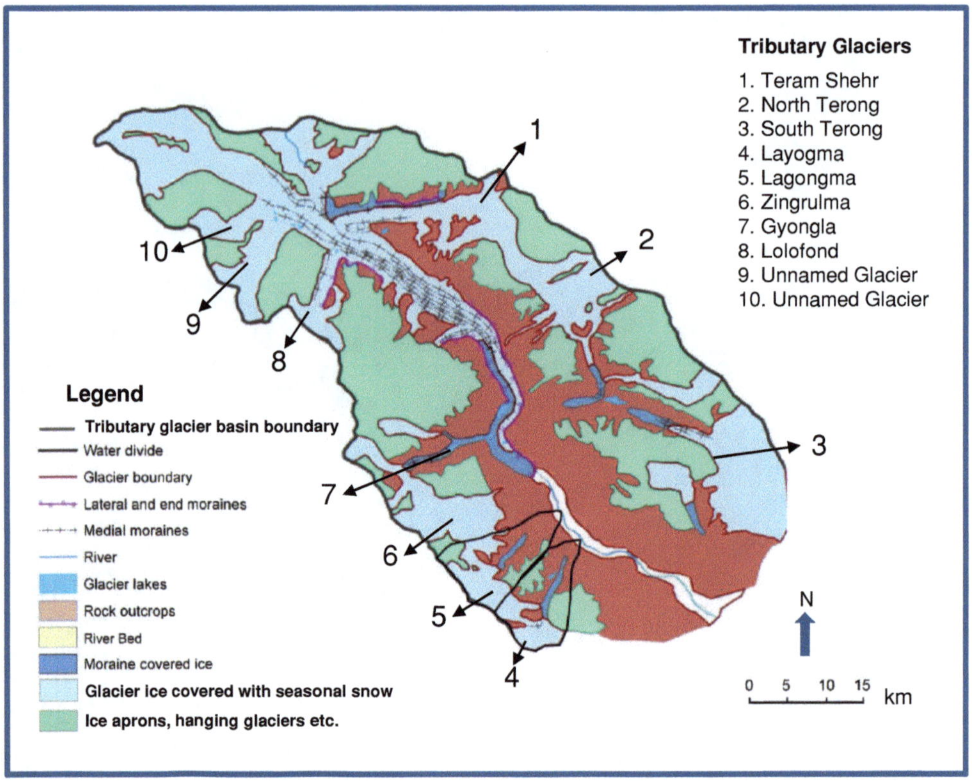

Fig. 3 Map showing different geomorphological features of the Siachen Glacier and Nubra Valley

The geological work done by the Siachen Glacier during the last glaciations has given rise to the development of a myriad of erosional and depositional glacial landforms, resulting in a truly unique and spectacular landscape (Fig. 3).

3.1 Erosional Features

Siachen, a compound glacier housed in a U-shaped glacial valley with steep valley walls on either side, is fed by a number of tributary glaciers with their own U-shaped valleys on both sides. On the left flank, it has three prominent tributary glaciers, namely Teram Shehr, North Terong and South Terong (Fig. 3). Teram Shehr is the longest tributary of Siachen having a length of ~32 km. North Terong and South Terong glaciers, which are now independent glaciers, occupy NNW-SSE trending valleys. On the right flank, there are five tributary glaciers namely Zingrulma (length ~6 km), Gyongla (~22 km), Lolofond (~11 km) and two unnamed glaciers in the north having lengths of ~12 km. There are numerous hanging glaciers with their individually carved glacial valleys. Because of such large number of tributary glaciers, there are many cirques, arêtes and glacier horns (Fig. 4). The highest of such horns in the Nubra Basin is a peak called "Sia Kangri" with an altitude of about 7,600 m a.s.l.

Generally, two types of crevasses, transverse and longitudinal, have been observed. The average gradient of the Siachen Glacier being low (~3°), not many transverse crevasses are noticed on the main glacier body. They are mostly confined to the areas of higher gradients and bends. They are also found at the confluences of various tributary glaciers and in the snout region. The longitudinal crevasses have been observed only in the snout region and are about 1 to 2 m in width. The tributary glaciers, such as Gyongla, Zingrulma, Lolofond and Teram Shehr, are highly crevassed because of relatively high gradient (~11°).

3.2 Depositional Features

The Siachen Glacier is characterized by the presence of extensive supraglacial lateral, medial and terminal or end moraines covering almost half of its total surface area. Although accumulation zone is relatively free of moraines, the ablation zone is largely covered with moraines, making Siachen a predominantly a debris-covered glacier.

The end or terminal moraines of Siachen have a length of ~4 km and a width of ~1.5 km (see Fig. 2). They mainly consist of unsorted, heterogeneous mixture of boulders of granite, granodiorite, granite gneisses, limestones, carbonaceous shales, slates and conglomerates. They are also characterized by a highly undulating and

Fig. 4 A typical cirque showing bergschrund. Multiple horns and arêtes are visible in the background

hummocky topography with a number of glacial lakes and supraglacial channels. The end moraines of Siachen coalesce with the end moraines of its tributary, the Gyongla, which has almost an E–W orientation.

The lateral moraines of the Siachen are seen on both the sides of the steep-sided valley. They consist of unsorted loose boulders of granites, granodiorites and granite gneisses with abundant rock waste.

Eight rows of medial moraines have been observed on the main glacier body. The western rows of medial moraines are entirely made up of pink granites and granite gneisses. The middle rows consist of sedimentary rocks, mainly limestones, dolomites and carbonaceous shales. The eastern rows of medial moraines consist mainly of granodiorites and granite gneisses. The differential melting of the glacier ice covered with medial moraines over a period of time has made the topography of the glacier-body highly undulating with the formation of number of morainic ridges (Fig. 5) and ice pinnacles.

Many active supraglacial streams and lakes are observed on the glacier body. The glacial lakes at some places have been formed as a result of whirling action of the supraglacial meltwater channels. A few of them have also been formed due to subsidence of the upper crust and caving in of the englacial cavities. The lakes thus formed vary in size from few meters to 30–40 m in diameter and are round to oblong in shape. The glacier lakes found in the snout region are shallow and smaller in size. They are interconnected by supraglacial channels/streams and at some places by englacial channels. Due to shrinkage of Teram Shehr and North Terong since their maxima, a few inactive glacier lakes, such as the one near the confluence of Teram Shehr Glacier with the main trunk glacier have also been formed.

Fig. 5 Numerous rows of medial moraines making topography of the glacier body highly undulating

The outwash plains of the proglacial stream of the Siachen Glacier are marked by extensive proglacial deposits of varved clays alternating with sand beds. These fluvio-glacial deposits are traversed by braided rivers. Tectonic activity during the late Quaternary period is evident by way of at least three episodes of uplift and formation of river terraces (Fig. 6). At places, wherever there are tributaries joining the main valley on both sides, these terraces are overlain by alluvial fans consisting of loose boulders near the valley slope and sandy to loamy soil in the middle portion and clayey soil on the periphery. Because of high fertility, some of these alluvial fans are cultivated and have human inhabitation on them (Fig. 7).

4 Retreat of the Siachen Glacier

Mountain glaciers located at higher elevations in the mid-latitudes and tropics like Himalaya are sensitive indicators of the changes in climatic conditions. The fluctuations in the climate are reflected in variations in the mass balance, glacier lengths, advance and retreat of the glaciers.

Majority of the glaciers in the Himalaya have had negative specific mass balance during the period from 1974–1975 to 1990–1991, coinciding with general rise in air temperature. Highly negative values of mass balance of the Siachen Glacier from 1986–1987 to 1990–1991 imply significant ice loss during this period (Bhutiyani 1999). Although no mass balance data are available for this glacier after this period, it is estimated that the glacier must have lost about 1.3 km^2 of its area since 1989 (Chander et al. 2012). The glacier has undergone retreat at a much lower rate, about 0.6 to 2.0 m/year between 1962 and 1998, as compared to other glaciers, which are comparatively shorter in length. This may be attributed to the fact that Siachen is a very long, compound and piedmont glacier with a low average gradient and its response time appears to be much longer than the other smaller glaciers (Bhutiyani 1999; Ganjoo and Kaul 2010).

Fig. 6 The braided channel pattern of the Nubra River on its outwash plain. A prominent river terrace is seen on the *left*

Fig. 7 View of an alluvial fan/talus cone with cultivated fields and human inhabitation on the periphery

5 Late Quaternary Glaciations

The understanding of past climatic changes serves as a benchmark against which predictive models of future climate can be evaluated. In recent years, many studies have been carried out to reconstruct the chronology of glaciations in the Himalaya and the Karakoram. Field evidence of remnant lateral moraines and striations on either side of the Nubra Valley up to its confluence with the Shyok River near village Deshkit (~90 km downstream of current position of snout) suggests that the Siachen Glacier might have extended up to the confluence and also fluctuated considerably during the late Quaternary. Studies based on field mapping, remote sensing and ^{10}Be terrestrial nuclide surface exposure techniques have indicated three glacial stages, namely, Deshkit 1, 2 and 3 (~45, ~85 and ~144 ka, respectively) which are synchronous with Milankovitch time-scales (Owen et al. 1998, 2008; Seong et al. 2007; Dortch et al. 2010). The Nubra Valley was extensively glaciated around 24 and 18 ka, a period corresponding with the Last Glacial Maximum (LGM) (Nagar et al. 2013). In addition, a number of north and south draining tributary glaciers may have contributed to the Siachen Glacier during the LGM. During the Little Ice Age (~1600 to 1850 AD) when Himalayan glaciers advanced marginally, the present position of the Siachen Glacier snout was insignificantly affected (Nagar et al. 2013). Recent warming trend has had a negligible effect on it and the glacier snout has remained, by and large, unchanged in the last millennium (Upadhyaya 2009; Ganjoo and Kaul 2010; Nagar et al. 2013).

6 Conclusions

Siachen, meaning a land with an abundance of roses, is the name of a long, compound and piedmont glacier in Karakoram Himalaya. The glacier is exclusive because it is the longest glacier in the Himalaya and the second longest glacier in the world outside the Polar Regions. Three major sets of fractures trending in WNW-ESE, NNW-SSE and ENE-WSW have dominantly shaped the glacial geomorphology of this part of Karakoram Himalaya. The Siachen Glacier is characterized by extensive supraglacial lateral, medial and end moraines, supraglacial streams and lakes and pro-glacial deposits. The area has witnessed three episodes of glaciations during the last ~145 ka, when the glacier may have extended up to the confluence of the Nubra with the Shyok River. If correct, this would indicate that the total length of the Siachen was nearly 165 km. The present position of the Siachen Glacier snout has not shown any detectable effect of twentieth century warming trend and it has remained largely unchanged in the last thousand years.

References

Bhutiyani MR (1999) Mass balance studies on Siachen glacier in the Nubra valley in the Karakoram Himalaya. J Glaciol 45(149):112–118

Chander S, Dewali SK, Sathish K, Snehmani, Bhutiyani MR (2012) Study of temporal variability of the Siachen glacier in upper Himalaya over the last decade using optical satellite data. In: International symposium on Cryosphere and climate change, Manali, India, 2–4 April 2012

Dortch JM, Owen LA, Caffee MW (2010) Quaternary glaciation in the Nubra and Shyok valley confluence, northernmost Ladakh. India Quat Res 74:132–144

Ganjoo RK, Kaul MN (2010) Is the Siachen glacier melting? Curr Sci 97:309–310

Ganser A (1964) Geology of the Himalayas. Wiley Interscience, London 289 p

Nagar YC, Ganju A, Satyawali PK, Juyal N (2013) Preliminary optical chronology suggests significant advance in Nubra Valley glaciers during the last glacial maximum. Curr Sci 105:96–101

Owen LA, Debyshire E, Fort M (1998) The Quaternary glacial history of the Himalaya. In: Owen LA (ed) Mountain glaciation. Quaternary proceedings, vol 6. Wiley, Chichester, pp 91–120

Owen LA, Caffee MW, Finkel RC, Seong YB (2008) Quaternary glaciation of the Himalayan-Tibetan Orogen. J Quat Sci 23:513–531

Seong YB, Owen LA, Bishop MP, Bush A, Clendon P, Copland L, Finkel R, Kamp U, John F, Shroder Jr (2007) Quaternary glacial history of the Central Karakoram. Quatern Sci Rev 26:3384–3405

Upadhyaya R (2009) Melting of Siachen glacier. Curr Sci 96:646–648

Ladakh: The High-Altitude Indian Cold Desert

Navin Juyal

Abstract

The arid landscape of Ladakh provides a rare opportunity to peep into the history of Earth surface processes. Because of the scanty rainfall, Quaternary landforms are better preserved in this part of the Himalaya. Glaciation, which is considered as the main driver of climate change, has undoubtedly played a key role in shaping the landforms. The alluvial fan dominated mountain slopes and valleys owe their genesis to the fluvial sculpturing and re-sedimentation of the glaciogenic sediments following the periods of glacier recessions. Lacustrine deposits, although important in terms of their climatic significance, constitute a minor component of the Quaternary landforms but provide important climatic data towards improving our understanding of temporal changes in moisture variability. In spite of the fact that the terrain is a cold-arid desert, sand dunes are scanty and limited to the river beds or the mountain flanks suggesting a precarious balance between the wind intensity and sediment supply. It can be suggested that the Ladakh Himalaya is a fascinating laboratory for Quaternary geomorphologists to unravel the history of landforms in a high, tectonically active and arid terrain.

Keywords

Cold desert • Glacial trough • Moraines • Alluvial fans • River terraces • Dammed lakes • Quaternary glaciations

1 Introduction

The barren and desolate landscape of Ladakh (or Ladhak) is a unique geological laboratory where the history of the Indian–Eurasian collision is preserved in vivid details. The oldest sedimentary sequence in Ladakh Himalaya belongs to the Tethyan Sedimentary Sequence (Zanskar Ranges) followed by the Indus molasses (between Zanskar Range and Ladakh Batholith). The Tethyan sediments pass gradually to the north in a large dome of greenschist to eclogitic (metamorphic) rocks. The Indus Suture Zone (ISZ) which broadly dictates the course of the Indus River in the region is represented by the occurrence of Ophiolite Mélanges and Dras Volcanic (Fig. 1). The northern boundary of the ISZ is demarcated by the predominantly granodioritic Ladakh Batholith which forms part of the plutonic remnants of the island arc that rimmed the Asian continent from Cretaceous to Eocene times. The emplacement followed creation of two major basins on the northern and southern side of the Ladakh Batholith. The basins were aggraded with coarse clastic continental sediments derived from the uplifted Ladakh Batholith in the north and the Tethyan Sedimentary Sequence (TSS) in the south (Indus molasses) with subordinate marine contribution. The molasses are post-collision deposits implying their deposition occurred sometime after the Eocene (Dézes 1999).

N. Juyal (✉)
Geosciences Division, Physical Research Laboratory,
Navrangpura, Ahmedabad 380058, India
e-mail: navin@prl.res.in

V. S. Kale (ed.), *Landscapes and Landforms of India*, World Geomorphological Landscapes,
DOI: 10.1007/978-94-017-8029-2_10, © Springer Science+Business Media Dordrecht 2014

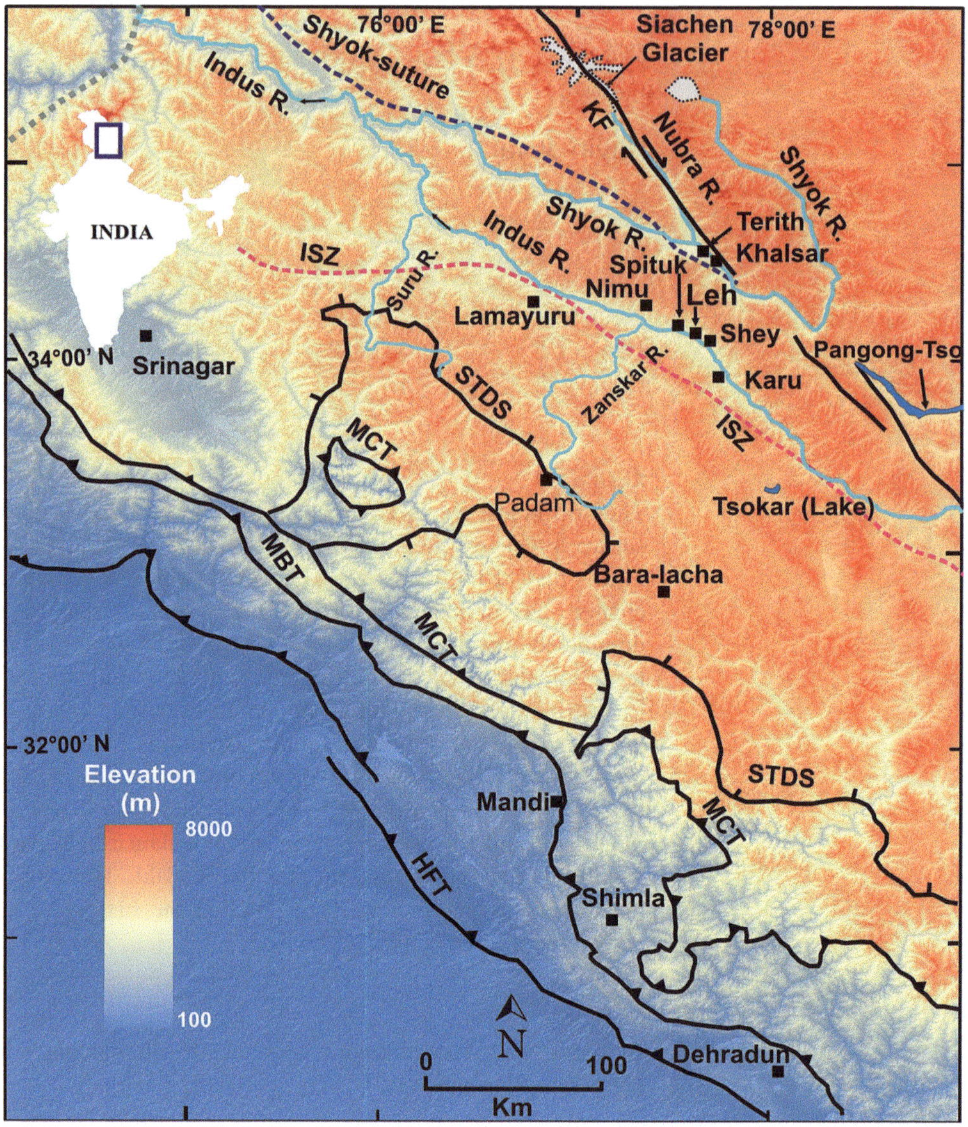

Fig. 1 Map of northwest Himalaya showing the major tectonic boundaries along with the sites mentioned in the text. HFT Himalayan Frontal Thrust, MBT Main Boundary Thrust, MCT Main Central Thrust, STDS South Tibetan Detachment System, ISZ Indus Suture Zone, KF Karakoram Fault

The Indian and Eurasian Plate collision ~50 Ma closed the Tethys Ocean and compressed the sediments deposited on the ocean floor into folded and faulted sequences. Even the pillow-shaped basaltic lava of the ocean floor and swarm of dykes were squeezed up (abducted). The northward convergence resulted in the stacking of the crustal slabs along the north dipping, and east–west striking terrain boundary thrusts which led to the evolution of present architectural framework of the Himalaya (Yin 2006). These are from north to south the South Tibetan Detachment System (STDS), which differentiate the TSS in the north from that of the Higher Himalayan Crystalline (HHC) rocks in the south. The Main Central Thrust (MCT) demarcates the tectonic boundary between the HHC in the north and the Lesser Himalaya (LH) meta-sedimentary rocks to the south. The rocks of LH are thrust over the Sub-Himalayan rocks along the Main Boundary Thrust (MBT). Finally, the Sub-Himalayan rocks which are composed of sediments essentially derived from the erosion of Himalaya are thrusted over the Quaternary alluvium of the Indo-Ganga Plains along the Himalayan Frontal Thrust (HFT) (Fig. 1).

2 Climate

Ladakh is a high-altitude (>3,000 m a.s.l.) desert lying north of the Higher Himalayan Ranges which act as a climatic barrier preventing the northward penetration of the Indian Summer Monsoon (ISM) creating desert-like conditions (Fig. 2). The decrease in moisture is reflected in the topography from the monsoon dominated lush green southern Himalayan slopes to barren and desolate landscape of Ladakh and Karakoram Ranges.

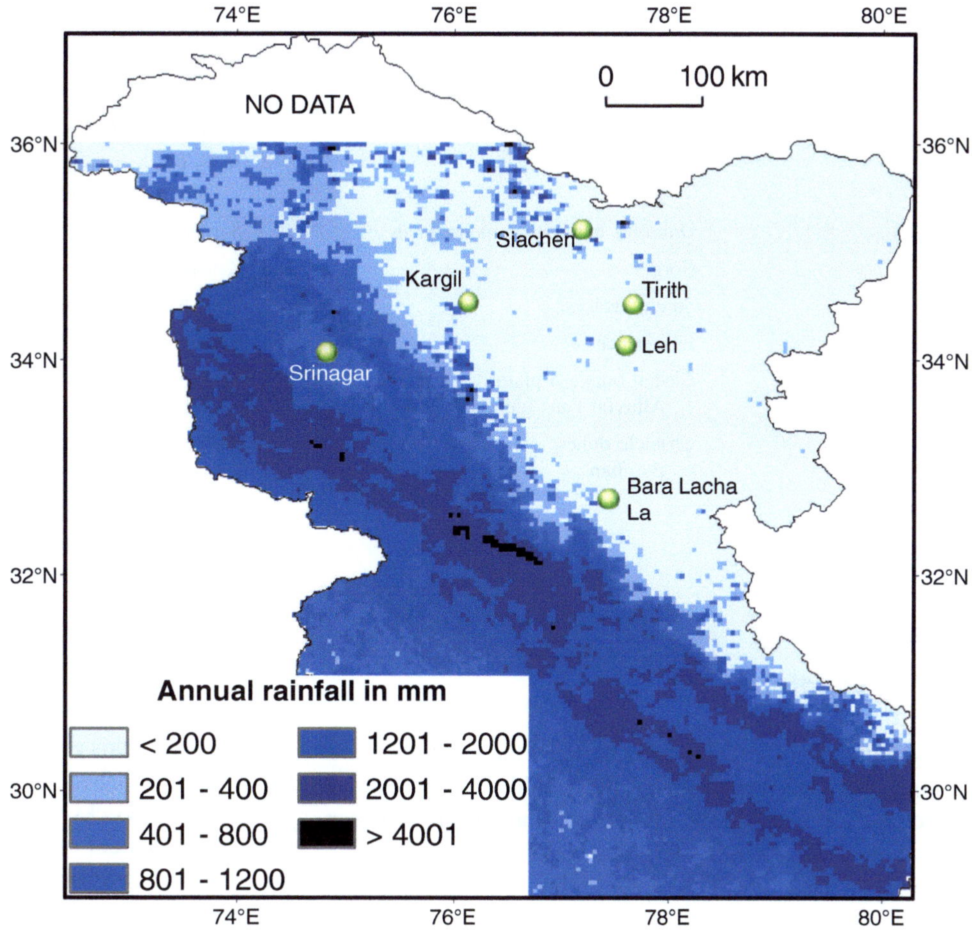

Fig. 2 Map showing the rainfall gradient in the northwest Himalaya. Average annual precipitation from TRMM satellite (*Data source* http://www.geog.ucsb.edu/~bodo/TRMM/)

The terrain receives considerable solar radiation due to its elevated topography. During the summer, temperatures are relatively high, although a large diurnal temperature range occurs, whereas the winter temperature plunges much below the sub-zero. There is an overall decreasing trend in the mean annual precipitation from 670 mm (at Dras) to ~100 mm at Leh (Holmes 1993). Mid-latitude westerlies are the major contributor of moisture as one moves northward towards the Karakoram Himalaya (Bhutiyani et al. 2010). Due to the overall arid climate, vegetation cover is scarce; limited to the areas proximal to glacial-fed streams and the cultivation is solely based on human irrigation system.

3 Drainage

The upper Indus River, along with the Shyok (Fig. 1), is the backbone of Ladakh. Traditionally, the source of the Indus River is considered to be the *Senge Khabab* or "Lion's Mouth", located near the sacred Mount Kailas, in Tibet. The river initially follows the northwest–southeast trending Karakoram Fault then turns westward and flows along the east–west trending suture zone (ISZ) in Ladakh. Considered to be the oldest river in Himalaya, it came into existence shortly after the collision of Indian–Eurasian Plates (>45 Ma) and remained confined in the suture zone since its establishment (Clift 2002). The river is joined by the Shyok, Shigar, Gilgit and Kabul Rivers from the northern Karakoram and Hindu Kush Ranges. In the vicinity of Nanga Parbat, the Indus has carved one of the deepest gorges in western Himalaya.

The nearly 500 km long Shyok River flows roughly parallel to the Indus River. The prominently longitudinal valleys of the two rivers are separated by the Ladakh Batholith. The Nubra River, the right-bank tributary of the Shyok, emerges from the Siachen Glacier and flows north–south to meet its parent stream at an obtuse angle (Fig. 1). This barbed drainage pattern is generally the result of river capture or back titling of headwaters.

From south, the Indus River is joined by the Zanskar River that originates in the Zanskar Range and flows northward. The tributary has deeply incised into the Indus Molasse forming a gorge.

Table 1 The salient landforms of the Ladakh Region

Geomorphic process	Landforms	Localities/sites
Glacial	Glacial troughs	Baralacha, Sarchu, Leh, Khardung, Nubra and Shyok Valleys
	Cirques	Tsokar Lake Basin, Northern Zanskar and Karakoram Ranges
	Moraines	Baga River (Baralacha) Yunam Tso (Sarchu Plain), Suru, Leh, Shyok and Nubra Valleys
	Drumlins and roches montonnées	Sarchu Plain, Baralacha Pass and Nubra Valley
Fluvial	Gorges	Zanskar, Suru and Indus Rivers
	Strath terraces	Near Nimu and Lamayuru
	Fill terraces	Sarchu Plain
	Glacial outwash plains Alluvial Fans (debris-flow dominated)	Sarchu and More Plains Between Karu to Nimu (Indus Valley)
Aeolian	Obstacle dunes Barchan	Between Shey and Sabu (Indus Valley) Hunder (Shyok Valley)
Lacustrine		Tso kar and Pangong Tso
		Baralacha La (Ynam Tso), Spituk (Indus Valley), Lamayuru, Khalsar (Shyok Valley)

4 Landforms

The landforms of Ladakh have attracted the earth scientists mainly for two reasons—(i) they hold key to resolve problems related to the dynamic relationship between Indian and Eurasian Plates; and (ii) they may help to unravel the causative factors controlling the mechanism of global climate change.

Ladakh displays a cold-desert landscape sculpted by Quaternary glaciations and cryogenic weathering, with superimposed effects of fluvial and mass movement processes (Table 1). Periglacial conditions exist at high altitude. Damming of rivers and dam-failure floods have been recognized as one of the major geomorphic processes in this terrain. First-order landforms include long ridges (Ladakh and Zanskar Ridges), prominent longitudinal valleys (Indus and Shyok Valleys) and fan belts (bajada). Glacial deposits are ubiquitous in the valleys. Hillslopes covered by frost-shattered debris and colluvium are striking features of the landscape. The strong control of lithology and tectonics is apparent in the landscape.

4.1 Glacial Landforms

Glacial processes have been the most important geomorphic activities operating in the region above 3,000 m altitude and are well represented by the erosional and depositional landforms. Features such as the "U" shaped valleys present the iconic image of the magnitude and extent of glacial sculpturing of the landforms in the region. Presently, these valleys are nearly vacated and where glaciers exist, they are confined to the valley headwalls (Fig. 3). The Khardung Glacier is located northwest of Leh town. A melt water stream that emanates from this glacier is the life line of Leh. In addition, presence of polished (desert varnish) striated boulders and whaleback bedrocks (roches montonnées) provide the evidence of the extent of past glaciers in the region. At places, glacially polished bedrock surfaces are engraved with hunting scenes (e.g. above Tirith and Deskit villages). Such engraving is typical of Neolithic culture and is dated to ~5 ka in Kashmir Valley (Pant et al. 2005) and implies that since the last 5 ka no major glaciation occupied these valleys (Fig. 4).

Moraines are the sedimentological expression of past glaciation. They can be found occurring between ~100 and 500 m above the valley floor and are represented by sharp-crested linear ridges which at times terminate with a curvilinear morphology. For example, in the Khardung Valley (north of Leh) a symmetrically paired lateral moraine can be seen around Ganglas village (Fig. 5a). Similarly, a detailed record of late Quaternary glaciation can be found preserved in the form of multiple lateral moraine ridges at Nubra and Shyok River confluence. At this location, four generations of lateral moraines, of decreasing altitude, are observed between 3,180 and 3,650 m a.s.l. (Fig. 5b).

4.2 Debris-Flow Dominated Alluvial Fans

In dry and elevated terrains, debris-flow dominated fans are ascribed to the re-sedimentation of glaciogenic sediments implying that the deposition essentially post-dates the events of glaciation. These deposits are formed between the elevated valleys and the flatter valley floor.

Fig. 3 East-west trending glacially carved "U" shaped valley near Kahardung La. Note the present Kahradung Glacier sticking to valley headwall

Fig. 4 Neolithic rock engraving depicting hunting scene can be found on glacially polished bed-rocks above Tirith and Deskit villages. The engraving suggests that after the recession glacier did not occupy this region at least during the last 5 ka

In the Indus River, around Leh, the fans which predominantly emanate from the Zanskar Ranges (southern flank of Indus River) are of coalescing type (bajada). Compared to the size of the debris-flow dominated alluvial fans, the valleys from where they emanate are quite small (Fig. 6). Thus, it can be speculated that the contribution of large volume of sediment from small valleys can only be explained by considering the availability of glaciogenic sediments following the recession of the valley glaciers. The lateral and vertical sedimentary facies show a wide variation in grain size; ranging from boulders to clay (Sant et al. 2011) and can be used to reconstruct the post-glacial climate and seismicity in the arid mountainous regions.

4.3 Fluvial Landforms

Although glaciers are effective excavators of sediments and rocks, their transportation capabilities are spatially limited. Compared to this, rivers not only erode the

Fig. 5 **a** Lateral and terminal moraines at the exit of Khardung Valley near Ganglas village. **b** Five degraded moraine ridges (marked as *I* to *V*) above Tirith village at the confluence of Nubra and Shyok Rivers. Neolithic rock engraving can be observed on surface *II*

Fig. 6 Debris-flow dominated alluvial fans from the northern Zanskar Range. These fans are the major source of sediments to the Indus River around Leh Valley

surface over which they flow but are efficient transporters. The fluvial landforms evolved by a combination of the temporal changes in the melt water discharge, infrequent flash floods caused by the lake out bursts and seismicity. Broadly, three major fluvial terraces can be found in the region—(i) fluvially modified debris flows/lake outburst terraces e.g. around Leh Town and Tangtse Valley (ii) valley-fill outwash gravel terraces e.g. around Sarchu and More Plains (Fig. 7) and the (iii) strath terraces e.g. at the junction of Indus and Zanskar Rivers (Fig. 8). Occurrence of more than one type of fluvial landforms can be ascribed to the sensitivity of fluvial system to spatial and temporal changes in climate and seismicity.

Fig. 7 Fluvial terraces in Sarchu Plain. The plain is filled by ~100 m thick outwash gravel indicating accelerated sediment flux during the period of deglaciation. The flat-topped terraces form six broad levels which we ascribed to the recessional limb of a major glacial lake outburst flood

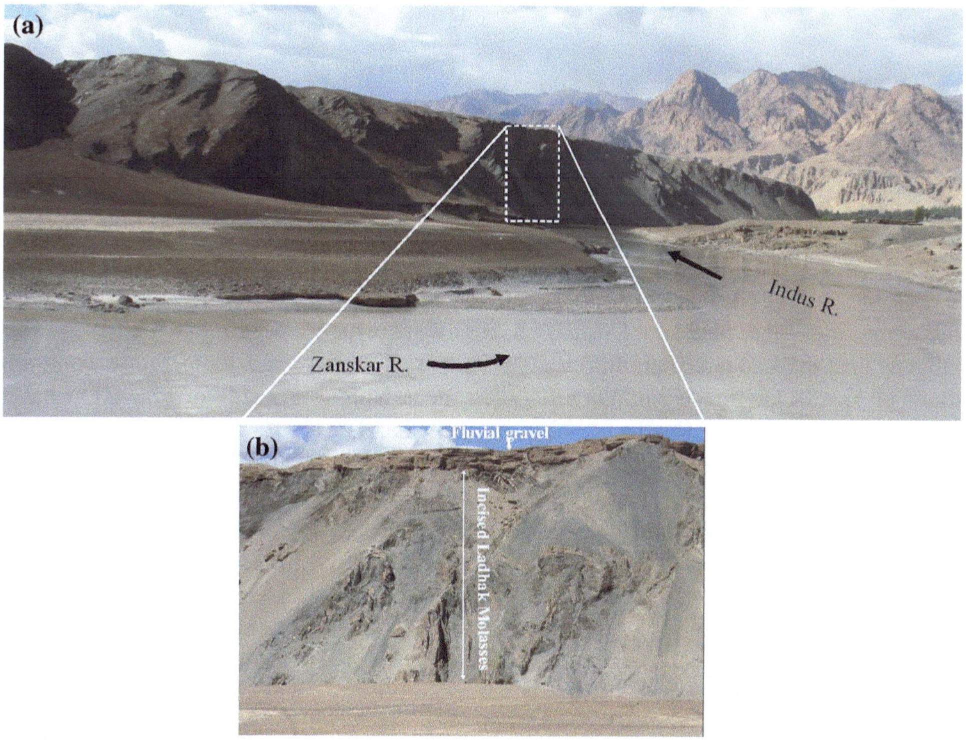

Fig. 8 **a** Strath terraces at the confluence of the Zanskar and Indus Rivers, near Nimu developed on the deformed Indus molasses. **b** The close-up of the strath terraces, note the thin fluvial gravel resting on the beveled Indus molasses

4.4 Sand Dunes

Being in the rainshadow, Ladakh experiences desert conditions. Usually, such areas contain aeolian features, such as sand dunes. In the Ladakh region, as compared to other landforms, these geomorphic expressions of aridity are less extensive. The sand dunes are represented by obstacle and barchan dunes, which can be observed proximal to the river valleys e.g. the Indus and Nubra-Shyok Rivers (Fig. 9a, b). The sediment is locally derived from the exposed river banks. Such dunes may or may not represent the role of climate parameters and cannot be always equated with phases of extreme aridity (climate). In the upper Indus Valley, well-preserved obstacle dunes can be located

Fig. 9 **a** Obstacle dune near Leh and **b** the barchans on the floodplain of Shyok River near Nubra-Shyok confluence

between Shey and Sabu on Upshi–Leh Highway, which abut against the Ladakh (Batholith) Range (Pant et al. 2005; Sant et al. 2011). In the Karakoram Valley, near Nubra-Shyok confluence (at Hunder), ~2–5 m high barchans dunes with their horns pointing towards the east (downwind) can be found. Lack of weathering suggests frequent mobilization and replenishment by fresh sediment supply (Fig. 9b) under hyper-arid conditions (intense wind activity) in Nubra–Shyok Valley.

5 Late Quaternary Climate

The timing and amplitude of palaeoglaciations represent important cornerstones of terrestrial palaeoclimatic research, because glaciers are arguably the most sensitive recorders of climate changes as they respond to the combined effect of snowfall and temperature. In recent years, attempts have been made to generate numerical ages on the moraine successions.

Multiple glaciation events have been recognized in the Ladakh Region. Interestingly, some workers believe the events did not coincide with the global Last Glacial Maximum (LGM) as expected, due to unfavorable moisture and temperature conditions. In the Zanskar Valley, three events of glaciation have been identified between 78 and 10 ka (Taylor and Mitchell 2000). Similar observation was made by Brown et al. (2002) in Khardung Valley near Ganglas village where they found that the most recent major glacial advance occurred ~90 ka and not during the global LGM. In comparison, at least five major glacial events, with progressive decreasing magnitude have been identified in Leh Valley (Owen et al. 2006). The oldest event is dated between 130 and 385 ka, whereas the youngest event is assigned early Holocene age. The decreasing magnitude is ascribed to the reduction in moisture flux due to uplift of the Himalayan Ranges to the south and the Karakoram Ranges to the west.

Further north in the Karakoram (Nubra-Shyok Valleys), Dortch et al. (2010) identified three glacial stages which they named as Deshkit-3 (144 ka), Deshkit-2 (81 ka) and Deshkit-1(45 ka). Deshkit-1 and 2 glacial stages seem to be influenced by the Indian summer monsoon precipitation. However, ages of 18 and 24 ka for moraines from Nubra Valley (eastern Karakoram) suggest that the area was

Fig. 10 The Lamayuru Lake deposits, popularly called as 'moon surface', provide one of the most spectacular sites in the arid landscape of Ladakh

extensively glaciated during the LGM also (Nagar et al. 2013). These inferences are at variance with those of the Dortch et al. (2010) but accord well with the recent climatic data which show that winter precipitation (mid-latitude westerlies) is the major source of moisture to the glaciers in Karakoram Region (Bhutiyani et al. 2010). Needless to say more dated sequences are required to resolve this important issue of the role of Indian summer monsoon versus the mid-latitude westerlies in modulating the Quaternary glaciation in the region.

Although moraines are direct manifestation of past glaciation, interpretation of glacial records based on age and position of moraines is subject to major temporal and spatial gaps. This problem can be circumvented using the sedimentary record of pro-glacial lakes. Consensus varies regarding the formation of the lakes in Ladakh Himalaya. Some workers ascribed it to tectonics, whereas others believe that they were formed following the recession of the valley glaciers (moraine dammed lakes). For example, lake deposits at Lamayuru are ascribed to the tectonically triggered debris slide $\sim$40 ka ago (Fig. 10). For Spituk Lake near Leh, dated to $\sim$50 ka, consensus varies in favor or against the role of tectonics. Similarly, the Tso Kar Lake located between Zanskar and Ladakh Ranges was considered to be formed by tectonic activity. However, the Khalsar Lake in the Shyok Valley (Karakoram) was attributed to damming caused by the terminal moraines of Nubra Valley. Except for the Tso Kar Lake, detailed history of past climate variability from the other lakes is still awaited.

The faulted Tso Kar Lake basin ($\sim$1000 km^2) has a closed drainage. Geophysical studies suggest that $\sim$500 m thick fluvio-lacustrine sediments were deposited in the trough (Wünnemann et al. 2010 and references therein). The study further suggests that basin hydrology was controlled by glacier and snowmelt water discharge during the last 15 ka. The contribution of monsoonal effective moisture was responsible for amplifying the water balance during wet periods. The Tso Kar Basin was under the influence of monsoonal precipitation during the late glacial and early Holocene epoch, due to the northward shift of the monsoon (ISM) domain. Presently, the periphery of the lake is an extensive sand covered area. Around four raised shorelines can be discerned implying steady decline in lake level since its formation (Pant et al. 2005).

6 Concluding Remarks

Regional geomorphology of this extraordinary landscape suggests that following the evolution of the present architectural framework of the Ladakh Himalaya, glacial processes dominated the terrain. This is eloquently manifested by the presence of innumerable "U" shaped valleys and wide spread occurrence of moraines. These landforms were subsequently modified by the interaction of fluvial and alluvial processes. The re-working of the glacial deposits by fluvial processes resulted in the formation of extensive debris flow dominated alluvial fans. In areas proximal to major structures, one can also observe well-developed strath terraces implying the tectonic overprinting on the landforms. Several relict lake successions are preserved in the region and although the evolution of these lakes is still being debated, they provide evidence of landform-modifying extreme events (dam-failure floods) as well as constitute an important archive for reconstructing still elusive high-resolution climatic history of this magnificent arid terrain.

References

Bhutiyani M, Kale VS, Pawar NJ (2010) Climate change and the precipitation variations in the northwestern Himalaya: 1866–2006. Int J Climatol 30:535–548

Brown ET, Bendick R, Bourlés DL, Gaur V, Molnar P, Raisbeck GM (2002) Slip rates of the Karakoram fault, Ladhak, India, determining using cosmic ray exposure dating of debris flows and moraines. J Geophys Res 107. doi:10.1029/2000JB000100

Clift PD (2002) A brief history of the Indus River. In: Clift PD, Karoon D, Gaedicke C, Criag J (eds) The tectonic and climatic evolution of the Arabian Sea region, vol 195. Geological Society, Special Publications, London, pp 237–258

Dézes P (1999) Tectonic and metamorphic evolution of the Central Himalayan domain in southeast Zanskar (Kashmir, India), vol 32. Mémoires de Géologie, Lausanne

Dortch JM, Owen LA, Caffee MW (2010) Quaternary glaciation in the Nubra and Shyok valley confluence, northernmost Ladakh, India. Quatern Res 74:132–144

Holmes JA (1993) Present and past patterns of glaciation in the northwest Himalaya: climatic, tectonic and topographic controls. In: Shroder JF Jr (ed) Himalaya to the sea: geology, geomorphology and the Quaternary. Routledge, London, pp 72–90

Nagar YC, Ganju A, Satyawali PK, Juyal N (2013) Preliminary optical chronology suggests significant advance in Nubra valley glaciers during the Last Glacial Maximum. Curr Sci 105:96–101

Owen LA, Caffee MW, Bovard KR, Finkel RC, Sharma MC (2006) Terrestrial cosmogenic nuclide surface exposure dating of the oldest glacial successions in the Himalayan orogen: Ladakh Range, northern India. Bull Geol Soc Am 118:383–392

Pant RK, Phadatare NR, Chamyal LS, Juyal N (2005) Reconstructing climate and seismic history of Ladhak-Karakoram region: potential of Quaternary landforms and sedimentary archives. Curr Sci 88:1789–1798

Taylor PJ, Mitchell AW (2000) The Quaternary glacial history of the Zanskar range, north-west Indian Himalaya. Quatern Int 65:81–99

Sant DA, Wadhawan SK, Ganjoo RK, Basavaiah N, Sukumaran P, Bhattacharya S (2011) Morphostratigraphy and Palaeoclimate appraisal of the Leh valley, Ladakh Himalayas, India. J. Geol Soc India 77:499–510

Wünnemann B, Demske D, Pavel T, Kotlia BS, Reinhardt C, Bloemendal J, Diekmann B, Hartmann K, Krois J, Riedel F, Arya N (2010) Hydrological evolution during the last 15 kyr in the Tso Kar lake basin (Ladakh, India), derived from geomorphological, sedimentological and palynological records. Quatern Sci Rev. doi:10.1016/j.quascirev.2010.02.017

Yin A (2006) Cenozoic tectonic evolution of Himalayan orogen as constrained by along strike variation of structural geometry, exhumation history, and foreland sedimentation. Earth-Sci Rev 76:1–13

The Vale of Kashmir: Landform Evolution and Processes

Rajinder Kumar Ganjoo

Abstract

The Vale of Kashmir is one of the most popular tourist destinations in India. Nestled between the imposing Pir Panjal Range and the towering Karakoram Ranges, the bowl-shaped intermontane valley is studded with numerous freshwater lakes and green meadows. The Jhelum River meanders through the 135 km long valley and exits the oval-shaped basin through a deep gorge near Baramula. The valley preserves a unique landscape known as Karewa. These thick lacustrine and fluvial deposits, capped by loess, form high terrace-like features and cover nearly half the valley. The floodplain of the Jhelum River and the multiple lakes constitute the remaining part of the valley. It is generally agreed that the Kashmir Valley originated from the draining of a huge lake (the *Karewa Lake*), which was formed as a result of tectonic upheaval and subsequent tectonic uplift of the Pir Panjal and Zanskar Ranges. None of the Karewa deposits provide evidence of glaciers having existed during their deposition within the valley. The uplift of Pir Panjal at ~4.5 Ma not only triggered the deposition of Karewa sediments in the valley, but the elevated ranges formed an orographic barrier to the incoming southwest monsoon winds and enhanced the dependence of the valley on winter precipitation.

Keywords

Intermontane basins • Tectonic geomorphology • Kashmir Valley • Pir Panjal • Jhelum River • Karewa Lake • Karewa deposits

1 Introduction

The Vale of Kashmir or the Kashmir Valley, famous for its captivating and exhilarating scenic beauty, is one of the most popular tourist destinations in India. Snow-clad mountains, sparkling lakes with *shikaras* (long wooden boats) and houseboats, lush green meadows, pine forests, silver-white streams, sprawling apple orchards and lush green rice fields attract more than a million Indian and foreign tourists every year. The bowl-shaped valley, nestled within the imposing Pir Panjal Range in the south and west, the towering Karakoram Ranges in the north, and the Zanskar Range in the east (Fig. 1), is drained by the Jhelum River, a tributary of the Indus River.

The picturesque Kashmir Valley is studded with lakes, such as Wular, Mansbal and Dal (Fig. 2). The first, a Ramsar site, is a remnant of a former bigger lake, and other two are of riverine origin and/or remnants of old beds of the river system. The Jhelum River originates from the eastern mountain girdle of the valley and flows through the Wular Lake. It exits the enclosed basin west of Baramula (Fig. 1) via a narrow gorge, cut across the Pir Panjal Ranges. The Liddar, Sind and Kishan Ganga are the main tributaries

R. K. Ganjoo (✉)
Department of Geology, University of Jammu,
Jammu, 180006, India
e-mail: ganjoork@rediffmail.com

V. S. Kale (ed.), *Landscapes and Landforms of India*, World Geomorphological Landscapes,
DOI: 10.1007/978-94-017-8029-2_11, © Springer Science+Business Media Dordrecht 2014

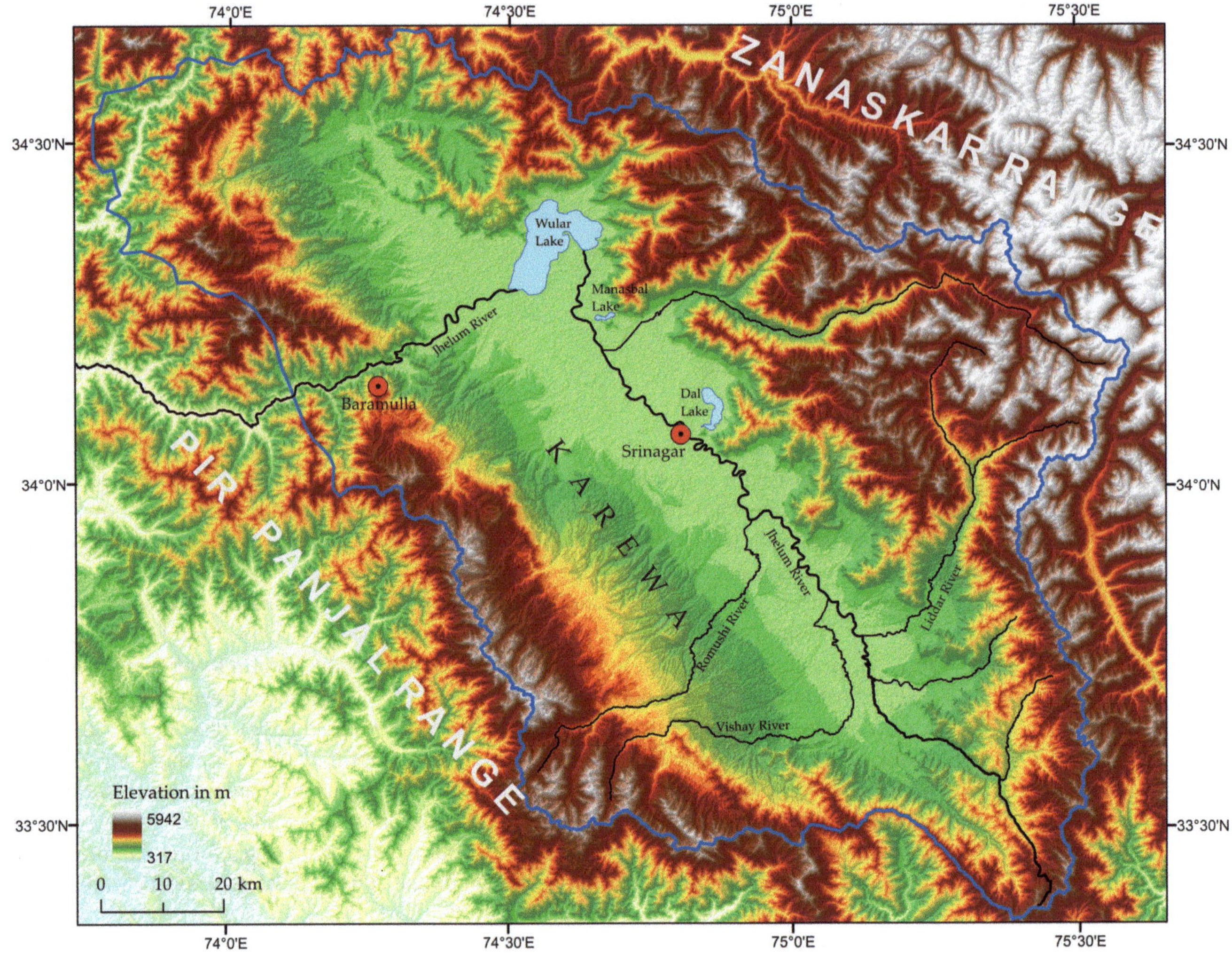

Fig. 1 The oval-shaped intermontane Kashmir Valley. The continuous *blue line* demarcates the watershed of the Kashmir Valley

from the north. The Romushi and Vishay Rivers are the main tributaries from the south.

The altitude of the valley at Srinagar is ~1,600 m a.s.l. As a result, the climate of the valley is generally cool in spring and autumn, mild in summer and cold in winter. Summer has little rain with cool nights. Emperor Jahangir (1569–1627 CE) of the Moghul dynasty, appreciating the scenic beauty and cool salubrious climate of the Kashmir Valley, called it as the "Paradise on Earth". The moderate climate inspired Muslim rulers to lay down beautiful gardens, namely Shalimar, Nishat and Chashm-e-Shahi along the periphery of the Dal Lake and at the foot of Zabarwan Hills, near Srinagar. Elsewhere, the valley is punctuated with green meadows namely Gulmarg (Fig. 3), Pahalgam, Sonmarg, etc. that are world known tourist resorts. In winter, Gulmarg is skiers' paradise (Fig. 4).

2 Geology of the Kashmir Valley

The oval-shaped Kashmir Valley is about 40 km wide and stretches approximately 135 km from NNW to SSE. The Jhelum River, which is closer to the northern side, divides the basin into two parts. The bowl-shaped Kashmir Valley is an intermontane fault basin that could be an "extensional rift basin" oriented parallel to the main orogenic belt with a tendency to sink with respect to the general rise of Himalaya (Agarwal and Agrawal 2005) (Figs. 5 and 6). The rocks, extending from southwest to northeast of the Kashmir Valley, comprise progressively complicated structures, whereas towards the plains of India there is a progressive uplift and folding. The bowl-shaped basin, part of the Kashmir Nappe (Fig. 6), consists of Precambrian basement overlain by a thick succession of fossiliferous Palaeozoic

Fig. 2 Picturesque Dal Lake with floating house boats and *shikaras*. The historical (18th century) Hari Parbhat Fort is seen in the background (Photo courtesy of Rakesh Chandra)

and Triassic rocks with thrusted contacts in the north (Zanskar Thrust) with Great Himalayan Range and in the south (Panjal Thrust) with Pir Panjal Range.

Almost 1.3 km thick succession of sand, mud and gravels belonging to Neogene to Quaternary age, fill the Kashmir Basin. The thick fill-deposits, locally known as Karewa or *Udra,* form elevated benches, hill-tops and plateaus (Figs. 5 and 7).

3 Major Landforms

The valley is an alluvium filled oval-shaped depression ($\sim$4,865 km^2) that was once occupied by a vast lake, known as the "*Karewa Lake*". The lake was gradually drained out, leaving flat-topped valley-fill Karewa (*Udra*) deposits (Fig. 7). The recent alluvium filling the bottom of the valley varies from 3 to 25 km in width along the Jhelum River. The valley can be divided into three sub-divisions: (i) the Jhelum Plain, (ii) the Karewas, and (iii) the Rimlands (Singh 1971).

The Jhelum Plain includes the floodplain extending from Khanabal to Baramula. The Jhelum River meanders sluggishly through the Jhelum Plain and overflows its banks in spring and autumn due to heavy rainfall and/or melt waters from the glaciers. The floods add fresh alluvium to the advantage of farmers and swamps beyond the Wular Lake. Flanking the floodplain is the Bahil tract that stretches between Kulgam and Shopiyan and forms the 'rice bowl of Kashmir'.

Flat-topped terrace-like features are prominently seen in the southern part of the vale. These distinct upland valley-fill deposits (Karewas) of lacustrine origin occur above the modern floodplain of the Jhelum River and its tributaries and are contrastingly infertile. Modern techniques of irrigation and acute shortage of cultivable land has compelled the local farmers to grow maize and fruits (apples, plums and almonds) on these uplands.

Fig. 3 Lush green meadows in foreground and snow covered mountain tops at Gulmarg in summer (Photo courtesy of Rakesh Chandra)

The mountains surrounding the valley that are blanketed with thick cover of conifer in the north and deciduous trees on the Pir Panjal, form the Rimland. The lush green meadows (Gulmarg, Pahalgam, Sonmarg, Yousmarg, etc.) are situated in these rimlands that are visited by shepherds along with their livestock for grass and clean water during summer. The Rimland rises above 3,500 m a.s.l. and is covered with snow till late April.

Triassic limestones of south Kashmir reveal some karstic features such as karren, springs, few caves and small conduits. Most of the springs are the places of worship by Hindus and termed as 'nag' such as Anantnag, Kokernag, Tulmul nag, etc.

4 The Karewa Deposits

The Karewa deposits in the Kashmir Valley cover about half the basin (Figs. 1, 5). The Karewa deposits on the Pir Panjal side have a much wider spread (12–15 km). These sediments were laid down in the tectonically created depression not only in response to the change in climate but also to various tectonic events that commenced in the Pliocene. The first phase is reflected in the Lower Karewa (=Hirpur Formation), followed by the deposits of the Nagum Formation (Upper Karewa), associated with the second phase. The entire Karewa Group is divided into Hirpur, Nagum, and Dilpur Formations (Fig. 5). The Hirpur Formation consists of gray to bluish-gray clay, light-gray sandy clay, fine to coarse-grained green to purple sand, conglomerate, lignite, and lignitic clay (Singh 1982). The Nagum Formation comprises of fine to coarse-grained greenish to purplish sand, gray and ochre sandy clay, ochre and cream colored marl and gravel. The lithologies of the Karewa do not support glacial origin for any of the deposits. The Upper Karewa deposits are exposed in the central and northeastern part of the Kashmir Valley and are capped by loess-palaeosol sequence termed as Dilpur Member. Loess exposed on the SW or Pir Panjal flank and on the NE or Himalayan flank is termed as 'older loess' and 'younger loess', respectively (Bhatt 1975, 1979).

Fig. 4 Snow-clad Gulmarg and waiting tourists/skiers in winter (Photo courtesy of B.S. Bali)

The age of Karewa Group has been a matter of debate. Opinions about the onset of its deposition vary from the Miocene—on the basis of diatom and ostracode analysis, through the Pliocene—on the basis of vertebrate fauna and cross-relationships with the stratigraphy of Siwaliks, to the late Pliocene—on the basis of correlation with glacial stage elsewhere. Termination of Karewa deposition may have occurred in the late Pleistocene or at the beginning of the Holocene. The 440 m thick section on Romushi River (Fig. 1) identifies Plio-Pleistocene magneto-stratigraphic sequence of Karewa Group from ~4.40 to 0.77 Ma (Basavaiah et al. 2010), supported by fission track dates (~2.6 and 2.12 Ma) of tephra layers (Fig. 8) from Pakharpura and Baramula. It is now thought that the emergence of Pir Panjal Range at ~4.5 Ma triggered the deposition.

Buried soils or palaeosols occur within the loess. The loess-palaeosol sequence (Dilpur Member) forms an important part of the Upper Karewa (Nagum Formation) that coincided with the deglaciation (~15 ka) and continued till ~10 ka, when the early Holocene humid phase set in. Neolithic habitational debris occurring above loess deposits has been reported from Burzahom and Puthkhah, indicating that the loess pre-dates the Neolithic settlement (Kusumgar et al. 1980).

5 Evolutionary History and Palaeoclimate

It is now generally agreed that the present-day topography of the Kashmir Valley originated from draining of a huge lake similar to those of the Kathmandu and Pokhara Valleys. The lake was most likely drained through an outlet west of Baramula, giving rise to the present gorge (De Terra and Paterson 1939). The resultant depression was gradually filled by sediments derived from the surrounding mountain ranges, particularly Pir Panjal. The former lake was created by the rise of Pir Panjal, which impounded the then existing drainage.

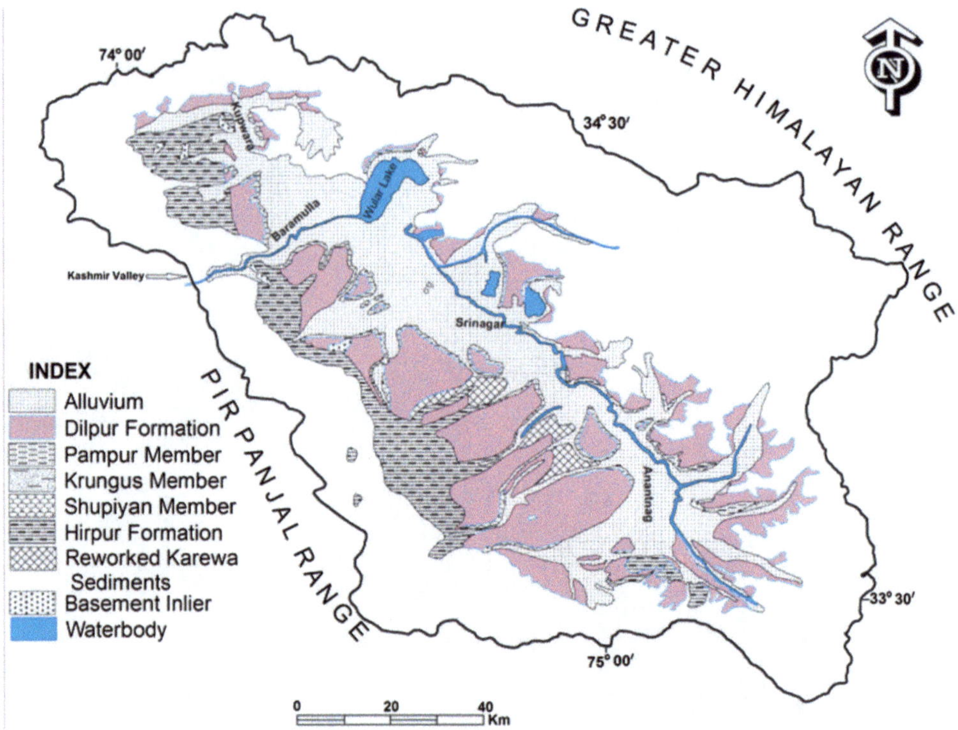

Fig. 5 Geological map of Karewas showing the geographical distribution of each member (after Bhatt 1982; Ahmad 2012)

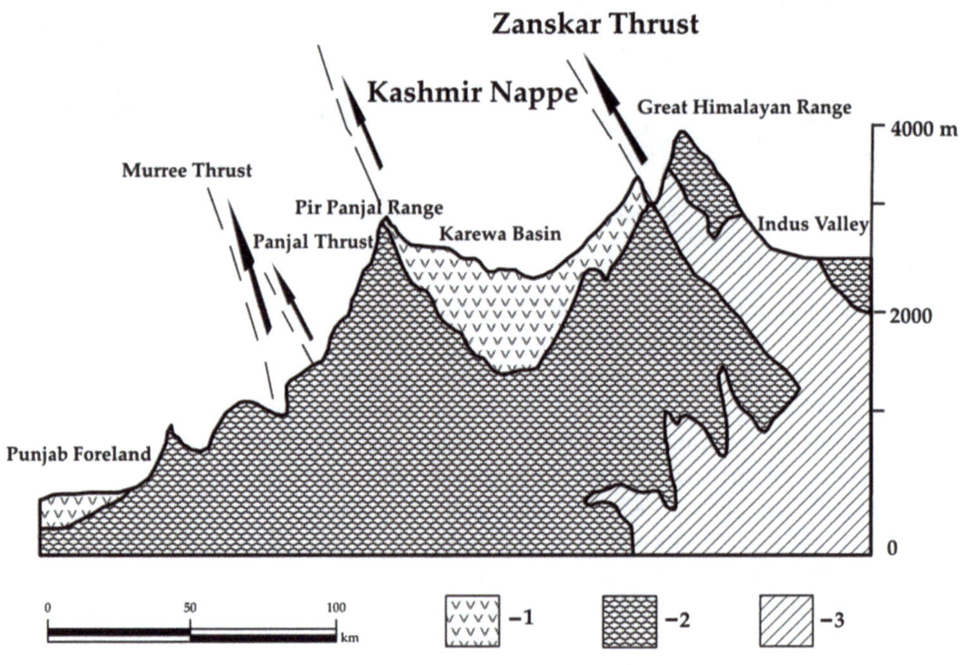

Fig. 6 Diagrammatic section across the Kashmir Himalaya, showing broad tectonic features (after Agarwal and Agrawal 2005). *1* Neogene-Quaternary deposits, *2* Palaeozoic and Mesozoic sequences, *3* Proterozoic basement

The sedimentation of Karewas was in a lake that was shallow, ~5–10 m deep and in some areas up to ~30 m, known as Hirpur Lake. The lake had marginal swamps, which are indicated by the presence of lignite and lignite mud deposits. The grey shales represent the deeper parts of the lake having fine grained carbonates. The extensive drainage system with subaqueous channels producing bar finger sand (elongated lenticular sand body) fed the Hirpur Lake. The progradation in the lake during the Upper Karewa (Nagum Formation) is represented by the development of alluvial fans grading laterally into braided river system (Singh 1982). The Shupiyan Member represents a major phase of lake regression

Fig. 7 General landscape of Karewas. The Karewas have been dissected into a multitude of steep sided ravines ranging in depth from 50–150 m (Photo courtesy of Rakesh Chandra)

from the southwestern part of the valley resulting in the development of braided river system, dominated by gravelly/conglomeratic facies. The thick conglomeratic deposits at the base of Karewa were laid by high energy stream flows, succeeded by fluvio-lacustrine deposits. The conglomeratic deposits are believed to have formed ~3.15 Ma. The Hirpur Lake was thus reduced to half the size and became shallower (~2–10 m). The reduced lake, known as Pampur Lake, supported extensive aquatic vegetation with no marginal swamps. The drainage system produced mostly Gilbert-type delta and delta sheet sand due to wave activity. The muddy facies thus formed are termed as Pampur Member (Fig. 9) of the Upper Karewa (Singh 1982). Thus, the erstwhile Karewa Lake after the initial deposition of the Lower Karewa (Hirpur Formation) started receding gradually from the SW side and shifted the basin more to the north and northeast where the Upper Karewa (Nagum Formation) was deposited.

The supply to the lake was mostly from the surrounding mountains and therefore the fossil record from Karewas represents a mixed community of mountainous and plain areas adjacent to the lake. Nevertheless, the presence of aquatic vegetation such as *Trapa, Typha, Nymphea* in the Karewa sediments suggests warm climate in the valley (Vishnu-Mittre 1965, 1979).

A sudden increase in the rate of sedimentation ~1.95–1.77 Ma (the Olduvai sub-chron) is attributed to the rising Pir Panjal Range that served as an orographic barrier to the incoming southwest monsoon and making the valley dependent on winter precipitation (Basavaiah et al. 2010). The glaciers during Quaternary were therefore, restricted to the montane valleys and did not extend to the Kashmir Basin. This inference is based on the re-interpretations of Karewa conglomerates from outwash gravels to tectonic alluvial facies (Holmes et al. 1989).

Available evidence indicates that the climate between ~4.40 and 3.0 Ma was warm and temperate with the influence of southwest monsoon, since the orographic barrier of Pir Panjal Range in the south was low. The rate of sedimentation within the basin was also low perhaps due to vegetated landscape and consequent low rate of erosion. The event of cold, temperate and dry phase between 3.0 and 2.2 Ma is recorded on the basis of the presence of *Arvicolid* (cold loving rodent) and pollen data. The fluvial source to the erstwhile lake between 4.40 and 1.95 Ma was from Pir Panjal Range. The fluvial activity was directed to NW during upper Matuyama and lower Brunhes chron (~1.07–0.77 Ma) suggesting the onset of Jhelum River flow in the Kashmir Basin (Basavaiah et al. 2010).

Fig. 8 **a** Karewa deposits (~70 m) exposed along Romushi River at Aglar. **b** A close up of the deposits showing volcanic ash (~2.4 Ma) horizon along the Romushi River section. The source of the ash is believed to be the Dasht-e-Nawar Volcano in Afghanistan (Photo courtesy N. Basavaiah). For more details refer to Basavaiah et al. (2010)

Fig. 9 **a** Upper Karewa tableland at Pampore having Pampur Member in the lower and Dilpur Member in the upper part of exposed section. Encircled is the human scale. **b** Close up of Pampur Member of Upper Karewa exposed near Pampore, Kashmir (Photo courtesy of I Ahmad)

Abundance of oak during the early Pleistocene period of the Lower Karewa indicates moist temperate climate. The presence of mixed oak woods and the pollen studies suggest that the average annual precipitation then over the valley was much higher than the present. The cold oscillation periods, inferred from the style of the Lower Karewa deposits, were responsible for reducing the forest cover and the spread of grasslands that were dominated by *Artemisia* and *Chenopodiaceae*—an alpine meadow landscape (Vishnu-Mittre 1965, 1979; Agrawal et al. 1985). The evidence of first such cold, temperate and dry event at ~3.4 Ma is substantiated by the presence of *Arvicolid*. The climate in the Kashmir Valley has been in transition from sub-tropical to cool to temperate with some variation in precipitation ~2.2 Ma, as indicated by the pollen record. The rainfall conditions, more particularly during the winter monsoon, enhanced around 0.77 Ma (Bronger et al. 1987; Basavaiah et al. 2010).

The development of landscape in the Kashmir Valley occurred prior to, concurrently with, and after Karewa deposition (Singh 1982; Bhatt 1979). The formation of Karewa Lake is a consequence of the existence of rudimentary Pir Panjal Range that was formed sometimes in Neogene and was responsible for blocking the south flowing drainage from already uplifted Great Himalaya Range. The uplift of the Pir Panjal Range pre-dates the initiation of sedimentation i.e. Hirpur Formation in the Karewa Lake. The transition from the structurally complex Lower Karewa (Hirpur Formation) to non-deformed Upper Karewa (Nagum Formation) indicates the occurrence of tectonic upheavals contemporary with the Karewa deposition, which are therefore, syn-Karewa. The existence of pronounced angular unconformity between the Hirpur and Nagum Formations on the Pir Panjal side of the Karewa deposits as observed in the field, substantiates it. Major change in the drainage system in some parts of the Karewa Basin subsequent to the deposition of Hirpur Formation further corroborates that the period of Karewa deposition was a key stage in the development of the landscape here. Presence of plant-fossil remains from Hirpur Formation situated at an altitude of ~3,350 m a.s.l, such as *Quercus glauce*, *Alnus*, *Cinnamonum*, fruits of *Traps patans*, and *T. bispinoas,* living species of which otherwise grow at altitudes ~1,400–1,800 m a.s.l further confirm the tectonic upheaval of Lower Karewa during the period of the sedimentation (Vishnu-Mittre 1965, 1979; Agrawal et al. 1985). The formation and gentle tilt of terraces formed subsequent to the deposition of Upper Karewa authenticate the post-Karewa tectonic upheaval in the region that resulted in the development of present-day landscape (Singh 1982).

6 Conclusions

The geomorphic, climatic and tectonic history of the bowl-shaped Kashmir Valley and its surrounding ranges is largely contained within the thick deposits within the basin, which today stand high above the floor of the valley and add to the uniqueness of the valley. Although the age and genesis of these Karewa deposits remain a contentious issue, multi-disciplinary studies strongly suggest that these deposits were laid down by fluvio-lacustrine, fluvial and aeolian processes from Neogene to Holocene, in response to the emergence of Pir Panjal. The bowl-shaped intermontane basin was created by tectonic uplift of the Pir Panjal Range around 4.5 Ma. The uplift impounded the then existing drainage and gave rise to a large lake, known as "Karewa Lake". The present large-scale morphology of the valley was created by draining of the lake water via an outlet west of Baramula across the Pir Panjal, giving rise to the present gorge, as well as by deposition of Karewa sediments. Archaeological evidence suggests that the deposition continued till pre-Neolithic times (early Holocene).

During the period of deposition of Karewa sediments, the climatic conditions have fluctuated from sub-tropical to cool to temperate. The main source of precipitation over the valley has also shifted from southwest monsoon to the mid-latitude westerlies after $\sim$2 Ma with the gradual rise of Pir Panjal. The changing climate conditions and tectonic pulsations have been instrumental in changing sediment accumulation rates in the Karewa Lake deposits from time to time. The present picturesque landscape of the Kashmir Valley owes its origin to these major tectonic and climatic events that occurred from Neogene to Quaternary.

References

Agarwal KK, Agrawal GK (2005) A genetic model of thrust-bounded intermontane basin using scaled sandbox analogue models: an example from the Karewa Basin, Kashmir Himalaya, India. Int J Earth Sci (Geol Rundsch) 94:47–52

Agrawal DP, Kusumgar S, Krishnamurthy RV (eds) (1985) Climate and geology of Kashmir—the last 4 million years. Current trends in geology, vol VI. Today and Tomorrow's Printers and Publishers, New Delhi, 247 p

Ahmad I (2012) Geochemical study of loess-paleosol Quaternary sediments of Karewa basin with reference to palaeoclimate of Kashmir valley, J and K, India. Unpublished PhD Thesis, University of Kashmir

Basavaiah N, Appel E, Lakshmi BV, Deenadayalan K, Satyanarayana KVV, Misra S, Juyal N, Malik MA (2010) Revised magnetostratigraphy and characteristics of the fluvio-lacustrine sedimentation of the Kashmir basin, India, during Pliocene-Pleistocene. J Geophys Res 115:1–17

Bhatt DK (1975) On the Quaternary geology of Kashmir valley with special reference to stratigraphy and sedimentation. Misc Publ Geol Surv India 24:188–203

Bhatt DK (1979) Lithostratigraphic subdivision of the Hirpur formation (Lower Karewa)—a critical review and modification. Himal Geol 9:283–391

Bhatt DK (1982) A review of the stratigraphy of the Karewa group (Pliocene-Quaternary), Kashmir. Man Environ 6:46–55

Bronger A, Pant RK, Singhvi AK (1987) Pleistocene climatic changes and landscape evolution in the Kashmir basin, India: palaeopedologic and chronostratigraphic studies. Quatern Res 27:167–181

De Terra H, Paterson TT (1939) Studies of the ice age in India and associated human cultures. Publication Carnegie Institute, Washington, 493 p

Holmes JA, Kingston, Street-Perrott FA (1989) The Quaternary glacial history of Kashmir, north-west Himalaya: a revision of de Terra and Paterson's sequence. Z Geomorph NF 76:195–212

Kusumgar S, Agrawal DP, Krishnamurthy RV (1980) Studies of the loess deposits of the Kashmir valley and ^{14}C dating. Radiocarbon 3:757–762

Singh IB (1982) Sedimentation pattern in the Karewa basin, Kashmir valley, India and its geological significance. J Palaeontol Soc India 27:71–110

Singh RL (1971) India: a regional geography. National Geographical Society of India, Varanasi, pp 347–389

Vishnu-Mittre (1965) Floristic and ecological reconsiderations of the Pleistocene plant impressions from Kashmir. Palaeobotanist 13:308–327

Vishnu-Mittre (1979) Palaeobotanical evidences of the environment of early man in northwestern and western India. Grana 18:167–181

Duns: Intermontane Basins in the Himalayan Frontal Zone

Sampat K. Tandon and Vimal Singh

Abstract

Duns are broad longitudinal intermontane depressions in the frontal parts of the Himalaya. Such basins are flanked by the outer Siwalik Hills in the south and the Lesser/Sub-Himalaya in the north. Duns consist of several landforms, prominent amongst which are relict surfaces, piedmonts, alluvial fans, floodplains, and river terraces. The geomorphic evolution of Duns is largely controlled by the combined influence of the basin margin structural regime as well as the tectonics of the structures that transect the Duns. Duns are formed in settings in which there is a forelandward migration of the Himalayan fold and thrust systems involving the progressive tectonic development of the Main Boundary Thrust (MBT) and the Main Frontal Thrust (MFT). Their evolution is connected primarily to the emergence of positive topography in the mid- to late Pleistocene associated with the hanging wall fold uplifts related to the MFT displacement, as well as the tectonic activity of the MBT on their northern flanks. Duns show variability within themselves and between them in length, area, slope and relief characters, because of the control exercised on these by the deeper structure of the Dun.

Keywords

Duns • Intermontane basins • Himalayan frontal zone • Landforms

1 Introduction

The ongoing convergence of the Indian Plate with the Eurasian Plate has, with time, resulted in the southward migration of the Himalayan front(s) and the development of new landscapes during the later phases of the Quaternary. Duns (also, Doons) are one of the major tectono-geomorphic elements that are associated with the southward migrating fronts of the Himalaya (Fig. 1). The word *Dun* is of Indian linguistic origin, and is used to describe broad synformal depressions that are bounded by the Outer Siwalik Hills in the south and the Lesser/Sub-Himalaya in the north. The significance of Duns is not only that they hold critical clues to the tectonics of the frontal zone and the Himalaya, but because of the accommodation space provided by them for the storage of Himalayan-derived sediments in the mid- and late Quaternary; thereby impacting, in some measure, the sediment flux from the Himalaya to the Ganga Plains. Because of the capacity of these depressions for the storage of sediment and hydrological fluxes, they provide a suitable environment for human settlements, for example Dehra Dun has several cities and industrial townships; with the population of Dehra Dun, the major city being more than half a million as per the last census in 2011. Because of the rapid settlement and human transformation of the Duns, several

S. K. Tandon (✉)
Department of Earth Sciences, Indian Institute of Technology, Kanpur 208016, India
e-mail: sktand@rediffmail.com

V. Singh
Department of Geology, Center for Advanced Studies, University of Delhi, Chhatra Marg, Delhi 110007, India
e-mail: vimalgeo@gmail.com

V. S. Kale (ed.), *Landscapes and Landforms of India*, World Geomorphological Landscapes,
DOI: 10.1007/978-94-017-8029-2_12,

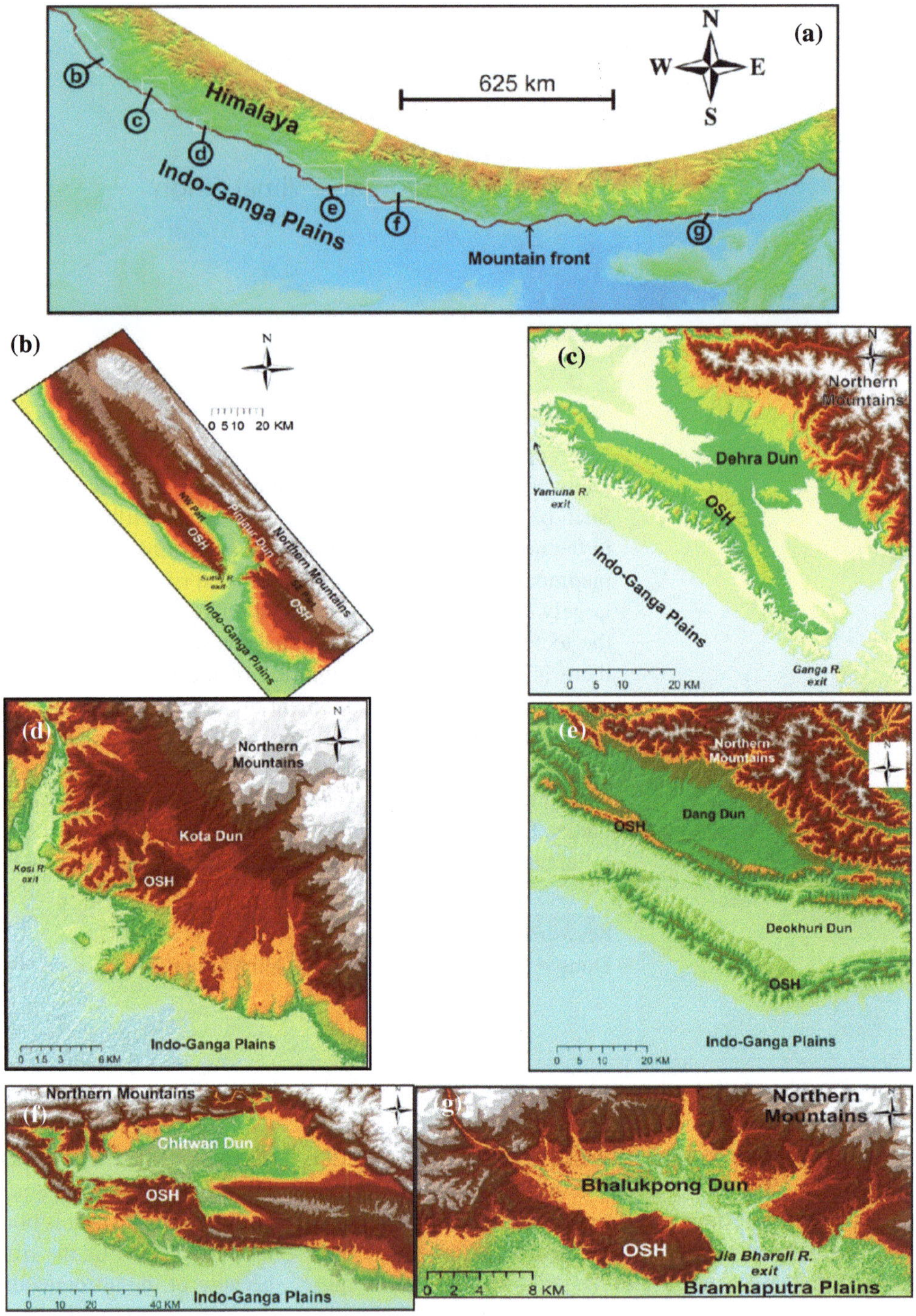

Fig. 1 **a** Map showing location of intermontane valleys along the Himalayan front. Digital Elevation Models of: **b** Pinjaur Dun, **c** Dehra Dun, **d** Kota Dun, **e** Dang and Deokhuri Duns, **f** Chitwan Dun and **g** Bhalukpong Dun. *OSH* Outer Siwalik Hills

environmental concerns have emerged, more notably the issue of limestone quarrying in Dehra Dun and the vulnerability of the industrial hubs in the Duns to earthquake risks related to the active tectonics of the various segments of the Main Frontal Fault (MFT). The MFT is considered the near surface equivalent of the Main Himalayan Thrust (MHT) and the deformation along the MHT during large earthquakes can be transmitted to the MFT making it susceptible to large earthquake hazard.

In recent years, much attention has been focused on the Duns including their tectonic geomorphology, active tectonics, drainage development, palaeoseismology of faults in the Duns and the associated earthquake risk, and the evolution of Dun landforms using sedimentological and OSL chronological methods (Singh et al. 2001; Malik and Matthew 2005; Delcaillu et al. 2006; Malik and Mohanty 2007; Thakur et al. 2007; Suresh et al. 2007; Singh and Tandon 2008, 2010; Barnes et al. 2011).

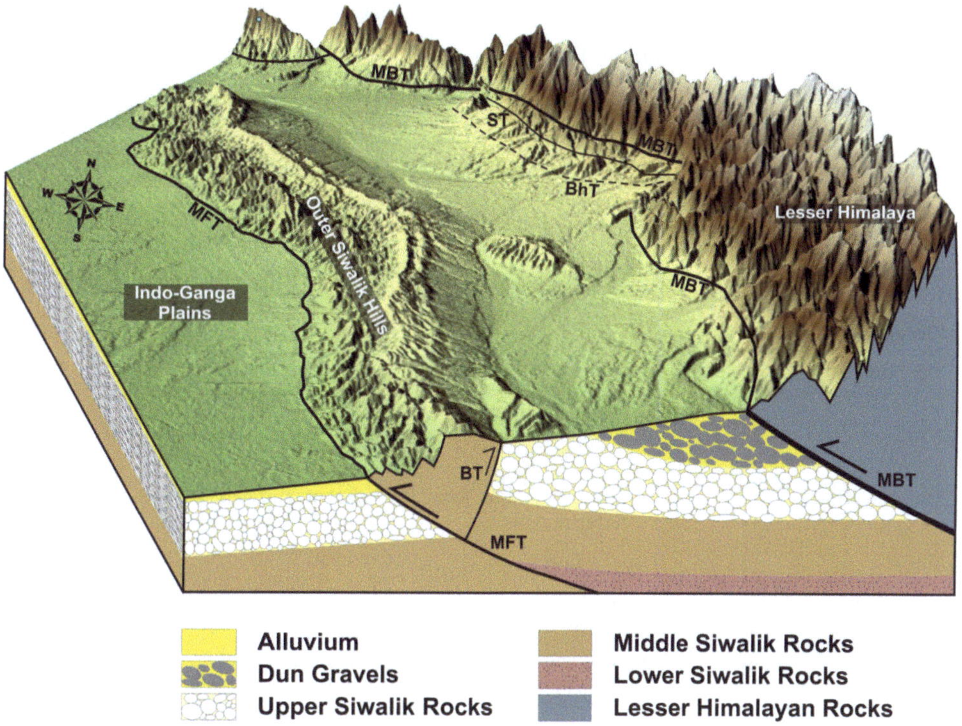

Fig. 2 Diagram showing cross-section structure and 3-D topography of the Dehra Dun (bed thickness not to scale). *MBT* Main Boundary Thrust, *MFT* Main Frontal Thrust, *BT* Bhimgoda Backthrust, *ST* Santaurgarh Thrust, *BhT* Bhauwala Thrust

2 Tectonic Setting

Major orogen strike-parallel tectonic discontinuities such as the Main Himalayan Thrust (MHT), the Main Boundary Thrust (MBT), and the Main Frontal Thrust (MFT) mark the boundaries between the Central Crystalline Zone (Higher Himalaya), the Lesser Himalaya, the Sub-Himalaya, and the proximal part of the Ganga Plains. In the Himalayan foothills, the present mountain fronts are associated with the MBT and the MFT; and in some segments in the Himalaya there are two mountain fronts, a southerly one associated with the MFT and a northerly one associated with the MBT or one of its many splays. The Duns occur where these two mountain fronts, one associated with the MBT and the other associated with the MFT are well developed; for example, the Pinjaur Dun, Dehra Dun, Kota Dun, Chitwan Dun, and the Dang Dun amongst others are all believed to occur in such a tectonic setting (Figs. 1 and 2). Nakata (1972) recognized the importance of these geomorphic features in the Himalayan frontal zone, and classified the Himalayan foothills into three structural types i.e., the Piedmont, Dun, and Re-entrant types with transitions between the piedmont and Dun types. He also highlighted the prominent differences in slope development between the northern and the southern part of a Dun, the former being generally more steep with geomorphic surfaces formed within the Duns at the foot of the south facing Himalayan slopes. More recently, the tectono-geomorphic implication of this observation has been emphasized by the interpretation that the slopes associated with the Siwalik Hills in the south are related to the hanging wall of the younger Himalayan Frontal Thrust (HFT), and the slopes of the northern part of the Duns are associated with the hanging wall of the MBT or one of its splays (Singh and Tandon 2008; Barnes et al. 2011).

3 Occurrence, Distribution and Characteristics of Duns

It is clear that a necessary condition for the initiation and formation of Duns is a fault related folding and consequent uplift that promotes the development of positive topography (outer Siwalik Hills); together with the pre-existing MBT related mountain front to the north this results in a broad synformal depression in the landscape bounded by two fault-related mountain fronts. As the HFT extends for thousands of kilometers along the strike length of the Himalayan front, and as it is segmented, its properties and characteristics are variable in different strike segments. Duns are likely to form in those segments of the HFT where there are frequent relatively larger slip events that lead to the formation and preservation of positive topography finally resulting in the outer Siwalik Hills. Therefore, it is expectedly observed that the spacing between adjacent Duns is quite variable (Fig. 1). Commonly, the formation of Duns results in drainage reorganization with major drainage exits being formed at the

Table 1 Dimensions, relief, geomorphic and geologic characteristics of three Duns

	Pinjaur Dun	Dehra Dun	Kota Dun[a]
Area in km^2	1,503	1,524	107
Length in km	150	100	21
Width in km	8–20	15–20	5
Elevation of the northern mountains	1,500 m	2,300 m	2,550 m
Elevation of the southern hills	320–780 m	410–900 m	440–700 m
Elevation of the valley floor	270–680 m	400–950 m	480–750 m
Structures within the Dun	Pinjaur thrust, Pinjaur Garden fault	Santaurgarh thrust, Bhauwala thrust	Dhikala thrust, Pawalgarh fault
Axial river and main drainages	Sirsa, Satluj, and Soan	Ganga, Asan, Yamuna and Bata	Dabka
Major landforms	Surfaces, piedmont, fans, terraces, hills	Piedmont, surfaces, terraces, hills	Fans, piedmont, terraces, hills
Shape	Longitudinal	Crescent	Spindle
Faults bounding the northern mountains	Barsar Thrust, Nalagarh Thrust	Main Boundary Thrust	Main Boundary Thrust
Rocks in the bounding northern mountains	Dagshai and Kasauli Formations	Lower Siwalik Subgroup, Lesser Himalayan sequences	Lower/Middle Siwalik Subgroup, Lesser Himalayan sequences
Rocks in the bounding southern mountains	Upper Siwalik Subgroup	Middle and Upper Siwalik Subgroup	Upper Siwalik Subgroup

[a] Data taken from Goswami and Pant (2007)

margins of the Duns, for example, the Dehra Dun is marked by the exits of the Ganga and Yamuna Rivers at its eastern and western margins, respectively.

Differences of up to one order in magnitude are noted in the area and length characters of some of the smaller Duns such as the Kota Dun and the larger Duns such as the Pinjaur Dun. Similarly, there may be considerable differences between different Duns in the respective elevation(s) of the southern bounding Siwalik Hills as well as the elevation(s) of the valley floors of the Duns (Table 1). It is also noted that the width may show significant variations both within a Dun, for example in the Pinjaur Dun and between Duns, for example between the Dehra Dun (15–20 km) and the Kota Dun ($\sim$5 km) (Fig. 1). Variations in the widths of the Duns most likely depend on the horizontal distance that the MFT has propagated along a flat before the fault related folding took place.

The Duns are commonly characterized by tectonically influenced drainage networks that show varying amounts of incision and aggradation in different reaches. Usually, the drainage systems from the northern slopes are larger and supply larger sediment volumes and support higher water discharges in comparison with the relatively shorter drainage systems originating from the southern slopes of the Duns (Fig. 3).

4 Landforms and Their Development in the Duns

A variety of landforms (Table 2) with different dimensions have been recognized in the Duns and these include piedmonts, talus wash surfaces, alluvial fans, floodplains, stream terraces, isolated hills and well-developed relict geomorphic surfaces (Fig. 3).

Broadly, landforms in the intermontane valley can be grouped into—(a) the landforms associated with the northern mountain front and the northern margin, (b) the landforms in the main basin (or Dun axial part), and (c) the landforms associated with the southern mountains and the southern margins. The landforms associated with the northern mountain front cover most of a Dun's area; it is because the northern mountains are associated with faults with longer tectonic histories and on which larger slip has occurred resulting in relatively higher elevations of the mountains and an adjoining Dun margin. Therefore, the streams originating on these mountains have larger catchment sizes and higher stream power (Fig. 3). The south bounding hills are associated with relatively younger faults on which less slip has occurred; hence the lower elevation of these mountains/hills. Due to this, the streams draining these mountains have relatively smaller catchment sizes and lower stream power. The

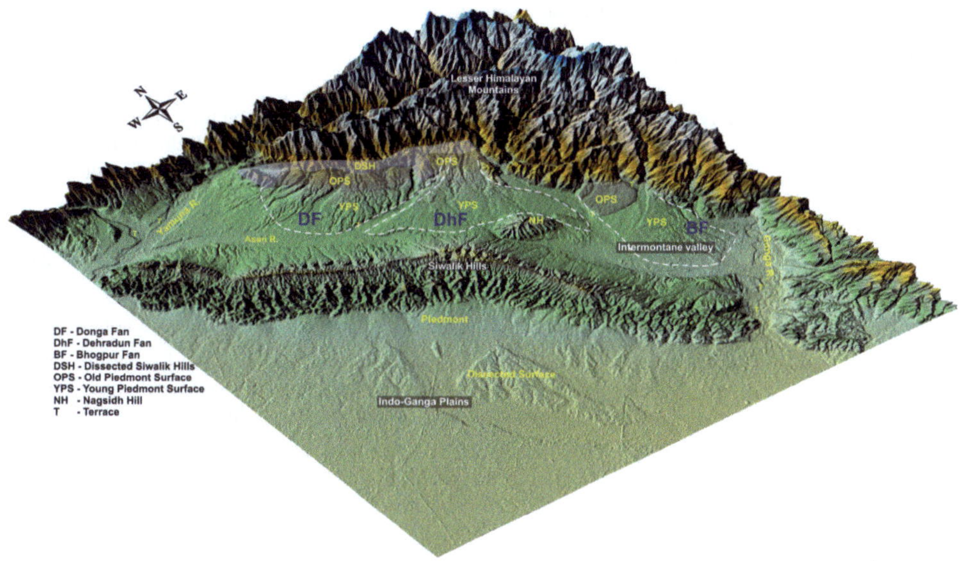

Fig. 3 Geomorphic units of the Dehra Dun showing their spatial distribution. Fan boundaries marked on the basis of previous work (Singh et al. 2001)

streams display various patterns along their length but generally they are straight for most part of their length. In some segments the streams are braided in nature and in some segments they are meandering; such changes occasionally coincide with the presence of a structure (either fault or fold axis). Wider stretches of the streams may reach up to a few hundred meters in width.

A large variability in the distribution of landforms between Duns is observed. The main factors governing this variability are the tectonic base level for all the streams set by the elevations of the mountain exits of the major rivers, and the local base level for the smaller streams (which is the main river that they join within a Dun).

Alluvial fans and piedmonts are the two common landforms that form in the Duns (Fig. 3) because of the break in slope across the mountain fronts, and progressive tectonic activity on the faults associated with the mountain fronts.

Relict geomorphic surfaces occurring at different levels are another important geomorphic unit; and these are described as a two dimensional fraction of land that can be defined by space and time (Ruhe 1975). These surfaces develop when the streams draining a Dun incise to adjust to local base level changes, causing erosion in the stream landscape and consequent degradation of the alluvial landforms to the extent that their original morphological characteristics are considerably modified or obliterated. For example, Nakata (1972) identified several geomorphic surfaces both in the Dehra Dun and the Pinjaur Dun that are developed locally, and also over relatively larger Dun areas. In the Pinjaur Dun (Fig. 4), the Kalka Surface (oldest), the Pinjaur Surface, Kiratpur Fan Surface and Jhajulla Surface (youngest) have been mapped by various workers (Nakata 1972; Singh and Tandon 2008), and are believed to represent the deposits of relict alluvial fans and piedmonts.

5 Duns: Sediment Storage and Sources

Duns act as sediment traps and hence impact the longitudinal sediment connectivity of some of the major Himalayan rivers such as the Ganga and Yamuna, and consequently their total sediment fluxes into the Ganga Plains. The potential of Duns to act as sediment traps is limited by their dimensions, and the ability of the rivers to cut through the folds that bound them in the south. Densmore et al. (2010) indicated that the Duns 'act as a sediment filter that is superposed between the sediment source in the Himalaya and the routing systems of the Ganga Plain'. They also showed that in the case of Dehra Dun, Holocene excavation of Dun sediments yields an additional sediment source term, which averaged over the last 10 ka, represents a small but significant percent of the present-day sediment load of the two rivers—Yamuna and Ganga.

6 Evolution of Duns

Recurrent fault related slip is a pre-requisite for the formation of Duns in order that new positive topography emerges in the form of incipient Siwalik Hills, and then this topographic build-up has to be sustained over time with the initiation and lateral propagation of more anticlinal/monoclinal folds in the terrain. Barnes et al. (2011) provided a detailed analysis of the development of Himalayan frontal fold topography, and showed from a study of the Chandigarh and Mohand anticlines that most growth in catchment size and relief is accomplished within 5 km of the fault tips. They suggested that "high slip rates, weak uplifting rocks, and rapid erosion may combine to quickly limit the topographic growth of the emerging folds and

Table 2 Geomorphic characteristics of Duns mainly from Pinjaur Dun, Dehra Dun and Kota Dun

Major Segments	Geomorphic Units	Description
Northern segment	Alluvial fan	Fans are noted in almost all the Duns and they vary in size from small (~14 km^2—Chopra Fan in Kota Dun) to large (52 km^2—Donga Fan in Dehra Dun). Slope of the fan surface varies from ~1° to ~14°. The fan surfaces are mostly under cultivation but in some areas they are densely vegetated. In most of the Duns the fans are incised.
	Piedmont	When the spacing between the streams draining the northern mountains is less it results in the development of piedmont, which is formed because of the coalescence of fans.
	Surface	Surfaces are present in most of the Duns; the southwestern part of the Pinjaur Dun shows four well developed surfaces. However, young Duns do not show such surfaces.
	River terrace	Terraces in the Duns develop due to incision by streams in response to tectonic activity along their northern mountain fronts, change in base level either due to breaking of a new thrust or incision by the main river. Number of terrace levels vary in different Duns from 3 to 5.
	Floodplain	Well-developed floodplains occur when a relatively large river flows through the Dun. For example, in the northwestern part of the Pinjaur Dun, where the Satluj River acts as the axial river it shows a well developed floodplain.
	Hills	Hills/residual hills are observed in the Dun, for example, the Nagsidh hill in Dehra Dun which reaches an elevation up to ~800 m. These are either residual or form because of the breaking of a new thrust within the Dun.
Axial part		Axial part of the Dun varies in its location from the central part of the Dun to the southern part of the Dun. Generally, the Duns are asymmetric (described in text). The axial river could be a major drainage e.g. the Satluj River in the Pinjaur Dun or very small like the Asan or the Sirsa Rivers. More than one axial river could also be present in a Dun if there is a drainage divide present in the Dun, for e.g., the Asan and the Suswa Nadi of Dehra Dun.
Southern segment		In the northwestern Himalaya, the outer Siwalik hills are asymmetric; as a result very small streams have developed on their northern slopes. In Dehra Dun, the streams draining the northern slopes of the outer Siwalik hills have formed talus wash covered surface. Occasionally, small fans are also present. However, in the case of the symmetric outer Siwalik hills, streams developing on their northern flank may develop a floodplain, for e.g., in the northwestern part of the Pinjaur Dun.

Fig. 4 Landscape view of a part of the Pinjaur Dun showing fan surfaces and fan-head incision in the foreground. Note the Himalayan Frontal Thrust (HFT) hanging wall related Siwalik Hills in the background. Photograph taken facing south. *E* and *W* represent east and west, respectively

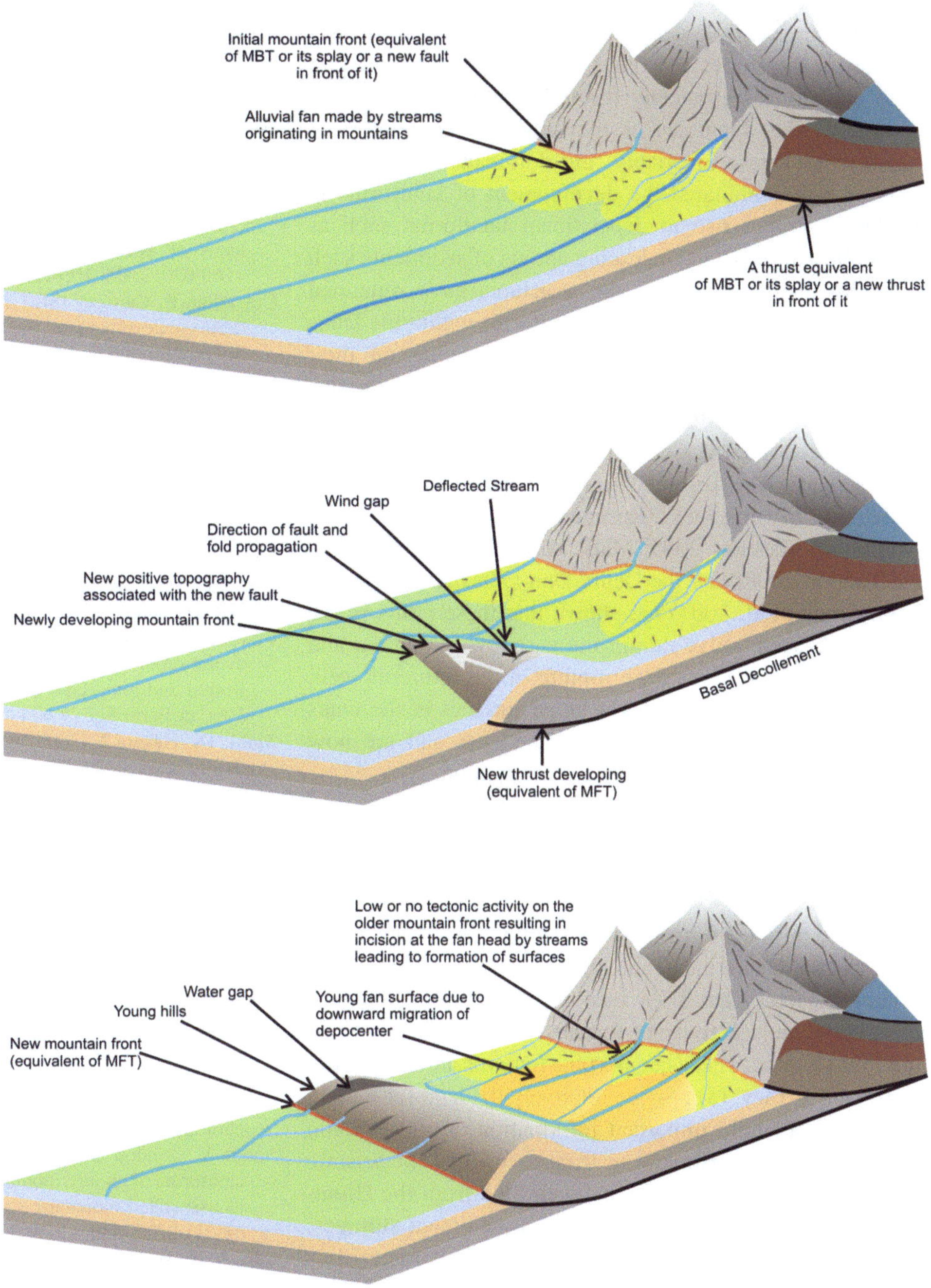

Fig. 5 A schematic model showing various stages of evolution of the intermontane valley. Note that after the new mountain front has developed, activity on the older fault (i.e., older mountain front) is reduced and the streams incise their older deposits to give rise to surfaces

disconnect their morphology from the displacement field". This implies that the growth of topography in these structures is limited by the strength of the rocks making up the folds.

Singh and Tandon (2010) showed that the Siwalik Hills bounding the Pinjaur Dun in the south consist of multiple fold structures with independent initiation histories, and subsequent growth patterns that involved their lateral propagation and linking to form significant positive topography of a few hundred meters. As a result of the formation of this new topography on the hanging wall of the MFT, considerable drainage re-organization results, determined to a large extent by the dynamics of the tectonically driven south-lying topographic barrier (Fig. 5). These changes are manifest in many forms including major changes in the loci of the major drainages coming from the northern slopes related to the MBT hanging wall, as well as some remarkable drainage deflections such as those observed in the Sirsa River in the Pinjaur Dun.

The latter phases of geomorphic evolution of landforms in a Dun are controlled by many factors such as the differences in the slopes and relief structures of the bounding north and south tectonically formed topographic domains, as well as the dynamics of the structural elements that transect the Duns. Invariably, the dominant relief of the northern part dominated the development of landforms in the Duns, giving rise to depositional landforms such as major alluvial fans and coalescing aprons of sediments such as piedmonts. As many of the structures have remained active in the late Quaternary, the landforms within the Duns show variable degrees of dissection depending on their proximity to these structures. These active faults have also influenced the formation of some other geomorphic features such as stream terraces, back tilted terraces, relict geomorphic surfaces, scarps and scarplets, and isolated hillocks made up of Siwalik rocks. Recent work on the Pinjaur Dun (Malik and Matthew 2005; Malik and Nakata 2003; Singh and Tandon 2008) and the Dehra Dun (Thakur and Pandey 2004; Singh et al. 2001) has shown that these faults in the Duns have been active during the past few millennia, and hold the potential for surface rupture in the future as well. Lastly, most of the flat and low slope surfaces in the Duns have come under various types of land use and are now witnessing considerable human transformation.

7 Conclusions

Duns were initiated as tectono-geomorphic landforms in the mid- to late Quaternary, and their later landscape evolutionary trajectory has been mostly controlled by the progressive tectonic activity of forelandward migrating thrust and fold systems. These geomorphic features are considered significant as they: (a) provide relatively flat land for human settlement, (b) hold important information on the tectonic deformation of the Himalayan front as well as for seismic risk assessment (c) influence sediment flux from the Himalaya because of their potential to act as sediment storages. Duns show variable dimensional and relief characteristics that possibly result because of the differences in the deeper basement related structures of the Duns. A variety of landforms such as piedmonts, alluvial fans, floodplains, stream terraces, relict geomorphic surfaces, and escarpments occur in the Duns; at a gross scale the geomorphic evolution of a Dun is guided by the combined controls exercised by the progressive and related tectonics of the basin margin structures. Duns act as sediment storages as well as sediment sources, but their role in sediment supply over different time scales is not understood.

References

Barnes JB, Densmore AL, Mukul M, Sinha R, Jain V, Tandon SK (2011) Inter-play between faulting and base level in the development of Himalayan frontal fold topography. J Geophys Res 116:F03012. doi:10.1029/2010JF001841

Delcaillau B, Carozza JM, Laville E (2006) Recent fold growth and drainage development: the Janauri and Chandigarh anticlines in the Siwalik foothills, northwest India. Geomorphology 76:241–256

Densmore AL, Sinha R, Barnes J, Jain V, Tandon SK (2010) Proximal storage and controls on sediment flux in large Himalayan river systems. In: EGU general assembly conference abstracts, vol 12, p 11276

Goswami PK, Pant CC (2007) Geomorphology and tectonics of Kota-Pawalgarh Duns, central Kumaun sub-Himalaya. Curr Sci 92:685–690

Malik JN, Mathew G (2005) Evidence of paleoearthquakes from trench investigations across Pinjore Garden fault in Pinjore Dun, NW Himalaya. J Earth Syst Sci 114:387–400

Malik JN, Mohanty C (2007) Active tectonic influence on the evolution of drainage and landscape: Geomorphic signatures from frontal and hinterland areas along Northwestern Himalaya, India. J Asian Earth Sci 29:604–618

Malik JN, Nakata T (2003) Active faults and related late Quaternary deformation along the northwestern Himalayan frontal zone, India. Ann Geophys 46:917–936

Nakata T (1972) Geomorphic history and crustal movement of the foothills of the Himalaya. Report of Tohoku University Japan, vol 2, 7th series (Geography)

Ruhe RV (1975) Geomorphology, Geomorphic Processes and Surficial Geology. Houghton Mifflin, Boston, MA

Singh AK, Prakash B, Mohindra R, Thomas JV, Singhvi AK (2001) Quaternary alluvial fan sedimentation in the Dehradun valley Piggyback basin, NW Himalaya: tectonic and palaeoclimatic implications. Basin Res 13:449–471

Singh V, Tandon SK (2008) The Pinjaur Dun (intermontane longitudinal valley) and associated active mountain fronts, NW Himalaya: Tectonic geomorphology and morphotectonic evolution. Geomorphology 102:376–394

Singh V, Tandon SK (2010) Integrated analysis of structures and landforms of an intermontane longitudinal valley (Pinjaur dun) and its associated mountain fronts in the NW Himalaya. Geomorphology 114:573–589

Suresh N, Bagati TN, Kumar R, Thakur VC (2007) Evolution of Quaternary alluvial fans and terraces in the intramontane Pinjaur dun, sub-Himalaya, NW India: interaction between tectonics and climate change. Sedimentology 54:809–833

Thakur VC, Pandey AK (2004) Late Quaternary tectonic evolution of Dun in fault bend/propagated fold system, Garhwal sub-Himalaya. Curr Sci 87:1–10

Thakur VC, Pandey AK, Suresh N (2007) Late Quaternary-Holocene evolution of Dun structure and the Himalayan Frontal Fault zone of the Garhwal sub-Himalaya, NW India. J Asian Earth Sci 29:305–319

The Chambal Badlands

Veena Uday Joshi

Abstract

Chambal Badlands of central India are one of the most extensive badlands in the world, and are one of the four severely dissected landscapes within the Middle Alluvial Ganga Plains (MGAP). This extensive dissected landscape with labyrinth of winding gullies has offered refuge to outlaws for centuries. Badlands or ravines generally but not exclusively occur in semi-arid and arid areas with erodible rocks. These areas, dominated by surface erosion by overland flow and gullies, are characterized by heavily dissected terrains with steep slopes and channels separated by sharp ridges. The gullies rapidly incise and extend headward. Evidence suggests that the evolution of the badlands along the Chambal River coincided with the incision of the river as a result of the strengthening of SW monsoon in the early Holocene. Lineament controlled block uplifts might have also affected these areas causing the streams to rejuvenate, inducing widespread gullying in the region. Evidence such as ruins of former settlements, and remains of temple foundations suggests that these badlands were formed and/or rapidly extended during the recent historical period. The possibility of further expansion of the badlands in response to human interference is expected in the future.

Keywords

Chambal • Badlands • Ravines • Gullies • Tectonics • Rejuvenation • Lineament • Monsoon

1 Introduction

Breaking the monotony of the vast Indo-Ganga Plains, stunning the eyes of the onlookers, emerge the Chambal Badlands with their awe striking appearance and spell-binding notoriety. Cascading from the lofty mountains of Himalaya in the alluvial Ganga Plains of India, lay two of the holiest rivers of India, namely, Ganga and Yamuna. But there are also some southern tributaries that originate from the Indian Craton and join the Yamuna and the Ganga with equal magnificence and vigor. The Chambal River is one such cratonic river that joins the Yamuna (Fig. 1) and forms a large alluvial valley at the junction of the northwestern lobe of the Vindhya Plateau and the southeastern fringe of the Aravalli Hills. From the source to its confluence with the Yamuna, the Chambal River is about 1000 km long, with a catchment area of $\sim$140,000 km^2, which is larger than that of any Himalayan tributary to the Ganga or Yamuna. The river drains over both the Deccan Traps basalts and Proterozoic Vindhyan rocks and contributes significant amount of sediments to the foreland basin (Sinha et al. 2009). The Chambal Badlands, one of the most classic badlands in the world and comparable in their magnitude and extent only with the Dakota Big Badlands of North America, occur predominantly along this river (Fig. 2). In the Indian context, these badlands have been famous for centuries for all the wrong reasons, rather than for their

V. U. Joshi (✉)
Department of Geography, University of Pune,
Pune 411007, India
e-mail: veenaujoshi@gmail.com

V. S. Kale (ed.), *Landscapes and Landforms of India*, World Geomorphological Landscapes,
DOI: 10.1007/978-94-017-8029-2_13, © Springer Science+Business Media Dordrecht 2014

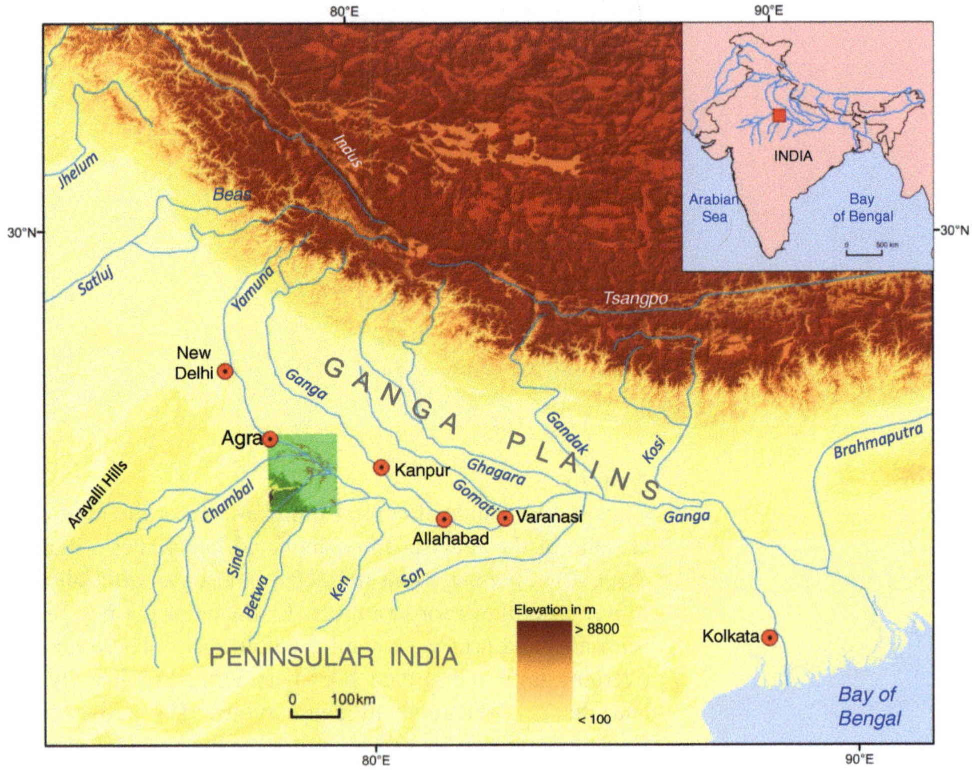

Fig. 1 Map showing the location of the Chambal River in central India. The *box* shows the location of Fig. 2. Circles represent major towns and cities

natural rugged beauty, dramatic landscape and rich faunal life. The name 'Chambal' is *synonymous with* banditry in the region. Some of the India's most infamous outlaws have operated from this region. But none of these can diminish the fact that the region is a display of nature's creative ability at its peak; a landscape so dynamic and spectacular, yet so difficult for the mankind to handle. That is the Chambal Badlands of India (Figs. 2 and 3).

Badland is a term applied to those landscapes, which are intensely dissected where vegetation is absent or sparse and is useless for agriculture. The term was first given to the arid, dissected plateau region of SW South Dakota by native Americans and fur trappers who found the area difficult to cross. Badlands develop in a wide range of materials and climate, but seem to be usually associated with arid and semi-arid environments and show a preference for unconsolidated and poorly cemented material or soft rocks (Harvey 2004). There are topographic variations in the badlands primarily due to differences in the sediment properties (Bryan and Jones 1997).

2 Geographical Setting

In India, ravines are found along many large alluvial rivers and occupy nearly four million hectares (ha) of land ($\sim$1 % of India's total land area) (Haigh 1984). There are four specific areas, known for severe ravine erosion in India: the Yamuna-Chambal Ravine Zone, the Tapi-Narmada-Sabarmati-Mahi Ravine Zone, the Chhotanagpur Ravine Zone, and the Siwalik-Foothills Ravine Zone (Sharma 1968). The largest among them is the Yamuna-Chambal Ravine Zone. The Chambal Ravines stretch for a length of 480 and 10 km wide belt along the banks of this cratonic river and occupy about half a million ha of land (Haigh 1984). These are deeply dissected ravines that spread extensively over the plain region.

The climate of the region is semi-arid, with an average annual rainfall of $\sim$900 mm with nearly 90 % of it occurring during the four monsoon months (June to September). Much of the summer precipitation is delivered with great intensity. Vegetation is mainly thorny acacia bushes which are very typical of such semi-arid landscapes (Fig. 4). Evergreen riparian vegetation is completely absent, with only sparse ground-cover along the severely eroded river banks and adjacent ravine lands.

The Chambal River passes through some of the most densely populated districts of three Indian states, namely, Madhya Pradesh, Rajasthan and Uttar Pradesh. The drainage area is rectangular in shape up to the junction of the Parvathi and Banas Rivers, while below the confluence the catchment becomes unusually narrow and elongated (Fig. 5a). Though the actual badland terrain supports very little population due to its ruggedness, there are several settlements within these badlands such as, Jiran, Kota, Bhilwars, Dholpur, Sawai Madhopur, etc. that carry large population.

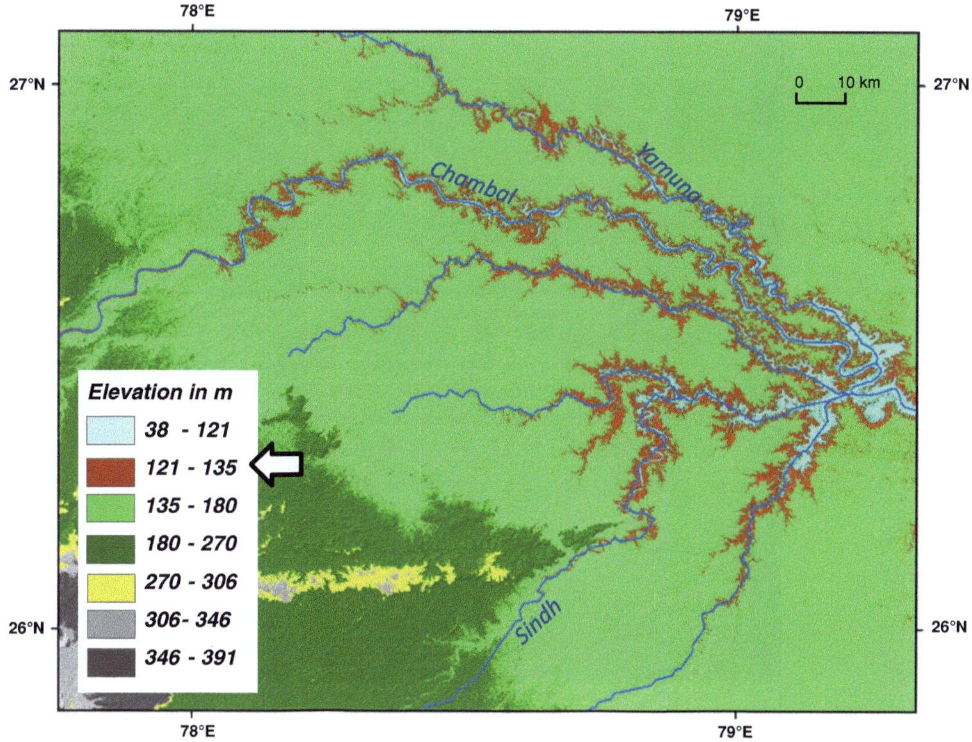

Fig. 2 Badlands along the Chambal and adjoining rivers. Near the confluence with Yamuna, the badlands developed along all these rivers coalesce to form an incredibly extensive ravine zone in this region. The *white arrow* represents the elevation category dominated by badlands

Fig. 3 Google Earth image of the Chambal Badlands (*dark* area). The Chambal River is seen on the *left*. The image covers approximately 40 km^2 area

The badlands are developed in ~50–60 m thick Quaternary alluvial deposits, which occur in a triangular-shaped area and are underlain by Bundelkhand Granite, the Gwalior Group, and the Vindhyan Supergroup rocks (Fig. 5b). The thickness of alluvium increases remarkably up to ~200 m near the confluence with Yamuna. The alluvial sequence consists of a highly oxidized polycyclic sequence of fine sands, silt and clay in varying proportions with occasional coarse channel sand and grit bodies. Calcrete (or *kankars*) and hardpan horizons are characteristically present in the sediment (Mishra and Vishwakarma 1999). These alluvial tracts exhibit remarkably homogeneous topography and the flatness of the surface is broken only by the ravine belts along the rivers.

3 Badland Morphology

The badland region is characterized by an intricate network of gullies and ravines that stretches for hundreds of kilometers not only along the Chambal River but also in the adjoining basins (Fig. 2). At many locations they attain a

Fig. 4 Two views of the Chambal Badlands. Gullies are v-shaped, very deeply incised, steep-walled and each channel is separated from the adjacent ones by sharp ridges. The difference in height between the gully *bottom* and inter-gully ridge *top* is between 20 and 40 m

width of as much as 10–12 km across. The entire badland zone demonstrates extensively and intensively dissected landscape having high drainage density; steep to sub-vertical scarps with narrow interfluves (Fig. 4). In the upstream reaches, gullies are narrow; depth is less than 1 m with a bank slope of between 45 and 80°. Further downstream, gullies change in their morphology and become wider (18–20 m), deeper with steeper bank slopes (>80°). Near the confluence with Yamuna, the gullies are of near-vertical slopes, very wide and deep with approximate width of 25 m and depth exceeding 45 m. Further beyond, the badlands of the Chambal, Yamuna, Kunwari, Sind, and Pahuj Rivers coalesce to form incredibly extensive badlands in this region (Fig. 2), the likes of which are very few in the world.

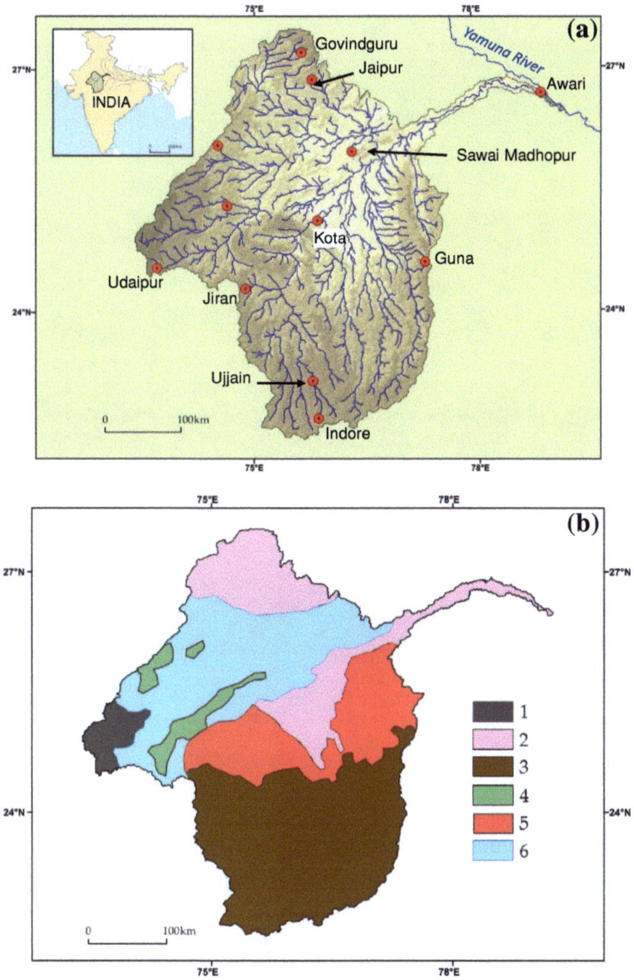

Fig. 5 **a** The rectangular-shaped Chambal drainage area. Note the remarkably narrower and elongated reach of the Chambal Drainage. **b** Geological map of the Chambal Basin *1* Aravalli Rocks, *2* Alluvium, *3* Deccan Traps, *4* Granite, *5* Vindhayan Rocks, *6* Gneisses

The badlands of Chambal resemble *Calanchi* landscape of Italy, except in its scale. A *Calanco* is a small hydrographic unit; horse-shoe shaped, with a tributary system in which each channel is separated from the adjacent ones by means of, more or less, sharp ridges with slope angles depending on the physical and mechanical properties of the bedrock. They often occur in colony to form *Calanchi*. These are heavily dissected terrain with steep, bare slopes and channels which rapidly incise and extend headwards (Alexander 1982). Chambal is the magnified version of *Calanchi*. Gullies are mostly v-shaped, very deeply incised and each channel is separated from the adjacent ones by sharp ridges (Fig. 4). The density of the gullies is so high that the interfluves appear like cones and the whole topography resembles unending colonies of hills. Morphology of the gullies varies from one location to the other and the variation is mainly attributed to the variation in their textural, stratigraphical and mineralogical properties. Wherever clayey horizon overlies the sediments, it gives rise to flat-topped hills with a steep scarp slope (Fig. 6).

The alluvial Himalayan Foreland exhibits the admixtures of sediments derived from two completely contrasting sources, namely the Himalaya and the Indian Craton. The Ganga and Yamuna Rivers bring down the Himalayan sediments and penetrate to distal foreland basin, whereas the cratonic rivers, especially Chambal and Betwa Rivers (Fig. 1), also transport huge volumes of sediments and deposit far north of the axial Yamuna River (Sinha et al. 2009). Himalayan sediments show high percentage of mica, illite and small percentage of chlorite and kaolinite but complete absence of smectite clays, whereas Chambal River transports high-grade metamorphic minerals from the Aravalli Hills and abundant smectite mostly derived from the Deccan Traps (Sinha et al. 2009). Both cratonic and Himalayan rivers have flooded over the entire region and reworked those sediments for prolonged period of time resulting in thorough churning and mixing of these sediments. The badlands have been formed on these reworked sediments from the two contrasting sources which eventually resulted in the variation in their morphological expressions. Smectite clays have high shrink-swell property and in sodium rich environment they disperse rapidly. Wherever the smectites are present in the alluvial sequence, the gullies have incised deeper and developed wider valleys and vice versa.

4 Evolution of the Badlands

Badlands are erosional forms that result from surface erosion by overland flow following heavy rainfall. Certain conditions favour the development of badlands, such as a semi-arid climate with a long dry season, available relief and easily erodible rocks (Harvey 2004). The climate over the Chambal Basin is semi-arid with a strong seasonal contrast. The Himalayan Foreland Basin stores much larger volumes of smectite rich sediments derived from the cratonic region than sediments derived from the Himalayan region (Sinha et al. 2009). Heavy monsoon rains for a short period and prolonged dry spells promote shrink-swell activity and crumbling of the alluvial sediments in the region. All these processes lead the area already vulnerable to badland formation.

Even though evidence, such as ruins of the former settlements, broken potteries and temples with their foundations almost entirely eroded away by the encroaching ravines, suggests that most of these ravines seem to have expanded rapidly in the last few centuries (Mishra and Vishwakarma 1999), there is neither documentary nor field evidence of a mass scale land use changes in this region in the historical and distant past that could have triggered the initiation and/or extension of badlands of such magnitude. Considering the

Fig. 6 When the clay horizon overlies the alluvial sediments, it gives rise to flat-topped hills with a steep scarp slope

scale and extent of the badland development along the Chambal River it appears that the process of badland formation has operated for a long time ($\sim 10^3$–10^4 years).

Base level is considered as one of the fundamental factors determining the badland morphology and development. The basally-induced incision and associated badland extension is generally goaded by climatic changes and tectonics.

Strengthening of SW monsoon during the early Holocene has been a well established fact that had left significant mark all across the subcontinent. As a response to the climate shift, the major rivers and their tributaries started incising their channels. Incision of the Yamuna River also began during this period, and the Chambal River which is a tributary to Yamuna responded without much time lag, leading to intensive bank erosion by bank gullies. The dispersive sediments readily collapsed and gully network expanded on the interfluves by further headward erosion. The badland formation, therefore, seems to have coincided with the intensification of the SW Indian monsoon at the end of the Last Glacial Maximum.

Although cratonic basins are considered to be tectonically relatively stable, one cannot completely rule out the possibility of tectonically-induced incision and dissection in the area. Abrupt shifting of the course of Chambal, as well as a prominent break along the longitudinal profile of the river, together with the dramatic increase in the thickness of alluvium have been attributed to a morpho-tectonic fault between Pinahat and Tikatpura by Mishra and Vishwakarma (1999). They have also inferred—(i) block uplift resulting in the sudden rejuvenation in relief and (ii) epeirogenic warping of the alluvial tract in the region. Although not well established, it appears that the evolution of the Chambal Badlands may be attributed to the combined effect of the strengthening of the southwest monsoon in the early Holocene and tectonic activity. Which of the two preceded the other or which of the two is really responsible for the badland development and expansion is difficult to ascertain? The debate will continue unless a good number of radiometric ages are acquired for this region.

5 Current Scenario

In the past few decades, the ravines of Chambal have become a national issue. The expansion of the ravines is causing serious threat to agriculture and irrigation projects in the lower Chambal Valley. The headward extension of the gullies constantly consumes the fertile agricultural land. The problem has worsened in the last few decades due to the reclamation of the gullied areas for agriculture at the peripheries of the ravines, which disturbs the surface and promotes renewed activities of gullies. Based on the analysis of the multi-date satellite imageries it has been demonstrated that between 1984 and 1998 there has been ~ 17 % increase in the ravine area (Pani and Mohapatra 2001). The expansion of ravines is destroying not only the agricultural land but also posing other problems. Inadequate food and fodder as a result of degraded habitats is posing stress on wildlife leading to their encroachment towards the cropland.

6 Conclusions

Badlands are the landscapes where the dynamic forces of erosion has exceeded the resistive forces of the land surface by magnitudes, to create myriad sculptures of labyrinth of winding gullies that dazzles the imagination. A combination of semi-arid climate with long dry season, sparse vegetation due to lack of moisture and unconsolidated sediments rich in smectite clays promote the development of such landscapes. Chambal Badlands of India, that spread across an area of half a million hectares along the Middle Ganga Alluvial Plain of India, is one such feature that has been notoriously famous for centuries in the country for reasons strictly not for any desirable causes. These ravines are feeding on the fertile alluvial lands and this is causing concern to the nation. Present understanding is that the evolution and/or expansion of the badlands coincided with the incision of the rivers following the strengthening of SW monsoon as well as lineament controlled differential block uplifts that have affected these areas during the Holocene Epoch. Due to its ruggedness, the interior of the ravine is virtually inaccessible but the peripheral areas have been reclaimed by farmers for agriculture in the last few decades. There is high possibility of further expansion of the badlands in the future primarily due to increased human activity.

References

Alexander DE (1982) Difference between ''Calanchi'' and ''Biancane'' badlands in Italy. In: Bryan R, Yair A (eds) Badlands geomorphology and piping. Geobooks, Norwich, pp 71–87

Bryan RB, Jones JAA (1997) The significance of soil piping processes: inventory and prospect. Geomorphology 20:209–218

Haigh MJ (1984) Ravine erosion and reclamation in India. Geoforum 15:543–561

Harvey A (2004) Badlands. In: Goudie A (ed) Encyclopedia of geomorphology. Routledge, London, pp 45–47

Mishra MN, Vishwakarma LL (1999) Morphotectonics of the Chambal and the Yamuna valleys in the Western Marginal Gangetic Alluvial Plains. Geological Survey of India. http://www.portal.gsi.gov.in/pls/gsipub/PKG_PTL_PORTAL_LINKS.pGetCaseStudyRegion?inpRegionId=35

Pani P, Mohapatra SN (2001) Delineation and monitoring of gullied and ravine lands in a part of lower Chambal Valley, India, using remote sensing and GIS. Proc ACRS, Singapore 1:671–675

Sharma HS (1968) Genesis and pattern of ravines of the lower Chambal Valley. Geogr Rev India 30:14–24

Sinha R, Kettanah Y, Gibling MR, Tandon SK, Jain M, Bhattacharjee PS, Dasgupta AS, Ghazanfari P (2009) Craton-derived alluvium as a major sediment source in the Himalayan Foreland Basin of India. Geol Soc Am Bull 121:1596–1610

The Kosi Megafan: The Best-known Himalayan Megafan

Rajiv Sinha

Abstract

The conical-shaped Kosi Megafan in north Bihar Plains (eastern Ganga Plains), eastern India is one of the most cited megafans in the world in the international literature. Named after its principal drainage, the Kosi River, this megafan has a distinct convex-up topography. Numerous palaeochannels on the fan surface represent the remnants of the earlier courses of the Kosi River through its westward journey during the last 200 years. Like other fans in the world, the Kosi megafan has formed due to high sediment flux, fluctuating discharge regime and the availability of large accommodation space. It is believed that the fan building process of the Kosi continues till date and this explains the unstable behavior of the river. The construction of a barrage close to the mountain exit and embankments on both sides of the river have certainly retarded the natural fan-building processes but the Kosi Megafan is still one of the most fascinating landforms on this planet.

Keywords

Kosi river • Megafan • Inland delta • Alluvial cone • Avulsion • Palaeochannels • River dynamics

1 Introduction

A megafan is a large (10^3–10^5 km^2), fan-shaped (in plan view) mass of clastic sediment deposited by a laterally mobile river system that emanates from the outer point of a large mountainous drainage network. Megafans form in alluvial plains and are characterized by a distinct morphology, specific tectono-climatic setting, unique hydrologic regime and complex sedimentary processes involved in their formation and preservation. As a result, megafans are quite localized (15°–35° N and S) in terms of their occurrence (Leier et al. 2005). Two most important characteristics of the megafans are (a) the principal drainage feeding the fan system which is generally quite dynamic and sweeps through the entire fan surface over a period of time, and (b) a typical distributary drainage network radiating from a mountain exit. The implicit controls for the formation of megafans would therefore include a source of abundant sediment supply such as an active orogen and a large accommodation space such as flat alluvial plain to allow for extensive sediment deposition as well as frequent migration of channels.

Geomorphic complexity of megafans has drawn enormous attention of researchers from all over the world but the Kosi megafan has remained an attractive example in most studies due to its large size, active Himalayan hinterland and exceptionally high sediment flux. The avulsive shifts of the Kosi River, dominantly westward (Fig. 2b), have been dramatic and the available records suggest a migration of over 150 km in 200 years (Gole and Chitale 1966; Wells and Dorr 1987). A recent eastward avulsion

R. Sinha (✉)
Department of Earth Sciences, Indian Institute of Technology, Kanpur 208016, India
e-mail: rsinha@iitk.ac.in

V. S. Kale (ed.), *Landscapes and Landforms of India*, World Geomorphological Landscapes,
DOI: 10.1007/978-94-017-8029-2_14, © Springer Science+Business Media Dordrecht 2014

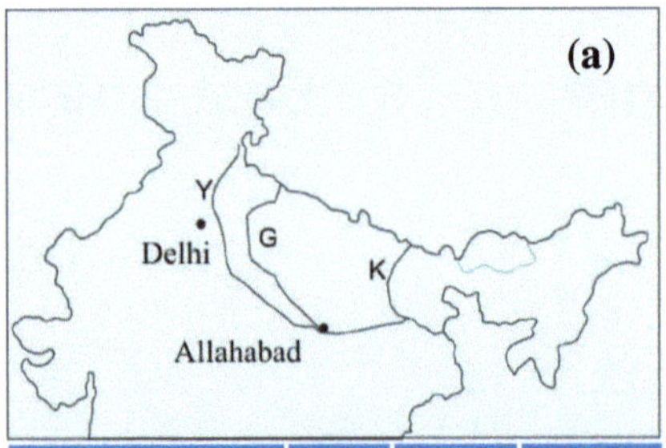

Parameter	Kosi	Ganga	Amazon
Catchment area (10^3 km^2)	**101**	1073	7180
Total length (km)	**1216**	2700	6518
Average Annual discharge (m^3/s)	**2236**	15,000	209,000
Annual sediment load (Mt/yr)	**43**	1670	1000
Sediment yield (Mt/y/km^2)	**0.43**	1.56	0.14

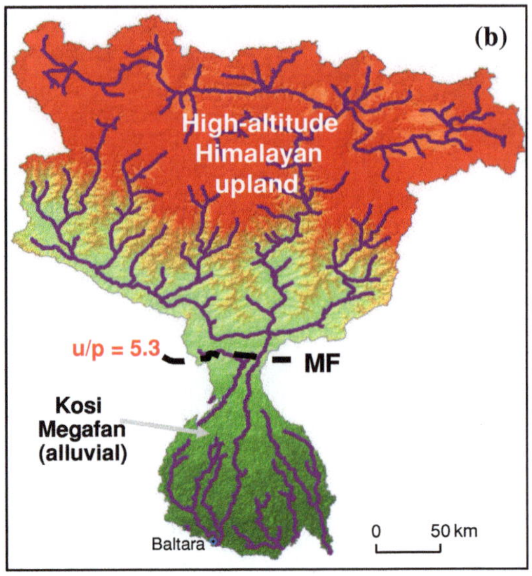

Fig. 1 Location map and characteristics of the Kosi Megafan. With a large upland to plains area ratio and a high sediment flux, the Kosi Megafan system has evolved a rapidly aggrading landform in the Himalayan Foreland. Mountain Front (MF) marks the sharp topographic break and separates the high altitude mountainous upland and the flat alluvial plains. G Ganga, Y Yamuna, K Kosi, u/p upland area to plains area ratio

(a sudden shift of channel to another location) in August 2008 resulted in a shift of ~120 km (Sinha 2009). This is the largest avulsive shift in a single event recorded so far anywhere in the world and this resulted in unprecedented floods in Nepal and north Bihar.

2 Location and Environment

The Kosi Megafan has a maximum length of 150 km from apex to the confluence to the Ganga and a maximum width of 120 km in its mid-fan region. With the total area of ~10,328 km^2 (fan area), the Kosi Megafan mostly lies in north Bihar Plains (eastern Ganga Plains) except for its apical part, which falls in Nepal. The principal drainage of the megafan, the Kosi, originates at an elevation of >5,500 m a.s.l. in Tibet and has a very large, tectonically active hinterland mostly lying in Tibet and Nepal Himalaya (Fig. 1b). The area of the mountainous/upland part of the Kosi drainage basin is more than 5 times larger than the part located in the plains (u/p ratio 5.3, Sinha and Friend 1994).

The Kosi River enters the plains at Chatra in Nepal and then flows for ~80 km in Nepal including the barrage at Bhimnagar before entering the north Bihar Plains in India. The river flows for ~160 km in north Bihar before meeting the Ganga River near Kursela and builds a large fan-shaped landform (Fig. 2a). The main channel of the Kosi in the alluvial reach is braided throughout and the thalweg keeps oscillating between the two ends of the 5–7 km wide channel belt as the river moves from upstream to downstream.

The Kosi is a typical monsoonal river and the monthly discharge of the river generally starts to peak in the month of June with the maximum in August/September. The average annual discharge (Q_{av}) of the Kosi is ~2,236 m^3/s. The average monsoon discharge computed for a period of 1985–2002 is 5,156 m^3/s that is ~5 times higher than the non-monsoon discharge of 1,175 m^3/s (Sinha et al. 2008). Such large difference between monsoonal and non-monsoonal discharge makes the river vulnerable to flooding as the shallow river sections cannot accommodate the excess discharge.

In terms of sediment flux, the average annual suspended sediment load for the Kosi at its most downstream station (Baltara) is 43 Mt/year (Sinha and Friend 1994) even though a large part of total sediment flux from the hinterland is intercepted at the barrage. For a total catchment area of ~101,000 km^2, modern annual sediment yield of the Kosi is 0.43 Mt/km^2/yr. Frequent avulsive movement of the Kosi River (Fig. 2b) has been attributed as *autocyclic* (resulting from processes within the system) and *stochastic* (non-deterministic behavior), which is typical of most alluvial fans across the world. However, the average avulsion frequency of 24 years (i.e. avulsion once in 24 years) for the Kosi is among the highest frequencies in the world compared to 1,400 years for the Mississippi River. High sediment flux and rapid aggradation within the embankments have been considered as the primary reasons for avulsion and flooding in this region (Sinha 2009; Chakraborty et al. 2010).

The Kosi River was embanked on both sides in 1955–1956 (Fig. 2a) and the Kosi Barrage became operational in 1963 as a part of flood protection measure and to provide irrigation for increasing agricultural productivity. However, the embankments have breached frequently and the August 2008 event was the 9th breach in the last ~50 years. Most of these breaches have resulted in large floods. Further development of irrigation and rail-road network in the megafan region has resulted in several

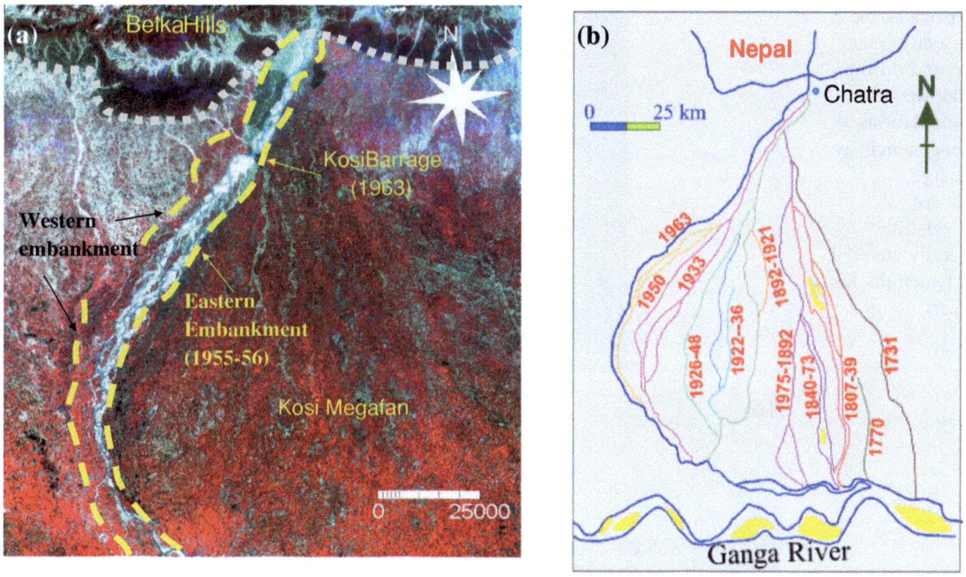

Fig. 2 **a** The Kosi Megafan has a typical conical shape and the fan surface is characterized by numerous palaeochannels as seen clearly on the satellite image. **b** Using historical maps, location of several palaeochannels corresponding to different time periods are shown (after Gole and Chitale 1966). The Kosi River was embanked by 1955–1956 and the barrage was operational by 1963. The river was thus confined and the natural fan building process was retarded. However, a large shift occurred in August 2008 and the river briefly occupied the central position of the fan. The river was put back within the embankments in January 2009 through engineering interventions

adverse effects viz. drainage congestion and waterlogging, rise of river bed level, and reduction in crop productivity due to reduced silt flux on the floodplains.

3 Landforms and Landform Diversity

The first systematic description of geomorphology of the Kosi Megafan was provided by Geddes (1960). Based on closely spaced contours, Geddes demonstrated that the Kosi Megafan (termed as 'cone') has a convex topography across the fan with a typical radial pattern of distributaries. Geddes recognized that the major factors controlling the size of the cone (megafan) are discharge and silt load and documented that "even in a normal season, the Kosi's sands may raise the belt of floodplain by a foot".

The Kosi Megafan has also been described as 'inland delta' (Gole and Chitale 1966) built by large sediment flux from the Himalayan orogen. Using topographic contours, the authors suggested that the Kosi has built a conical delta with contours running almost circumferentially with the center situated in the vicinity of Belka Hill, west of Chatra (Fig. 2b). The dynamic behavior of the Kosi River was attributed to delta cone building process. Since the radial distance from the apex to the toe of the base plane of the delta is maximum in the central part, the cone building, and hence, the shifting of the river is expected to be slower compared to the eastern and western edges of the cone. This was suggested to be the main reason why the north-south course of the Kosi River during 1807–1879 was more stable. The rate of movement of the Kosi River and cone building are therefore attributed to rate of sediment accumulation on the cone and the deficient river slope. The authors considered that the cone building activity of the Kosi is still incomplete and this explains the unstable behavior of the Kosi River compared to the adjoining Gandak River which has also built a megafan. Although the authors contested the earlier idea of a sudden eastward shift (Wells and Dorr 1987), the August 2008 avulsion (Sinha 2009) proved this to be true.

The term "inland delta" is no longer used to describe large fans. The term 'megafan' was first used by Gohain and Parkash (1990) who also provided a detailed geomorphic description of the landforms on the Kosi Megafan surface. They distinguished four distinct geomorphic surfaces (L1–L4) on the Kosi Megafan, well separated by elevation differences. The active channel course with low bars forms the lowest surface (L1) and the remaining three surfaces are parts of the active floodplain. The lowest, nearly barren floodplain surface (L2) is 0.5–0.9 m higher than the water surface during the low flow period and is flooded every year. The next sparsely vegetated surface (L3) is ~1 m higher than the lowest floodplain and is submerged during high flows of annual floods. The highest floodplain surface (L4) is 0.5–0.8 m higher than L3 and comprises the islands and banks covered by grasses, tall trees which get flooded during the peak discharge.

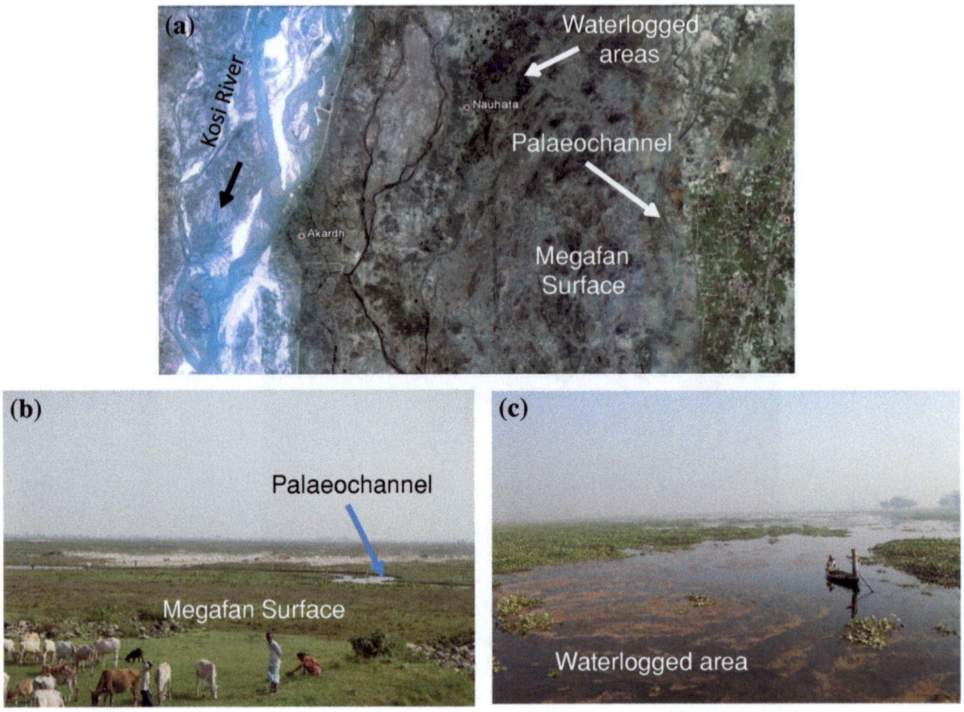

Fig. 3 **a** Characteristics of the megafan surface as seen on the Google Earth image and during field visits. **b** The flat megafan surface has minor undulations due to several palaeochannels as seen on ground. **c** Large waterlogged patches are conspicuous on the satellite images and are generally covered by algae and water hyacinths for most parts of the year

The fan surface as seen on satellite image (Fig. 3a) is characterized by numerous '*dhars*' (small channels) representing palaeochannels of the Kosi River which are partially activated during the monsoon season (Fig. 3b). Large waterlogged patches, locally called "*chaurs*", on the megafan surfaces (Fig. 3a, c) are conspicuous and most of them are low-lying depressions which are fed by rainwater and groundwater seepage. Some of these waterlogged patches in the lower reaches and close to the embankments are very large which are related to seepage along the embankment but may partly represent accumulation of floodwater after overbank flooding.

The transverse profile of the megafan is broadly convex-up but highly irregular due to numerous palaeochannels dissecting the megafan surface; the eastern part of megafan is lower than the western part (Chakraborty et al. 2010). The longitudinal profile of the megafan is concave-up with a general slope from north to south, being steeper in the north (55–75 cm/km) and flatter in the south (6 cm/km). Like many other megafans across the world, the presence of radiating channels on the fan surface reflects a divergent slope attributed to convex-up lateral geometry. Based on the paleochannel network and their discordant relationship, Chakraborty et al. (2010) identified three accretionary lobes on the Kosi Megafan although their relative chronology is not clear. An important conclusion of this work was the negation of the idea of continuous westward movement of the river during the pre-embankment period. The authors instead suggested, based on the analysis of historic maps, a random and oscillating shifting of the river except for a prolonged position in the axial part of the river as documented earlier by Gole and Chitale (1966).

4 Evolutionary History

Limited literature on the geologic and geomorphic evolution of megafans suggests three important conditions to form such large-scale landforms: (a) abundant sediment flux (b) adequate relief and (c) large accommodation space. Although several Himalayan rivers form a point-source dispersal system, only a few of them such as the Kosi has formed a megafan apparently due to enormous sediment flux from a large upland area and a flat alluvial terrain which provided accommodation space for a positive topography to build up over time. Leier et al. (2005), in a compilation of 33 megafans across the globe, also suggested that rivers with moderate to large drainage basins with moderate to high relief and large but fluctuating discharge are more likely to form megafans.

Megafans usually form at the mountain-plain interface in tectonically active areas. A simple geometric model was proposed by Horton and DeCelles (2001) for the evolution of megafan (Fig. 4). Initial stage (A) involves the uplift of the mountain thrust belt and several alluvial fans are deposited along the topographic divide associated with the small stream having small catchment area. The second stage (B) starts when the small stream with low stream power cannot cross the downslope structure. Consequently, the catchment area increases due to headward erosion and river piracy and

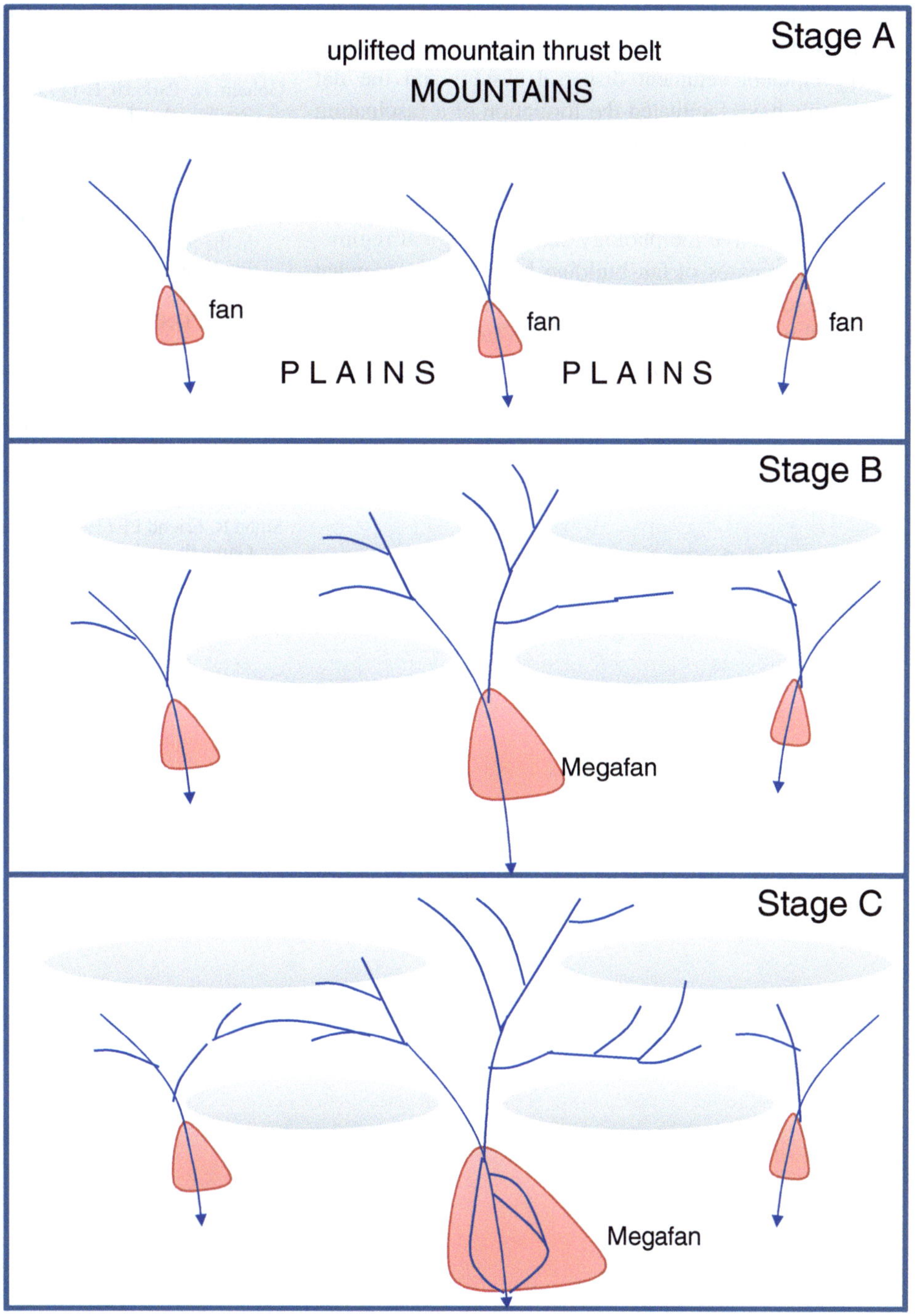

Fig. 4 A schematic diagram showing stages in the evolution of megafan. *Stage A* involves the uplift of the mountain thrust belt and deposition of several alluvial fans along the topographic divide associated with the small stream having small catchment area. In *Stage B* one of the small streams increases the catchment area by headward erosion and river piracy and megafan starts to form in the mountain thrust belt. The final stage (*Stage C*) involves further increment of catchment area and integration of drainage network in the mountain thrust belt resulting in large-scale megafan deposition

megafan starts to form in the mountain thrust belt. The final stage (C) involves further increment of catchment area and integration of drainage network in the mountain thrust belt resulting in large-scale megafan deposition. Our knowledge on the evolution of the Kosi Fan is severely handicapped because of the lack of sub-surface stratigraphic data which hold the clues for long-term geomorphic development of such unique landforms. However, all the available records suggest that the Kosi Fan formed during late Quaternary period but more precise estimates would have to wait for absolutes dates of fan deposits.

5 Conclusions

The Kosi megafan is one of the most cited megafans in the world. Large size, dynamic river regime and very high sediment flux are some of the most significant characteristics of the Kosi Megafan system. Tectonically active Himalayan hinterland and highly variable discharge regime have resulted in a very efficient sediment dispersal system and the flat alluvial plains have facilitated the formation of a fascinating landform in the Himalayan Foreland. Described as 'inland delta' and 'alluvial cones' by earlier workers, the 'megafan' term is fairly well accepted now for these large-scale landforms with distinctive morphology and hydrological regime. The natural processes of fan building have been somewhat retarded due to large-scale interventions on the Kosi River in the form of barrage and embankments. Long-term evolutionary history of the Kosi fan remains unexplored.

References

Chakraborty T, Kar R, Ghosh P, Basu S (2010) Kosi megafan: historical records, geomorphology and the recent avulsion of the Kosi River. Quatern Int 227:143–160

Geddes A (1960) The alluvial morphology of the Indo-Gangetic plain: its mapping and geographical significance. Inst British Geogr Trans Pap 28:253–276

Gohain K, Parkash B (1990) Morphology of the Kosi megafan. In: Rachoki A, Church M (eds) Alluvial fans: a field approach. Willey, Chichester, pp 151–178

Gole VC, Chitale VS (1966) Inland delta building activity of Kosi River. J Hydraul Div ASCE 92:111–126

Horton BK, DeCelles PG (2001) Modern and ancient fluvial megafans in the foreland basin system of the central Andes, southern Bolivia: Implications for drainage network evolution in fold-thrust belts. Basin Res 13:43–61

Leier AL, DeCelles PG, Pelletier JD (2005) Mountains, monsoons and megafans. Geology 33:289–292

Sinha R (2009) The great avulsion of Kosi on 18 August 2008. Curr Sci 97:429–433

Sinha R, Bapalu GV, Singh LK, Rath B (2008) Flood risk analysis in the Kosi river basin, north Bihar using multi-parametric approach of analytical hierarchy process (AHP). Indian J Remote Sens 36:293–307

Sinha R, Friend PF (1994) River systems and their sediment flux, Indo-Gangetic plains, northern Bihar, India. Sedimentology 41:825–845

Wells NA, Dorr JA (1987) Shifting of the Kosi River, Northern India. Geology 15:204–207

The Sikkim-Darjeeling Himalaya: Landforms, Evolutionary History and Present-Day Processes

Leszek Starkel and Subir Sarkar

Abstract

The Sikkim-Darjeeling Himalaya contains the stretch of tectonically active Eastern Himalaya. It is composed of three main tectonic units: Higher Himalaya, Lower Himalaya and Siwaliks separated by thrusts, but joined by great fluvial system of the Tista River. The Higher Himalaya with relief up to 2,000–4,000 m was uplifted by about 2,000 m in the Quaternary rising above the snowline. Its mountain massifs previously had fluvial relief which later had been totally transformed by glacial and cryonival processes. The Lower Himalaya dissected 1,000–2,000 m locally with remains of mature relief fragments are continuously in the forest belt. The Siwaliks in this part of the Himalayan Range are reduced to a narrow belt, which blende with Lower Himalaya. Along the Frontal Fault it rises above 1,000 m directly over the alluvial plains of the Sub-Himalayan foredeep, which is still active and split into blocks of various tectonic tendency. Steep edge of the Himalaya is exposed to high monsoon rainfalls and is afflicted by heavy downpours and continuous rains. Following deforestation this marginal part and also the southern Sikkim area have been affected by large-scale mass movement processes and floods. Only the Tista and several large rivers are transporting heavy sediment loads down to the Brahmaputra River. The smaller streams dissecting the mountain edge form a belt composed of fans.

Keywords

Uplifting mountains • Morphoclimatic vertical zones • Monsoonal rainfall • Effects of deforestation

1 Introduction

Darjeeling (in West Bengal) and Gangtok (in Sikkim) are the two most popular hill resorts in northeastern India. Surrounded by snow-capped mountains, verdant hills and valleys and vast slopes with tea gardens, these two Himalayan hill resorts attract thousands of tourists every year, particularly during the hot Indian summer.

The Sikkim-Darjeeling Himalaya occupy a 100 km long segment of the highest mountain barrier on the globe, at the southern edge of the elevated Tibetan Plateau, separating the tropical-monsoonal lands from the central Asian cold and arid elevated mountains and basins. The Tista River occupies

L. Starkel (✉)
Institute of Geography and Spatial Organisation, Polish Academy of Sciences, św. Jana 22, 31-018 Kraków, Poland
e-mail: starkel@zg.pan.krakow.pl

S. Sarkar
Department of Geography and Applied Geography, North Bengal University, Darjeeling 734013, India
e-mail: subirsnbu@yahoo.co.in

V. S. Kale (ed.), *Landscapes and Landforms of India*, World Geomorphological Landscapes,
DOI: 10.1007/978-94-017-8029-2_15, © Springer Science+Business Media Dordrecht 2014

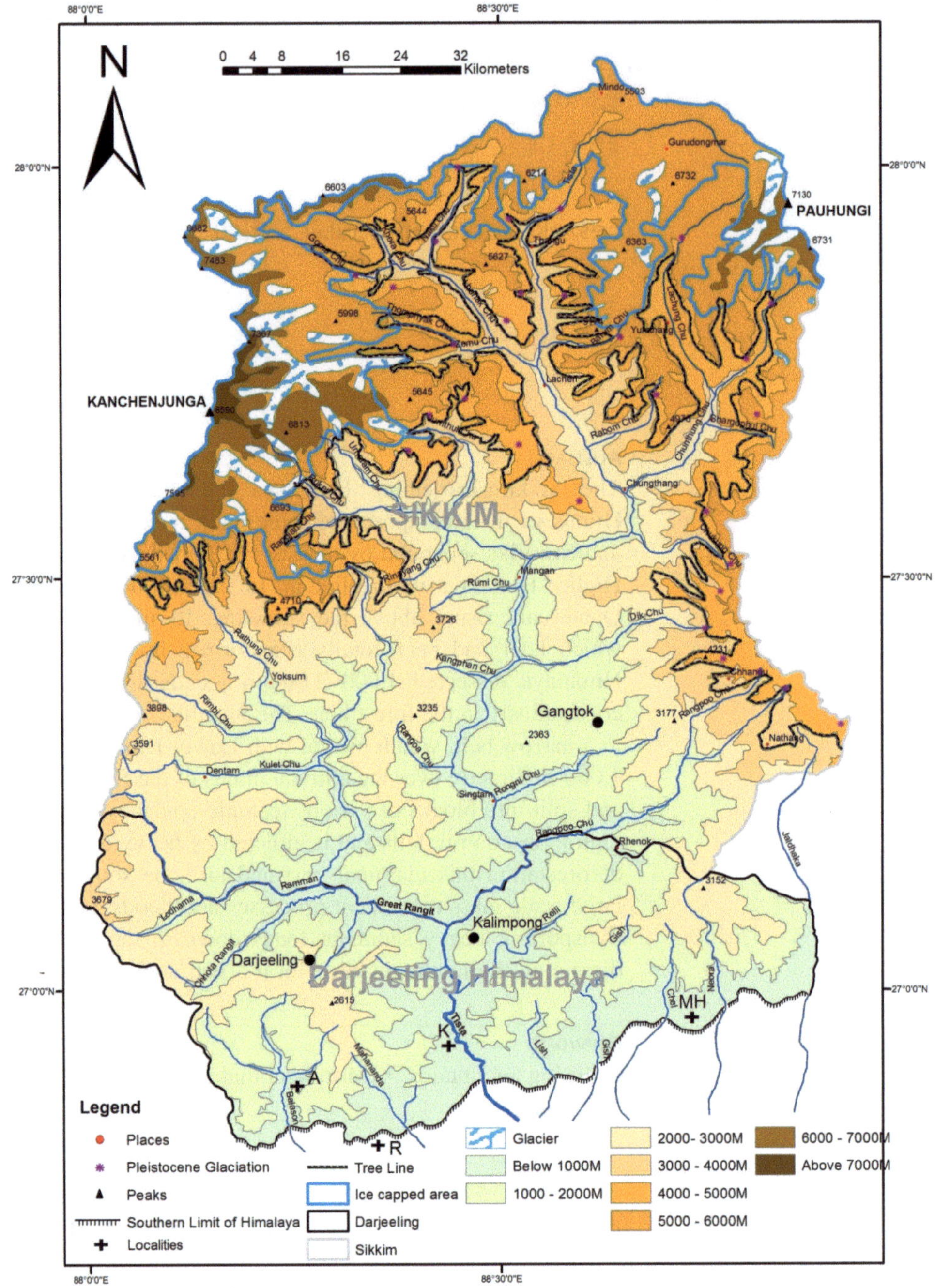

Fig. 1 Elements of the Sikkimese-Darjeeling Himalayan landscape. *A* Ambootia landslide valley, *R* Rangamati River level, *K* Kalijhora site with dated late Pleistocene terrace, *MH* Mission Hill with Gorubathan Surface

a wide depression between Kanchenjunga (8,590 m a.s.l.) in the west and the Pauhunri (7,130 m a.s.l.) in the east (Fig. 1). Kanchenjunga is the third highest peak in the world.

2 Geology and Vertical Morphoclimatic Zonations

East-central part of the Himalaya consists of three main units overthrust from north towards south and is underlain mostly by metamorphic rocks. Mountains are still rising at a rate between 0.5 and 5.0 mm/year (Valdiya 1998). The rising Himalaya attain an altitude of over 8 km a.s.l. and the subsiding foredeep is filled by Neogene-Quaternary sediments up to a depth of 10–12 km below sea level. The northern highest block of High Himalaya, made up of resistant metamorphic rocks, is separated by the Main Central Thrust (MCT) from the Lower Himalaya which is elevated only 2,000–3,500 m a.s.l. This part is built of rocks of various resistance and forms an extensive syncline. In its northern part the beds are inclined towards the south and towards the north in the southern part. This unit is

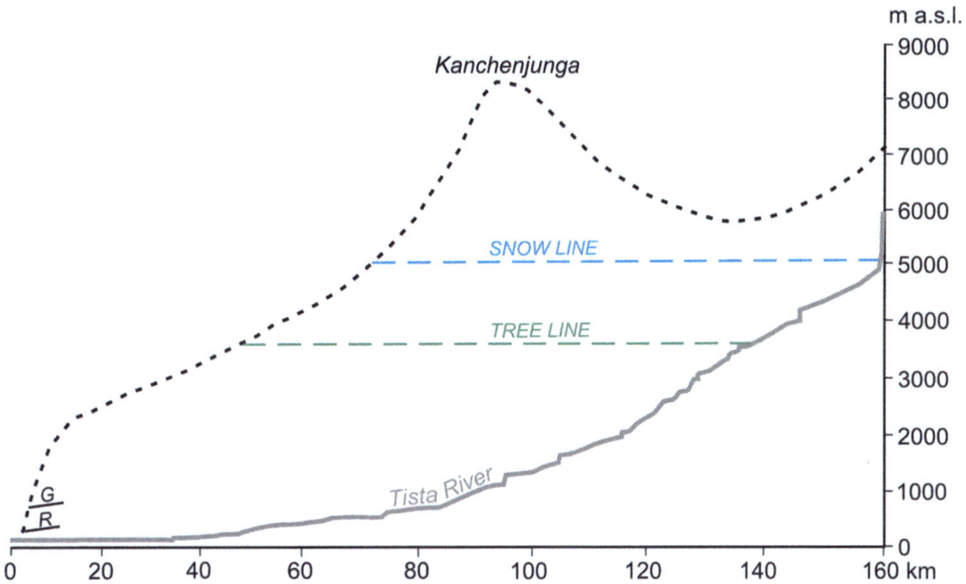

Fig. 2 Simplified orographic S-N transect along longitudinal profile of the Tista River with several knickpoints and main watershed. Position of snowline and upper tree line indicated. *R* Rangamati River level, *G* Gorubathan Surface

overthrusted along Main Boundary Thrust (MBT) over the Neogene-Quaternary Siwalik beds which are only several hundreds of meters thick. The steep inclined sandstones and conglomerates of Siwaliks are separated by a distinct Frontal Tectonic Line from horizontally bedded alluvia of the Ganga-Brahmaputra Plains. In the eastern part, the Siwaliks are less prominent than that of the western Himalaya.

The effect of the tectonic differentiation is reflected in the presence of vertical morphoclimatic zones, and associated processes and relief. The forest belt, reaching the upper tree line at ~3,500 m a.s.l., is dominated by fluvial processes, the sub-alpine zone is characterized by cryogenic processes, and with the snowline at ~4,850 m a.s.l., the zone of permanent ice and glaciers often descends to 4,000 m a.s.l. (Figs. 1 and 2) in the eastern Himalaya. The highest part is sculpted by glaciers throughout the year. These high ridges are characterized by sharp peaks, wide glacial cirques and deep troughs with steep walls (Fig. 3). The characteristics continue downstream indicating greater extent of glaciers in the past.

The river valleys in the High Himalaya are incised to a depth of 2,000–3,000 m and in the Lower Himalaya to a depth of 1,000–2,000 m. Narrow ridges, convex or straight slopes, frequently undercut by lateral erosion, are noticed in most valleys. As a result, mature convex-concave profiles are very rare and may develop only due to different resistance of rocks when quartzites or sandstones are exhumed forming hogbacks or cuestas and softer beds are removed.

3 Geomorphic Processes

The type and intensity of processes are related to morphoclimatic vertical zonation as well as to the rainfall gradient connected with the mountain barrier. The area above the snowline located at 4,800–5,000 m a.s.l. forming part of High Himalaya occupy about 2,000 km^2 (>20 % of the catchment). Mountain glaciers cover only ~270 km^2. Many of these glaciers are 10–20 km long and are now slowly retreating (Fig. 3). In the cryogenic belt, a long snowmelt season plays an important role in denudation. In the forest zone, the leading process is chemical weathering. At lower elevations, depending upon the slope gradient and lithology, the regolith thickness varies from 1–2 m to >5 m. High percent of sandy and silty fraction facilitates deep infiltration of water into soils which are therefore susceptible to mass movements (Froehlich et al. 1990; Starkel and Basu 2000).

The intensity of degradation depends on the intensity and frequency of heavy rainfall. The Sikkim-Darjeeling Himalayas are exposed to moisture-laden air masses moving directly from the Indian Ocean through a wide gap between the Deccan Plateau to the west and the Meghalaya Plateau to the east. The highest annual rainfall of 5,000–6,000 mm is recorded exactly at the base of the steep front of the Darjeeling Hills, which rise from 100–200 to 1,000–1,500 m a.s.l. The northern lee slope of the first ridge receives less than half of the amount received by the southern flank. The rainfall gradually declines below 1,000 mm further north (Starkel 1972; Starkel and Basu 2000; Starkel et al. 2008). One or more heavy downpours (200–400 mm in several hours) recur once every decade when debris flows occur on steeper slopes. Continuous rains of above 1,000 mm, covering greater areas, are recorded 2–3 times in a century when cyclonic systems penetrate up to the central Sikkim along transversal depression of the Tista Valley. The last such major event occurred during 2–5 October 1968, and recorded 200 mm of rainfall in 4 h on saturated soils, causing widespread devastation. Even in the forest, 1–2 % of the

Fig. 3 Gurudongmar Glacier with frontal moraines in the northeast Sikkim

surface was eroded. The great Ambootia Landslide (Fig. 1), covering an area of 56 ha, was triggered during this event (Froehlich et al. 1992). The water level in several streams rose by 10–15 m and the flows carried boulders several meters in diameter and in some valleys aggradation reached up to 10 m (Starkel 1972). During this event, the water level of the Tista River rose by 20–25 m and the discharge reached ~18,150 m^3/s (Starkel and Basu 2000). The smaller creeks dissecting the mountain front (like Lish and Gish—tributaries of Tista) with braided channels and fans (Fig. 4) have risen at the rate of 2–3 m after heavy rains within a decade (Starkel et al. 2008).

The other zone of higher precipitation is the southern slope of the Kanchenjunga massif above the upper timberline, where major part of annual precipitation is represented by snow, which falls during the whole year. It is reflected by the glacial tongues extended further down.

Valley deepening in the mountain sections has mostly occurred during heavy rains and floods. Proceeding upstream their role diminishes. Considering the evidence from other parts of the Himalaya it may be expected that the role of earthquakes and rapid snowmelts are important too. This inference is supported by frequent occurrence of great semi-circular niches on the upper slope sections, which look similar to the great rockslide at Ambootia (Froehlich et al. 1992).

4 Evolutionary History

Continuous uplift and overthrusting during the Neogene and Quaternary are reflected in the gradual rejuvenation and very leisurely maturation of the relief. These processes have proceeded with varying rates following phases of tectonic uplift and climate change.

The most substantial transformation occurred in the present-day glacial zone of the High Himalaya, which have been uplifted by at least 2,000 m during the past 2–3 million years (as evidenced by floristic and faunistic findings), when former fluvial V-shaped valley heads were transformed by glacial exaration into wide cirques and U-shaped glacial troughs surrounded by arêtes.

In the deepened river valleys there were at least two breaks, reflected by the presence of two levels with flattened ridges; the upper at 500–600 m above the present river bed and the lower one ~200–300 m above the river bed (Starkel 1972). The gradual valley deepening probably coincides with distinct rocky steps in the longitudinal profiles

Fig. 4 The apex of the alluvial fan of Jainti River at the steep front of Himalaya

of the mid-upper courses of the Tista and other rivers at elevations 1,300–4,000 m a.s.l. (Fig. 2) (Mukhopadhyay 1982).

The remnants of old fans located on strata built of Siwalik beds, are preserved on both sides of the Frontal Tectonic Line (Nakata 1972; Basu and Sarkar 1990). The remnants of Gorubathan Surface at about 700 m a.s.l. and 380 m above the river, with rounded boulders of 2 m in diameter covered by iron crust and probably deposited by debris flows, are observed near the outlet of Chel River at Mission Hill (Fig. 1). Another patch of lower Rangamati level (150–200 m above the river) is observed near the outlet of Balasan River to the west of the Tista River.

Such steps indicate that the dissection of the scarp along the Frontal Fault by smaller streams has taken place concomitantly with the uplift. The lack of coarse gravel along the Tista River might indicate that these sediments were eroded or deposited farther downstream in the plains.

East of Tista, the narrow Siwalik Belt almost disappears and steep mountain front retreats towards the north. Probably the Frontal Fault has not developed up to now. Elongated hills bordered by parallel fault scarps, build by stratified gravels and dated between 30 and 60 ka BP (Starkel et. al. 2008) appear above the younger piedmont fans in this section.

The lowering of upper tree line and snow line by about 1,000 m during cold stages of Pleistocene resulted in the increase in frost activity and extension of glaciers. The remains of frost weathering are still visible near the present upper tree line. Much better indicators are the moraines descending 20–40 km downstream of the present-day glaciers up to ~2,500 m a.s.l. at the steeper slopes of Kanchenjunga Massif and only to ~3,000 m in the NE part of Sikkim (Mukhopadhyay 1982). Probably these differences are connected not only with the elevation of ice fields but also with the exposure of slopes to precipitation. The headwaters of the Tista Valley are in the rainshadow. The extension of glaciers reached the lowest elevation during the last glaciation when about 20 % of Sikkim was under ice.

The last cold stage is reflected in the fluvial deposition along the Tista River. A section of 60 m high terrace, built of sand and fine gravels and ~40 m thick, is exposed at Kalijhora, 5 km upstream of a narrow outlet from Himalaya

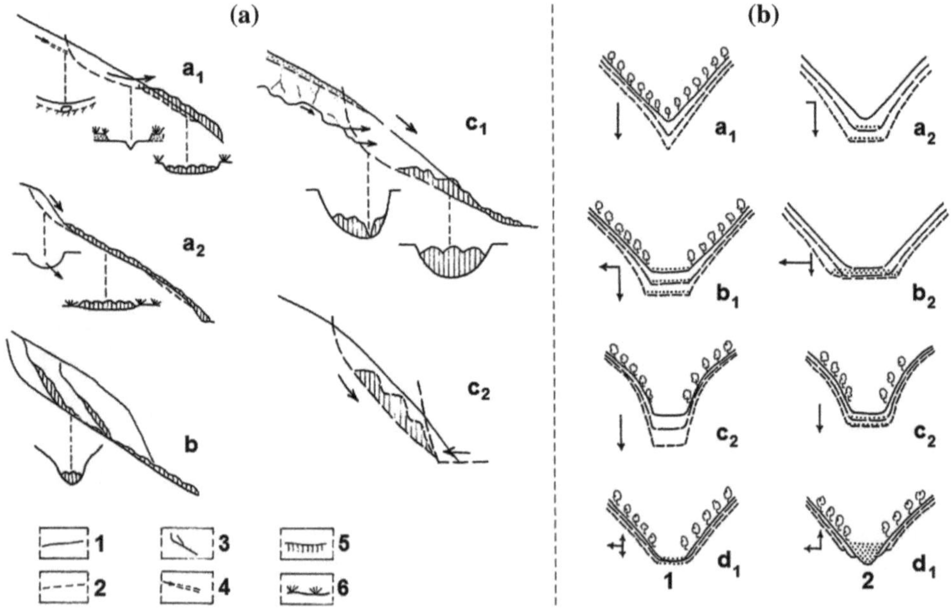

Fig. 5 Typical transformations of slopes and river valleys in the Darjeeling Himalaya (after Starkel 1972). **a** Types of mass movements during heavy rain in October 1968, a_1—mudflow and piping tunel, a_2—slump—mudflow, *b*—mudflows directed by slope gullies, c_1—deep rock slides, c_2—slope undermined by river. **b** Tendencies of transformation of valley cross-sections before and after deforestation (*1* and *2*) *a*—upper valley reach, *b*—middle reach, *c*—outlet of Tista from the mountains, *d*—outlet of smaller river (with progressing aggradation). *1* slope profile before movement, *2* new profile, *3* joints, *4* piping tunel, *5* sliding—mass, *6* soil under tea bushes

at Sivoke (Bluszcz et al. 1997). The deposits were laid down between ~47 and 17 ka (TL date). This indicates that the delivery of fine material from the periglacial zone and from glacial meltwaters has been very high. Probably the whole mountain reach of the Tista Valley had been filled by glacio-fluvial deposits. New phase of erosion is documented by next generation of piedmont fans, connected with increased monsoonal activity from ~14 ka.

5 Role of Human Activity

The deforestation of the lower Darjeeling Himalaya started when its southern part up to Great Rangit River was ceded to the East India Company in 1836 and in 1850–1860s when tea plantations were founded. Then followed road construction and finally in 1881 the construction of the Darjeeling railway. In the 20th century, the forest clearance continued, although some areas were partly replaced by parallel cultivation of fast growing Japanese cedar (*Cryptomeria japonica*). The density of roads increased with the growth of cities. At present, land use pattern in this part of hills in West Bengal is dominated by forest (38 %) followed by tea plantations (23 %) and over 20 % is under other agricultural activities. Darjeeling and the two other big cities (Kurseong and Kalimpong) located on the slopes are growing rapidly partly due to increasing touristic activities (their population increased from 0.5 to ~2 millions in the last 60 years).

The southern part of the Sikkim State started to develop rapidly in the last few decades, both in terms of agriculture (gardens, tea plantations) as well as in tourist services, as visible in the construction of numerous roads and multi-storey hotels.

All these activities have a great impact on the acceleration of geomorphic processes, both on the hillslopes as well as on the valley floors. Their role is reflected in the mean values of runoff and sediment load. Measurements from the mid-20th century in small catchments (~150 km^2) have shown that in the Neora River Catchment, which experienced 31 % deforestation, a minimum specific runoff of 2.4 m^3/s/km^2 was recorded, whereas in the neighbouring Gish Catchment, which experienced 58 % deforestation, the recorded runoff reached only 0.08 m^3/s/km^2.

The human impact is particularly evident after heavy rains. During heavy downpours, debris flows and mudflows pose a serious threat to roads and buildings on steep slopes, especially in the cities (Starkel and Basu 2000; Basu 2006). Undercuttings along the roads, frequently associated with exploitation of coal layers, further accelerate mass movements.

Fig. 6 Mud-debris flows dissecting slopes occupied by tea gardens near Darjeeling after the continuous heavy rain in October 1968

Regional catastrophic rainfall events, like the October 1968 event, may change the evolution of slopes and river channels (Starkel 1972). At tea gardens, 10–26 % of slope surfaces were transformed by debris flows or mudflows, which deepened existing gullies or created new shallow chutes (Figs. 5 and 6). After each such event the density of dissection is increasing. Taking into consideration the fact that after the establishment of tea gardens three such events (in 1899, 1950 and 1968) have occurred, it may be concluded that during the past 150 years, the density of partition of slopes could have increased by 2–3 times.

Debris flow deposits of several meters thick on the valley sides often contribute to the delivery of coarse sediments (2–5 m) to the river channels. Removal of such big blocks is not possible and thus new shallow channel has been formed through the washing away of fine particles and gradual sinking of big blocks. Thus, many smaller streams in the lower Darjeeling Himalaya are exhibiting a tendency toward aggradation, instead of downcutting characteristic so common in uplifting mountains (Fig. 5). Several smaller streams dissecting along the steep front of Himalaya, draining partly deforested areas, are building alluvial fans which are growing upstream also due to loss of flows by infiltration at the apex of the fans. Therefore, aggradation is progressing upstream (Starkel et al. 2008). Only the Tista River with differentiated regime has preserved natural erosional tendency of a river draining young mountains (Fig. 5).

In southern Sikkim, especially in Gangtok and other cities, the most active process is landsliding, especially concentrated along roads and on steeper slopes.

6 Conclusions

The Sikkim-Darjeeling Himalaya represents a part of the globally highest mountain barrier at the margin of the elevated Tibetan Plateau, which is still rising and overthrusting. The effect of uplift is the formation of morphoclimatic vertical zones resulting in the shift of former fluvial-process-dominate relief to the cryogenic and glacial ones. The contrast between marginal part and interior of mountains is greater due to the blocking of humid air masses by the steep mountain front. Enhanced human activities in the last two centuries have been responsible for accelerated degradation and rapid mass movements, and in the marginal zone the overloaded rivers have inhibited the deepening of valleys (with the exception of the Tista River), so characteristic of the active mountain belts in a collisional setting.

References

Basu S (2006) A comprehensive study of landslides and related phenomena in the Darjeeling Hills and their control. D. Litt. Thesis, University of Calcutta, p 245 (unpublished)

Basu SR, Sarkar S (1990) Development of alluvial fans in the foothills of the Darjeeling Himalayas and their geomorphological and pedological characteristics. In: Rachocki AH, Church M (eds) Alluvial fans: a field approach. Wiley, Chichester, pp 321–333

Bluszcz A, Starkel L, Kalicki T (1997) Grain size composition and age of alluvial sediments in the Tista valley floor near Kalijhora, Sikkim Himalaya. Studia Geomorphol Carpatho-Balcanica 31:159–176

Froehlich W, Gil E, Kasza I, Starkel L (1990) Thresholds in the transformation of slopes and river channels in the Darjeeling Himalayan, India. Mt Res Dev 10:301–312

Froehlich W, Starkel L, Kasza Z (1992) Ambootia landslide valley in the Darjeeling Hills, Sikkim Himalaya, active since 1968. J Himal Geol 3:79–90

Mukhopadhyay SC (1982) The Tista Basin. A study in fluvial geomorphology. Bagchi and Company, Calcutta, pp 308

Nakata T (1972) Geomorphic history and crustal movements of the Foothills of the Himalayas. Science Rep Tohoku Univ Ser VII Geogr 22:39–177

Starkel L (1972) The role of catastrophic rainfall in the shaping of the relief of the lower Himalaya (Darjeeling Hills). Geogr Polonica 21:103–160

Starkel L, Basu S (eds) (2000) Rains, landslides and floods in the Darjeeling Himalaya. Indian National Science Academy, New Delhi, p 168

Starkel L, Sarkar S, Soja R, Prokop P (2008) Present-day evolution of the Sikkimese-Bhutanese Himalayan Piedmont. Prace Geograficzne IG i PZ PAN 219:122

Valdiya KS (1998) Dynamic Himalaya. Universities Press, India, p 178

The Brahmaputra River in Assam: The Outsized Braided Himalayan River

Jogendra Nath Sarma

Abstract

The Brahmaputra is a very large braided river, which flows though a narrow intermontane valley in Assam with low gradient. It is the fourth largest river in the world in terms of average discharge, but its discharge is mainly contributed by its tributaries. The river is ranked second in sediment load. The river is the backbone for economy of Assam State and Bangladesh. Characteristic fluvio-geomorphic features of the river are channels of various orders, channel bars, side bars, unit bars, crevasse splays, natural levees, swamps, palaeochannels, etc. Floods, bank erosion, and natural dam-burst floods are major hazards of the river. The Brahmaputra River is an antecedent river as it is older than the Himalaya that are rising since the Miocene times. The sediments from the Himalaya and the southern hills were deposited in the Assam Valley giving rise to full development of the braided pattern in recent times.

Keywords

Assam • Brahmaputra • Braiding • Large floods • Sediment load • Bank erosion • Avulsion

1 Introduction

The Brahmaputra is one of the largest braided rivers in the world. It is an international river as it flows through Tibet in China as the Tsangpo (Yarlung Zangbo), India as the Siang/Dihang and Brahmaputra and Bangladesh as Jamuna. The river flows both as single thread and braided channel through Tibet, but it becomes totally braided in India and Bangladesh. The most distinguishing characteristic of the Brahmaputra River in Assam, India, is its channel size vis-à-vis its valley width, and large unstable and dynamic braids. The Brahmaputra Valley in Assam is about 35–90 km wide, whereas the width of the Brahmaputra River channel itself varies from 1.1 to 18.6 km. Hence, the outsized river occupies a large part of its valley (Fig. 1).

The river originates in Tibet, where climate is cold and dry (annual rainfall ~400 mm) temperate steppe variant. Flowing easterly for about 1,600 km with moderate gradient (1.6 m/km), the river then takes a sharp bend and flows southward through deep bedrock gorges cutting across the Higher and Lesser Himalaya and the Siwalik Hills with high gradient (4.3–16.8 m/km) over a series of precipitous cataracts and falls before entering the plains near Pasighat in India (Fig. 2a). About 600 m wide river in the Himalaya becomes 10 km wide just 12 km downstream of Pasighat due to sudden decrease in the channel gradient to 0.27 m/km. The river has a wide braided course throughout Assam with an average gradient of 0.16 m/km, except for three locations where it becomes narrow as it cuts through hard granite and gneissic rocks. Downstream, in Bangladesh, it is about 12 km wide. The total length of the Brahmaputra is about 2,880 km and the contributing drainage area up to the

J. N. Sarma (✉)
Department of Applied Geology, Dibrugarh University, Dibrugarh, 786004, Assam, India
e-mail: jnsdu@yahoo.com

V. S. Kale (ed.), *Landscapes and Landforms of India*, World Geomorphological Landscapes,
DOI: 10.1007/978-94-017-8029-2_16, © Springer Science+Business Media Dordrecht 2014

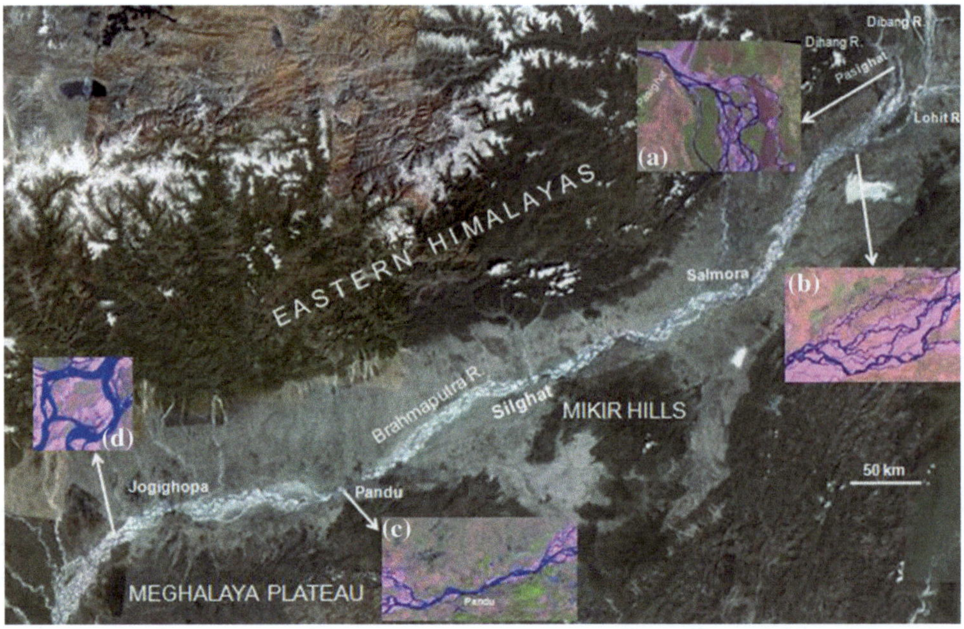

Fig. 1 Satellite image of the Brahmaputra River through Assam valley. *Insets* **a** single channel of the Dihang River enters into the plains near Pasighat and immediately develops braiding, **b** anabranching-cum-braided channel on the north of Rohmoria, **c** narrowest reach at Pandu with wide braided channel at both the ends, **d** a single large channel gets symmetrically divided into two smaller channels near Dhubri. R represents river

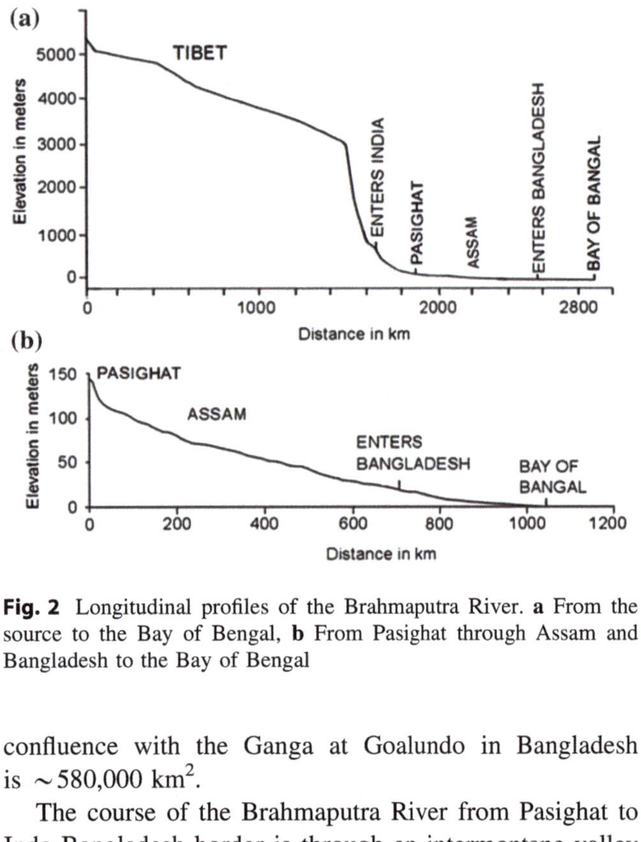

Fig. 2 Longitudinal profiles of the Brahmaputra River. **a** From the source to the Bay of Bengal, **b** From Pasighat through Assam and Bangladesh to the Bay of Bengal

confluence with the Ganga at Goalundo in Bangladesh is ~580,000 km^2.

The course of the Brahmaputra River from Pasighat to Indo-Bangladesh border is through an intermontane valley locked from three sides by hills and plateaus, whereas it flows through alluvial and coastal plains of Bangladesh with further low gradient (0.079 m/km) (Fig. 2b). Although it is navigable over the Tibetan Plateau, it is not so in the Assam Valley due to heavy siltation. The river is the backbone for the economy of Assam and Bangladesh. Its basin has the highest biodiversity hotspots in India. Although ravaged by annual floods since time immemorial, the fertile soils and pleasant climate of Brahmaputra Valley have attracted people from various corners to settle here permanently.

2 Channel Hydrology

The Brahmaputra is the fourth largest river in the world in terms of average flow discharge at its mouth with a flow of 19,830 m^3/s (Goswami 1985), although the river ranks 22nd in terms of drainage area. Hence, discharge per unit drainage area is amongst the highest in the world. At Pandu, near Guwahati, the flow in the Brahmaputra yields 0.0306 m^3/s/km^2. The mean annual flow of the Brahmaputra at Pasighat is ~5,869 m^3/s, which increases to ~17,030 m^3/s at Jogighopa, situated about 610 km downstream, implying that the discharge increases progressively downstream. About 68 % of the discharge of the Brahmaputra is contributed by its tributaries, such as Dibang (6.6 %), Lohit (7.9 %), Subansiri (7.9 %), Jia Bhareli (4.9 %), Manas (5.5 %), Sonkosh (2.8 %), and Burhi Dihing (1.9 %). The mean monthly discharge is highest in July (18.9 % of the total annual discharge) and lowest in February (1.7 % of the total annual discharge). The annual mean rainfall in Assam is ~3,000 mm.

3 Flood Hazard

The Brahmaputra is well-known for its severe and devastating floods. The highest flood discharge (73,000 m^3/s) was recorded in 1962 near Guwahati. The average annual flood

Fig. 3 Google Earth image of a part of the Brahmaputra River around Kaziranga showing features of the Brahmaputra River channel and adjacent valley. The channel is about 8–10 km wide. *1* second order channel, *2* third order channel, *3* mid-channel bar, *4* side bar, *5* sand, *6* bars with vegetation (islands, *chars*), *7* water body (swamp), *8* tributary, *9* abandoned channels, *10* natural levee. Flow of the river is from *right* to *left*

discharge at Guwahati is about 50,000 m^3/s with a recurrence interval of 2.6 year. The bankfull discharge is ~35,000 m^3/s, which occurs every year. Large floods with flows of the order of 70,000–100,000 m^3/s have a return period of 100 years. More than 10,000 km^2 of land (~12 % of Assam), is annually affected by floods. The flash floods of the Himalayan tributaries contribute huge peak discharges leading to flooding of the plains.

4 Sediment Load

Among the world rivers, the Brahmaputra is ranked second in sediment load that amounts to one billion tonnes of clastic sediment and 100 million tonnes of dissolved matter annually. At Pandu, the river transported an average annual suspended sediment load of ~402 million tonnes during 1955–1979 (Goswami 1985). About 70 % of the sediment of the Brahmaputra is derived from the Higher Himalaya. Using a sediment budget, Goswami (1985) estimated an overall aggradation of the 607 km Assam reach of the Brahmaputra from Pasighat to Jogighopa during 1971–1979 as 16 cm. The north bank Himalayan tributaries contribute enormous volumes of sediments; the chief contributions are from Beki and Jia Bhareli as 305 and 297 million tonnes, respectively.

5 The Channel Reaches: Width and Braiding Index

The Brahmaputra River channel in Assam is a classic example of braided river (Fig. 3) The river displays three types of reaches—(1) braided (island/bar)—83 % (2) anabranching-braided—13 %, and (3) narrow single channel reach (node)—4 %. The longest anabranching channel is the Kalang Suti (165 km).

The Brahmaputra flows in a single channel for three short rocky reaches at Silghat, Pandu and Jogighopa (width 1–2 km) (Fig. 4). At Pandu (near Guwahati) the river is narrowest (1.1 km), where the water depth varies from 18 to 27 m. The channel becomes wider downstream of the constricted reaches, for example, ~20 km downstream of Pandu it widens to 18.6 km.

Larger braid channels are between 0.6 and 1.5 km wide, and the depth varies from 2 to 9 m during dry season. Usual anabranching channels display sinuous thalweg

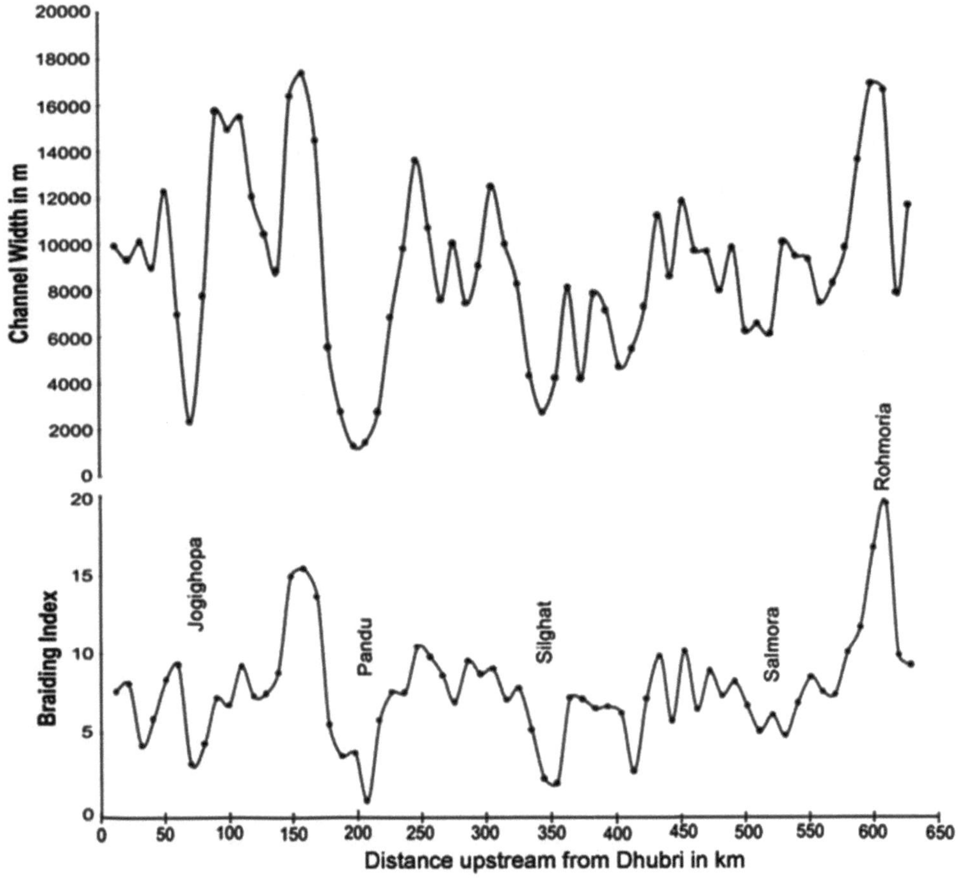

Fig. 4 Variation of channel width (w) and Braiding Index (BI) of the Brahmaputra in Assam from Dhubri to Kobo. Note that both w and BI are low at Jogighopa, Pandu, Silghat and Salmora and high at Rohmoria and downstream of Pandu. $BI = 2(\Sigma L_I)/L_r$, where ΣL_I is the length of all the islands and/or bars in the reach, and L_r is the length of the reach measured midway between the banks of the channel belt

with alternate bars. Their length ranges from 22 to 35 km, and the width varies between 265 and 622 m. The smaller braid channels ($\sim$100 m wide) occur on top of the bars.

The width of the Brahmaputra River channel has shown a constant increasing trend. The mean width of the river measured at 64 equally spaced segments in 1916–1925 was only 5,949 m, which increased to 7,455 by 1966–1972 and then to 9,012 m in 2009. The braiding index of the Brahmaputra River in Assam in different segments varies from 1.21 to 17.33, with an average between 7 and 8. It has shown minor changes with time. The channel width and braiding index are directly interrelated (Fig. 4).

6 The Channel Bars and Islands

Four types of bars are observed in the Brahmaputra, namely mid-channel braid bars, lateral or side bars, tributary mouth bars and unit bars (Figs. 3 and 5). Mid-channel bars are generally elongated, rhombic or triangular in shape with the longer axis aligned parallel or perpendicular to the direction of the main channel. Tributary mouth bars are small in size. Islands (*chars*) are diamond shaped, they have mature vegetation, settlements and are relatively stable. About 70 % of the channel bars have length between 0.8 and 1.0 km. Side bars are both elongated and triangular in form and are attached to the banks. The unit bars are rhombic or lobate and emerge only during low flow stage. At low flow stage, large areas of the bar tops ranging in length from a few tens of meters to 18.4 km, are exposed.

The Majuli Island (26° 57′ 2″ N and 94° 10′ 11″ E) of Brahmaputra River is the largest inhabitated river island in the world. But Majuli is not a braid bar island, it was a piece of land called Majali on the southern bank of the Brahmaputra till 1750, when the Brahmaputra partly avulsed southward to capture one of its *parallely* flowing tributary, the Dihing, converting the land area in between into an island, which is now called Majuli. In 1917, the island had an area of 751 km^2, but in 2011, it was reduced to 492 km^2 because of erosion by the Brahmaputra. Majuli is perennially devastated by floods.

7 Floodplain Features

The alluvial plain of the Brahmaputra varies in width in different portions of the Assam Valley, the widest part across Sivasagar is 95 km and the narrowest part across

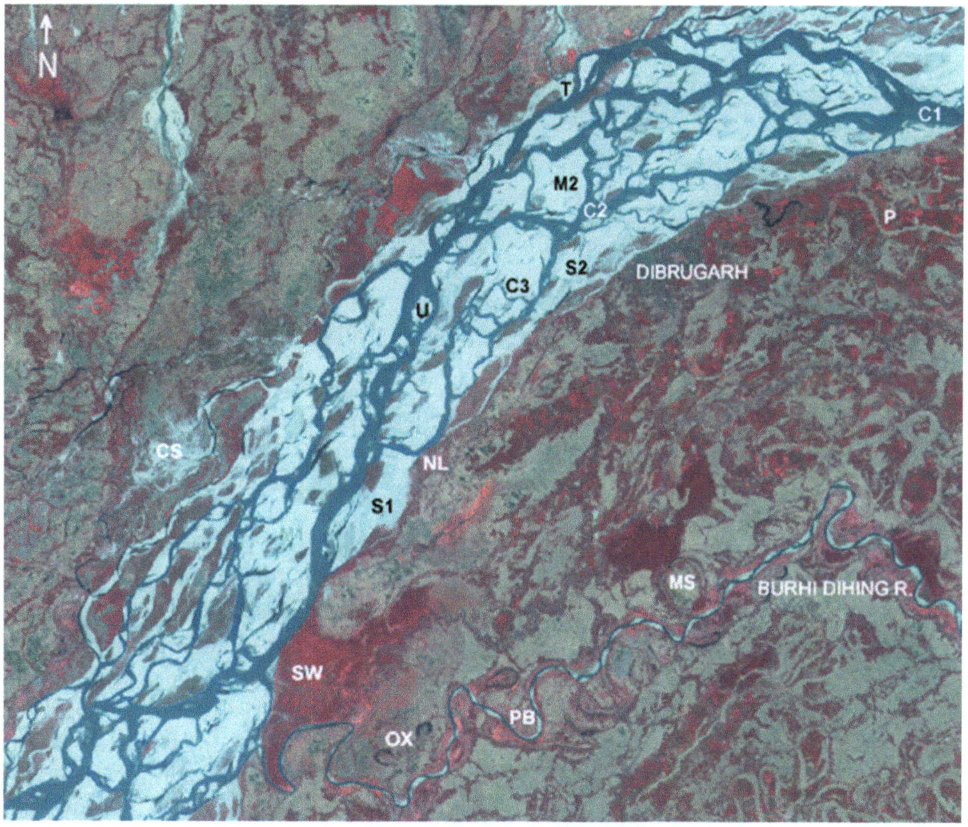

Fig. 5 Indian Remote Sensing 1D satellite FCC image (December 2002) of the Brahmaputra River near Dibrugarh showing Brahmaputra main channel and floodplain features. Large (C1) and small (C2) second order channels, third order channel (C3), mid-channel bars (M1 and M2), tributary mouth bar (T), unit bar (U), side bars (S1 and S2), swamp (SW), natural levee (NL), crevasse splay (CS), palaeochannel (P), and ox-bow lake (OX), point bar (PB), meander scroll (MS) and the meandering tributary Burhi Dihing River. Length of the reach of the Brahmaputra in the image is about 56 km

Silghat is 35 km, with an average width of about 75 km. The valley width to channel width ratio is 7.6 across Dibrugarh in the upstream, 7.5 across Dhubri in the downstream, 4.2 across Silghat due to narrow valley, 4.5 at Palasbari due to widest channel, and 64.5 at Guwahati due to narrowest channel.

Natural levees occur as wedges along the banks, breaching of the same during flood results in crevasse-splay deposits. Large swamps are developed behind the levees on the floodplain. All of the south-bank tributaries are meandering rivers. Meandering tributaries develop ox-bow lakes, meander scars, point bars, channel bars, side bars, point-bar islands, natural levees and deferred tributaries (Figs. 3 and 5). The north-bank tributaries show development of braiding near the foothills of the Himalaya. The interfluves comprise many palaeochannels (former old channels, now abandoned), which were formed due to piracy of the headwaters (Fig. 5). Some misfit streams now occupy the palaeochannels. The Kaziranga National Park (26° 40′ 0.1″ N and 93° 20′ 59″ E) is forest-edged riverine grassland inhabited by the world's largest population of one-horned rhinoceroses having an area of 430 km^2. There are about 204 swamps within this park; most of these swamps represent earlier abandoned river courses.

8 Fluvial Processes

8.1 Bank Erosion, Bank Materials and Causes of Bank Failure

High rate of bank erosion of the river results from frequent failure of its banks (Kotoky et al. 2005). Bank materials in the Brahmaputra River are mostly fine sand and silt with clay fraction less than 5 %. Shear failure in the upper bank materials seems to be by far the most widespread mode of bank failure. It is caused by currents undercutting the upper bank levee materials during high flows, due to which large blocks of natural levee sediments are sheared off into the river. Failure can also occur by over-steepening bank materials (Fig. 6a) because of the migration of the main flow closer to the bank during falling flood stages or where the flow approaches the outer bank at an angle. Large scale slumping is observed during falling stages of the river and highly saturated clayey silt will liquefy and tend to flow towards the channel causing bowl-shaped shear failure of the overlying silty deposits (Fig. 6b).

Between 1912 and 1996 the net bank erosion of the Brahmaputra was ~870 km^2 (Sarma and Phukan 2006). During this period, higher amount of shift in the bankline

Fig. 6 **a** Erosional scarp due to over-steepening bank materials at Rohmoria. **b** Bowl-shaped shear failure on silty bank materials at Nimati Ghat, near Jorhat (Reproduced from Kotoky et al. 2005, with permission)

was between 60 and 130 m/year. In Bangladesh, both the banks have shown a shift of ~90 m/year (Thorne et al. 1993).

8.2 Landslides-Dam Failure Floods

Being situated in a tectonically active zone, occurrence of landslides is frequent, which, in turn, are responsible for great morphological changes on Tsangpo, Dihang, Dibang, Subansiri and Lohit Rivers and their tributaries. Failure of landslide-dammed lakes on these rivers was responsible for flood havocs downstream. For example, the natural dam across the Subansiri, caused by the 1950 Assam earthquake, burst four days later and a 6 m high wave swept across many villages and caused 532 deaths. Breaching of a natural dam formed across Yigongzanbu, a tributary of the Tsangpo, on 10th June 2000 washed away all large bridges of the Siang and flooded Pasighat.

8.3 Avulsion

When a river abandons its earlier course and forms a new channel it is called avulsion. It is the most important change shown by the Brahmaputra. The Brahmaputra avulsed to join the Dihing, thereby creating Majuli Island, around 1750 CE. A large bend of the Brahmaputra was straightened between Pandu and Hazo around 1532 CE. Dibru-Saikhowa reserved forest (RF) was earlier situated on the southern bank of the Brahmaputra and Lohit Rivers. In 1995, an avulsion of the Lohit near Saikhowaghat diverted its flow to join a small river, the Dangori, which captured the Dibru and through it joined the Brahmaputra River (Sarma 2005). Thereafter, Dibru-Saikhowa RF has been converted into an island similar to the Majuli, comprising an area of ~322 km^2.

As per the 1779 map of Rennel, the Brahmaputra flowed southwesterly along the base of the Garo Hills through Jamalpur, Nasirabad, Maimansingh to meet the Meghna River at Bhairab Bazar. The Brahmaputra avulsed into its present course along the Jamuna to the south probably around 1787 (Bristow 1987) . The former course of the Brahmaputra was about 10 km wide and is presently occupied by the Old Brahmaputra, a small meandering river of about 200 m in width, which is active only during the flood season.

9 Evolutionary History

The Brahmaputra River is an antecedent river, as it is older than the Himalaya. The present valley is believed to have taken shape after the rise of the Himalaya during late Miocene. There is evidence to suggest that the river flowed due south, before it occupied the present Assam Valley sometimes in the late Miocene following capture of Tsangpo-Irrawaddy in mid Miocene (Fig. 7). The southern course of the river was blocked by the rise of the Barail Range. As the Himalaya achieved their full height, the rivers draining the southern slopes of Himalaya vigorously eroded the rocks producing enormous amount of sediments. Simultaneously, a long, narrow furrow was formed in front of the mountain. Into this furrow now flowed the Brahmaputra towards west. With the Brahmaputra deflected west, a new drainage network evolved in Assam. The south-flowing independent rivers of long-standing such as the Subansiri, Manas, Jia Bhareli, Sonkosh, etc. now joined the Brahmaputra as tributaries and on the left bank the river received streams rising in the newly uplifted Patkai and Naga Hills and Meghalaya (Shillong) Plateau. Immediately after traversing Assam the course of the Brahmaputra is determined by a fault zone (Dhubri Fault) trending north-south along the border of the Garo Hills, formed in the early

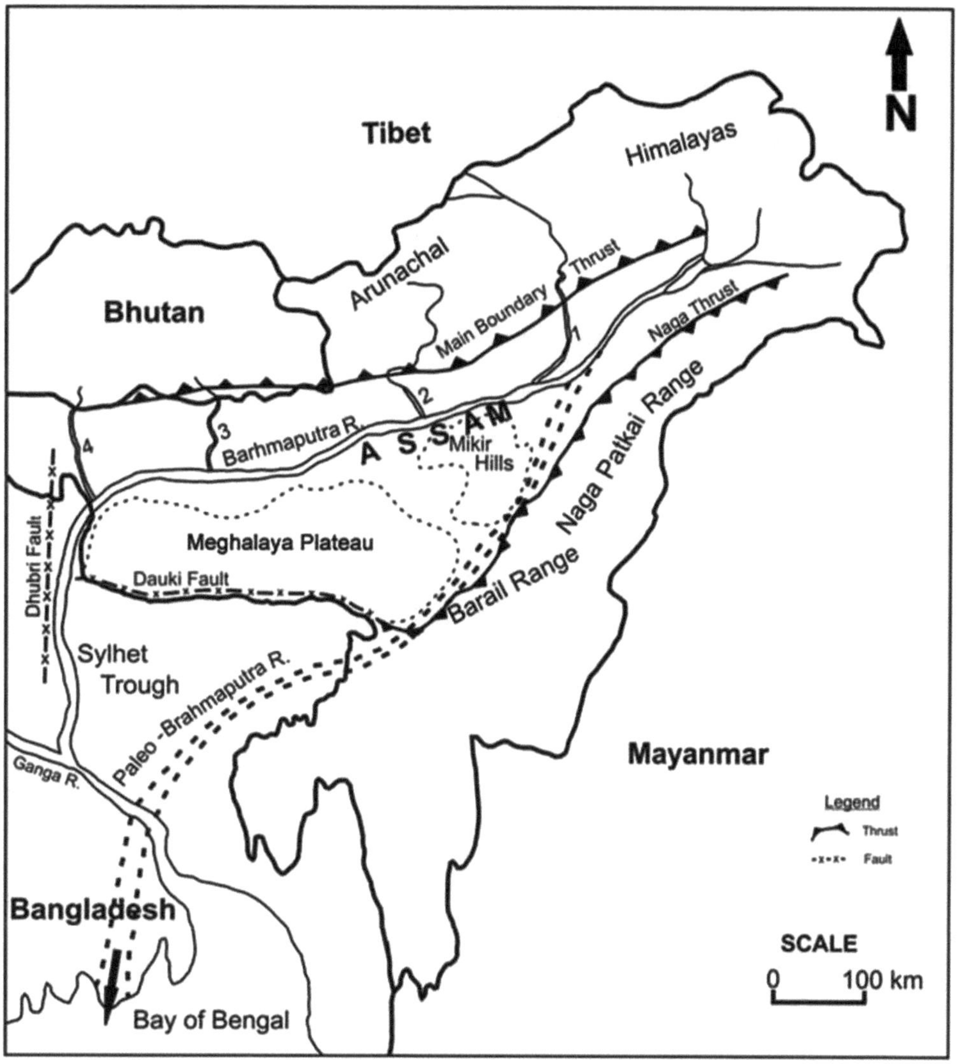

Fig. 7 The present course of the Brahmaputra River and probable ancient southerly course of the paleo-Brahmaputra, prior to rising of the Barail Range. *1* Subansiri, *2* Jia Bhareli, *3* Manas, *4* Sonkosh

part of Miocene. The sediments from rising Himalaya and the southern hills were deposited, raising the valley floor rapidly. By Quaternary, the valley floor was raised to such an extent that the river was unable to carry its entire load to the sea. Consequently, the river started depositing its sediments within the channel, giving rise to full development of the braided pattern in recent geological past.

10 Conclusions

The Brahmaputra is one of the largest braided rivers in the world. It is an outsized river because it is up to 18.6 km (average 9 km) wide and occupies most of its 35 to 90 km wide valley. Remarkable variation of slope of the channel from 16.8 to 0.16 m/km occurs as the river enters Assam Valley thereby intensifying development of braiding pattern. Geological evidence suggests that the Brahmaputra has occupied the elongated Assam Valley since about late Miocene times. Hydrologically, the river has some unique characteristics. The discharge per unit drainage area in the Brahmaputra is the highest in the world. Further, the Brahmaputra is ranked second in terms of sediment load. The high discharge and immense sediment load are responsible for the dynamic nature of channel. Changes in the braid bar morphology and bankline migration are common during the monsoon season. The Brahmaputra has a very high rate of bank erosion and bank migration. Avulsion of the main Brahmaputra and its large tributaries had been giving rise to the creation of very large islands, such as Majuli, and new river like the Jamuna in Bangladesh. The river is highly flood-prone and frequently experiences large floods. More than 12 % of the geographical area of the floodplain is annually affected by monsoon floods of the Brahmaputra and its tributaries. Each year its floods provide water for agriculture and fisheries on which some millions of people and a variety of wildlife depend. Not surprisingly the mighty river is known as the lifeline of Assam (India) and Bangladesh.

References

Bristow CS (1987) Brahmaputra River: channel migration and deposition. In: Ethridge FG, Flores RM, Harvey MD (eds) Recent development in fluvial sedimentology. Soc Eco Paleo Miner Special Pub 39:63–74

Goswami DC (1985) Brahmaputra River, Assam, India: physiography, basin denudation and channel aggradation. Water Resour Res 21:959–978

Kotoky P, Bezbaruah D, Baruah J, Sarma JN (2005) Nature of bank erosion along the Brahmaputra River channel, Assam, India. Curr Sci 88:634–640

Sarma JN (2005) Channel form and morphology of the river Brahmaputra in Assam, India. Geomorphology 70:226–256

Sarma JN, Phukan M (2006) Bank erosion and bankline migration of the river Brahmaputra in Assam, India, during the twentieth century. J Geol Soc India 68:1023–1036

Thorne CR, Russell APG, Alam MK (1993) Planform pattern and channel evolution of the Brahmaputra River, Bangladesh. In: Best JL, Bristow CS (eds) Braided rivers. Special Pub. Geol. Soc. London 75:257–276

The Meghalaya Plateau: Landscapes in the Abode of the Clouds

Pawel Prokop

Abstract

The Meghalaya Plateau is a distinct horst located on the northeastern flank of Peninsular India. It is one of the rainiest, most tectonically active areas in the world and hosts the richest types of karst phenomena in India. The combination of its regional uplift, associated river incision, headward erosion and chemical weathering in varied lithology, has facilitated the development of diverse landforms. This has resulted in deep valleys with magnificent waterfalls contrasting with mature undulating hills and karst topography over short distances. Meghalaya is a region of great scenic beauty and is an interesting tourist destination, in addition to its importance as an area for investigating the interaction between climate, tectonics and erosion.

Keywords

Plateau • Rainfall • Tectonics • Karst • Structure-controlled relief • Headward erosion

1 Introduction

Tectonically uplifted and seismically active areas with high rainfall are characterized by the complex nature of water circulation, weathering processes and landscape evolution. These, together with a variable lithological composition favour diverse geomorphology. The Meghalaya Plateau (also Shillong Plateau) is located on the northeastern flank of the Peninsular India and enjoys extreme conditions on the global scale; the region takes distinction of being one of the rainiest and most tectonically active places in the world, as well as hosting the richest area of karst within the Indian subcontinent.

Meghalaya, literally the "Abode (alaya) of the Clouds (megha)" in Sanskrit, forms a plateau that is the first orographic barrier encountered by the humid southwesterly monsoon winds on their way from the Bay of Bengal to the Himalaya. The massif, built up of metamorphic and intrusive basement rocks, and overlain in the south by a sedimentary complex, is a distinct geomorphological unit stretching 300 km E-W (Mazumder 1986). The rapid uplift of the plateau some 8–14 Ma ago (Biswas et al. 2007; Clark and Bilham 2008) enhanced the orographic rainfall. Intensified weathering and erosion in the warm and humid climate, caused development of great landform diversity. The southern escarpment receives $\sim$12,000 mm of rainfall annually (recorded in Cherrapunji) and has the contrast over short distances of rugged terrain with deep valleys and that of mature undulating hills and karst topography (Starkel and Singh 2004).

P. Prokop (✉)
Department of Geoenvironmental Research,
Institute of Geography and Spatial Organization,
Polish Academy of Sciences, św. Jana 22, 31-018,
Kraków, Poland
e-mail: pawel@zg.pan.krakow.pl

V. S. Kale (ed.), *Landscapes and Landforms of India*, World Geomorphological Landscapes,
DOI: 10.1007/978-94-017-8029-2_17,

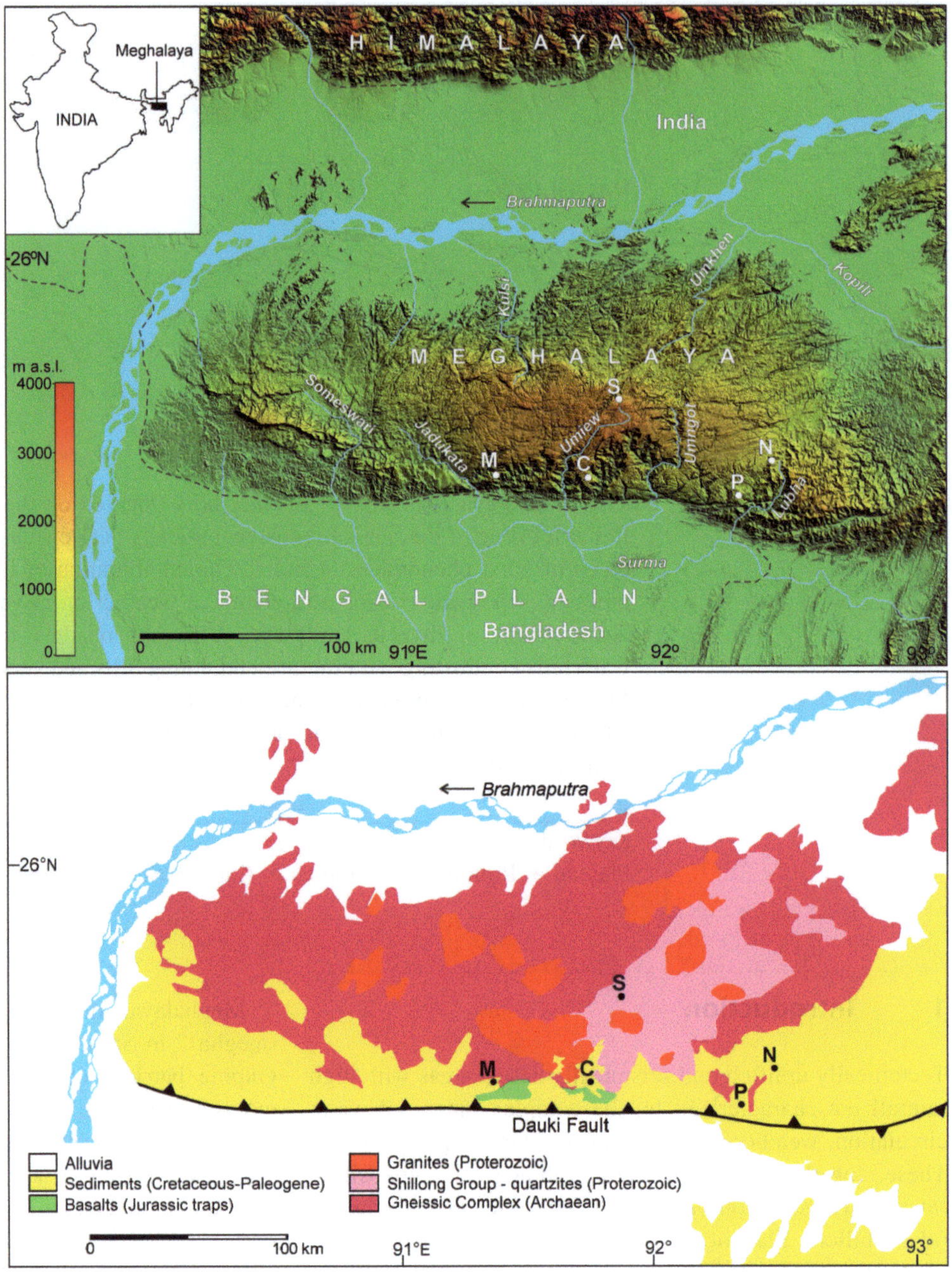

Fig. 1 Digital hypsometric coloured and shaded relief map of the Meghalaya (*upper panel*) and its geology (*lower panel*) (compiled on the basis of GSI 1974 and Mazumder 1986). *C* Cherrapunji with Mawmluh and Mawsmai caves, *S* Shillong, *M* Maw Tynhiang cave, *P* Synrang Pamiang cave, *N* Nongkhlieh ridge with Liat Prah cave

2 Geographical Setting

The Meghalaya Plateau is the northeastern extension of the Indian Peninsular Shield built up of metamorphic and igneous rocks overlain in the south by a sedimentary complex. This "pop-up" horst (raised fault block) is elevated nearly two km above the Bengal Plain in the south and the Himalayan foredeep in the north (Fig. 1). The highest part (1,964 m a.s.l) of the plateau near Shillong is the main water divide between Brahmaputra (Assam, India) and Meghna (Bangladesh) Rivers. Its southern escarpment, related to the Dauki Fault, is about 1,200–1,500 m high and much steeper than the northern slope. In its middle sector, the escarpment is dissected into several spurs by canyons up to 1,000 m deep.

The climate is subtropical monsoonal with the warm rainy season spanning from June to October and the dry cool season from November to May. The mean annual air temperature is closely related to elevation and varies from 24 °C in the foothills to 16 °C at the highest elevations. The annual rainfall distribution pattern is strongly controlled by the southern escarpment of the Meghalaya Plateau. It varies from

Fig. 2 Gently rolling topography of mature plateau with metamorphic and igneous rocks west of Shillong

6,000 mm in the southern foothills to 11,000–12,000 mm in Cherrapunji, which is located on one of the spurs (Fig. 1). Rainfall then decreases with distance from the southern edge of the plateau to 2,200 mm in Shillong and only 1,600 mm in the Brahmaputra Valley. About 80 % of the annual rainfall occurs between June and September.

Wide variations in rainfall, temperature and relief have contributed to a great diversity of forest flora, which constitutes the major land cover. Tropical and subtropical evergreen forests occur up to approximately 1,000 m a.s.l. and in deep valleys the forests can reach altitudes of 1,600 m a.s.l. The upper surface of the plateau is deforested and overgrown by grass with scattered subtropical pines. Deforestation, in the last few hundred years has contributed significantly to the acceleration of runoff and the stripping of the saprolite, which survives only partially on slopes. Therefore, the land use system influences to some extent the intensity of contemporary geomorphological processes.

3 Landform Diversity

3.1 Mature Plateau Top

The upper (top) surface of the Meghalaya Plateau has a mature, structure-controlled landscape with relief energy of 50–150 m (Fig. 2). A rectilinear drainage pattern in the NE-SW direction, affected primarily by exposed fractures, is a distinctive feature. Lithological contrasts between areas of quartzites, granites and sedimentary rocks affect the landform diversity. Near Shillong, quartzites build residual ridges separated by plateaus and shallow valleys. Ridges usually have peaks higher than 1,900 m a.s.l. with the quartzite beds standing nearly vertical close to the contact with the granites. The granite area in this part of the plateau shows little vertical differentiation with deeply weathered hills rising to heights of 1,800–1,850 m a.s.l. (Mazumder 1986). The relief, which is dissected by the wide valley floors sometimes with

Fig. 3 Structure controlled river bed cut in resistant ferruginous sandstones on the Cherrapunji spur

meandering rivers, is the product of protracted weathering and planation. It is possible that some parts of the plateau are the remnants of an extensive Gondwana palaeosurface, later covered by Cretaceous-Palaeogene marine sediments and exhumed in the late Cenozoic. Notable exceptions to the described topography exist closer to the headward erosion zones, where the relief is enhanced in the vicinity of the incised reaches of the major rivers (Migoń and Prokop 2013).

The sedimentary complex is less weathered and built up of almost horizontally bedded sandstones and conglomerates overlain by limestones. Over the spur surface near Cherrapunji, the main drainage network is represented by parallel streams in the NE-SW direction. The drainage network density exceeds 15 km/km^2; therefore, the slopes and interfluves are densely dismembered. Sequences of beds that are more resistant form several irregular scarps on the spur surface, which are especially distinctive between 1,400 and 1,500 m a.s.l. They are also expressed in the form of steeper slope segments and small waterfalls and rapids on the consequent stream flows. The faster retreating parts of the slopes are stabilised by block sliding from the more resistant beds. The horizontal bedding and formation of an iron crust on the rocky channel floors facilitates the lateral expansion of the river channels, which are mainly cut in the bedrock (Fig. 3). They are disproportionately wide to the catchment's size and are adapted to large discharges during periods of heavy rain. Over the less resistant complexes of beds are developed more gentle undulated hills (Starkel and Singh 2004; Soja and Starkel 2007). Deforestation in the past has caused soil degradation on the slopes and the development of a resistant surficial layer armoured in coarse gravels or the exposure of the bedrock. The formation of a protective layer in conditions of high rainfall has led to the establishment of a new stable system with high overland flow but low soil erosion, which reaches only 0.21 kg/m^2/year (Starkel and Singh 2004).

3.2 Rejuvenated Valleys

The south facing escarpment of the Meghalaya Plateau is dissected into several spurs by a system of valleys cut into the basement of the metamorphic and igneous rocks. The valleys have been incised by headward erosion, which has exploited the lines of the NE-SW faults and joints. Within the sedimentary complex, valleys up to 800–1,000 m deep

Fig. 4 Dissected southern margin of the Meghalaya plateau with headward eroded amphiteatral valleys within horizontally bedded sedimentary complex near Cherrapunji

are reminiscent of canyons or deep gorges (Fig. 4). Some of them terminate at amphitheatre valley heads with waterfalls developed on precipitous sandstone faces up to 300 m high (Starkel and Singh 2004). Scarp retreat is accomplished by spring sapping, rock fall and major slope failures. The tendency of slope retreat is conditioned by the occurrence of the alternating sandstone, limestone and conglomerate beds of various resistances. The slopes below are densely dissected by high orographic rainfall and are littered with big sandstone boulders derived from massive rock falls.

Deeper towards the north of the plateau where only basement rocks occur, the upper parts of the valley slopes change from vertical to very steep. Hundreds of shallow landslides are triggered on their slopes by heavy rainfall and frequent earthquakes every year. The natural recovery of their scars is fast in the warm and humid climate with dense vegetation. Debris flows concentrated in channels and ephemeral stream beds, transport large boulders and tree trunks, which frequently generate drainage obstructions.

The longitudinal profiles of the rivers form a relatively smooth curve with low gradients in the downstream reaches that gradually increase upstream. The rejuvenated sections of the V-shaped valleys with waterfalls change upstream to flat-bottomed shallow valleys with straight or even meandering channels. Major knickpoints signify valley rejuvenation whereas minor knickpoints are rock or joint controlled. The mature relief of non-rejuvenated upper river courses is inherited from the previous stages of evolution.

4 Limestone and Sandstone Karst

Karst processes related to structural and morphological conditions, are developed within the 300 km long and up to 30 km wide strip of sedimentary deposits along the southern margin of the Meghalaya Plateau. Over 1,300 caves with almost 400 km of cave passages have been explored to date and new caves are surveyed every year (Gebauer 2008). The Meghalaya hosts the longest cave (Liat Prah, 30 km), the deepest cave (Synrang Pamiang, 317 m) and the highest cave passage density near the Nongkhlieh Ridge (125 km of cave passages in an area of 30 km^2) in India (cf. Fig. 1). The caves range in size from small hillside openings to vast interconnected subterranean systems of many chambers

Fig. 5 Speleothems in Mawsmai cave, near Cherrapunji

and galleries. Many of these caves have impressive river pathways, deep shafts, canyons and keyhole passages mixed with huge fossil passages. A wide variety of speleothems have been recorded in the cave systems; some of which have been successfully investigated and dated for reconstructions of past Indian monsoon variability (Berkelhammer et al. 2012).

The most accessible surface portion of the Cherrapunji spur is built up of an almost 150 m-thick limestone-sandstone complex of littoral facies. The chemical denudation in the area of the Cherrapunji spur is a dominant process but is differentiated in space. Several occasional measurements indicate that the dissolved load is of the order of 200 tonnes/km^2/yr in the area of the sandstone, whereas it could be four to five times higher in area of the limestone (Starkel and Singh 2004).

The valley floors in this region are characterised by rising limestone mesas of 100–150 m in height with steep scarps and typical karst topography, such as sinkholes and poljes. Several small streams disappear in ponors. This water reappears in the karst springs and along the steep escarpment of the spur at the contact of the limestones and ferruginous sandstones. The whole limestone band, located 2 km southwest of Cherrapunji is crossed by underground passages belonging to two caves. Mawmluh Cave, which is the 4th longest in the Indian subcontinent, is 7 km long with five entrances. Mawsmai Cave is only 1 km long, 4.5 m wide and 15 m high with three entrances (Fig. 5). This cave

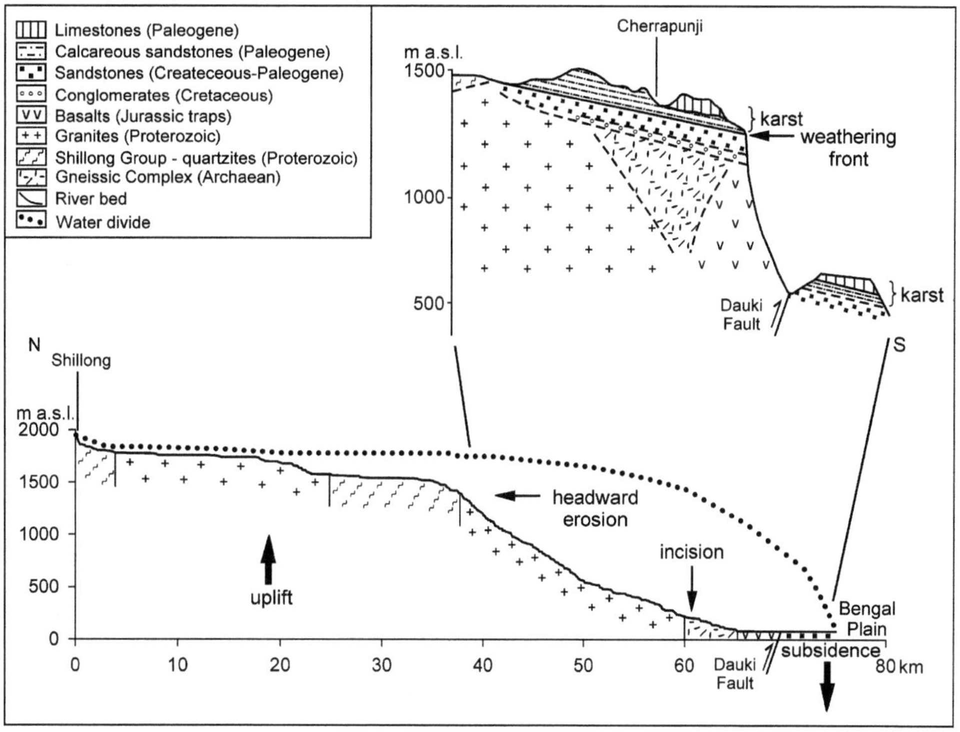

Fig. 6 Simplified diagram showing the escarpment evolution between Shillong and Cherrapunji along tectonically active Dauki Fault

has a large section of fossil passages and two streams that join inside. It is also the only cave in the region adopted for sightseeing.

Despite there being more than a thousand explored limestone caves in the Meghalaya region, only a few sandstone caves have so far been described. An interesting example of active sandstone karstification is Krem (cave) Maw Tynhiang, which is located west of the Cherrapunji spur. It is one of the longest sandstone caves in the world (3.16 km) but has not yet been fully explored. Some relic caves, Krem Pubon and Krem Lymbit, around the Cherrapunji spur are developed mostly in soft calcareous sandstones, which are sometimes intermixed with well-cemented ferruginous beds of sandstones and conglomerates. No or limited speleothem deposits inside the caves indicate a limited carbonate content in the sandstone (Breitenbach et al. 2010).

5 Long-Term Landscape Evolution

The Meghalaya Plateau is a basement block formed by gneisses and quartzites with intruded granites of Precambrian and upper Proterozoic age (Fig. 6). The surface was subject to long periods of planation to form a continental-scale surface of low relief, characteristic of much of the Indian Shield, prior to the fragmentation of Gondwana. The breakup of Gondwana by the end of the Jurassic caused the eruption of basalt traps and the formation of the W-E trending Dauki Fault (Mazumder 1986). During the late Cretaceous-Palaeogene transgression, sediments showing facies transition from littoral marine to coastal-deltaic covered the surface of the present plateau. In the Cenozoic, the plateau has been subjected to compressional forces in N-S and E-W directions resulting from the collision of the Indian Plate with the Tibetan and Burmese Plates, respectively. In effect, major uplift probably started in the Miocene (ca 8–14 Ma) with rates ranging from 0.4 to 0.53 mm per year (Biswas et al. 2007; Clark and Bilham 2008). Earthquakes from microlevel to $M > 8$ (noted in 1897) express seismic activity and indicate the ongoing rise of the plateau (Bilham and England 2001).

Concurrently with the plateau uplift, along the Dauki Fault, the removal of sedimentary cover and headward erosion of the escarpment was activated. Further upstream, erosion focused on and exploited pre-existing zones of structural weakness, mainly aligned NE-SW, which has led to the development of deeply incised fracture-aligned valleys (Mazumder 1986). Their steep slopes are intensively deepened and widened by debris flows and landslides. Within the sediment cap, scarp retreat by spring sapping, rock falls and major slope failures has been responsible for the formation of spurs separated by deep amphitheatre-like valleys with waterfalls. The presence of sandstone and limestone beds of various resistance has created a very characteristic step-like topography of ridges and slopes with retreating escarpments, as well as wide flat valley floors.

When headward erosion reached the metamorphic and igneous area, transformation of the deeply weathered preexisting plateau topography was initiated. Headward erosion enhanced local relief and consequently, the weathering systems through etching and stripping (Migoń and Prokop 2013). Away from the main drainage lines, the remnants of the poorly differentiated plateau survived as level surfaces.

Simultaneously with mechanical erosion, high monsoonal rainfall and the warm climate support chemical weathering and the circulation of underground water within the sedimentary rocks. The dominant process of solution in the karst landform development was influenced by a set of tectonic fractures, as well as the frequent contact of soluble limestones with more resistant sandstones and basement rocks. Many of the caves are aligned and developed along a rectilinear set of tectonic faults and joints, which were initially widened under phreatic conditions and then enlarged under vadose conditions later on.

6 Conclusions

The southern slope of the Maghalaya Plateau shows that under conditions of heavy rain, the continuous tectonic uplift and lithology still play a leading role in the evolution of the present-day relief. The transformation of the primary planar surface into a more differentiated relief is an example of relief rejuvenation subsequent to plateau uplift, associated with river incision and various types of erosion operating in the bedrock with varying resistance to degradation. Therefore, apart from the spectacular landscape of the contrast between young valleys and mature plateau, the Meghalaya region offers an abundance of smaller-scale active and relic geomorphological phenomena over relatively short distances.

The location of Meghalaya between the Bay of Bengal and the Himalaya makes it a key location for investigating the interaction between climate, tectonics and erosion at a regional and even global scale. The high rainfall over the southern slopes is crucial in the flood processes noted in northern Bangladesh. A variety of speleothems in caves provides the opportunity to study monsoonal rainfall regimes archived in stalagmites over a timescale of hundreds to thousands of years.

References

Berkelhammer M, Sinha A, Stott L, Cheng H, Pausata F, Yoshimura K (2012) An abrupt shift in the Indian monsoon 4000 years ago. In: Giosan L (ed) Climates, landscapes, and civilizations. Geophysical Monographs, vol 198. American Geophysical Union, Washington DC, pp 75–87

Bilham R, England P (2001) Plateau 'pop-up' in the great 1897 Assam earthquake. Nature 410:806–809

Biswas S, Coutand I, Grujic D, Hager C, Stöckli D, Grasemann B (2007) Exhumation and uplift of the Shillong plateau and its influence on the eastern Himalayas: new constraints from apatite and zircon (U-Th-[Sm])/He and apatite fission track analyses. Tectonics 26:TC6013

Breitenbach SFM, Donges JF, Kharpran Daly B, Kohn T, Kohn T (2010) Two sandstone caves on the southern edge of the Meghalaya Plateau, India. Cave Karst Sci 37:49–52

Clark MK, Bilham R (2008) Miocene rise of the Shillong plateau and the beginning of the end for the Eastern Himalaya. Earth Planet Sci Lett 269:337–351

Gebauer HD (2008) Resources on the speleology of Meghalaya state, India. Part 1: overview. In: Laumanns M (ed) Berliner Höhlenkundliche Berichte 33, Selbstverlag des Speläoclub Berlin, Rangsdorf

GSI (1974) Geology and mineral resources of the states of India, part IV, Meghalaya. Geol Surv India 30:69–90 (Miscellaneous Publication)

Mazumder SK (1986) The Precambrian framework of part of the Khasi Hills, Meghalaya. Rec Geol Surv India 117:1–59

Migoń P, Prokop P (2013) Landforms and landscape evolution in the Mylliem granite area, Meghalaya plateau, North-East India. Singap J Trop Geogr 34:206–228

Soja R, Starkel L (2007) Extreme rainfalls in eastern Himalaya and southern slope of Meghalaya plateau and their geomorphic impacts. Geomorphology 84:170–180

Starkel L, Singh S (eds) (2004) Rainfall, runoff and soil erosion in the globally extreme humid area, Cherrapunji region, India. Prace Geograficzne 191, IGiPZ PAN, Warszawa

The Sundarbans and Bengal Delta: The World's Largest Tidal Mangrove and Delta System

Kimberly G. Rogers and Steven L. Goodbred Jr

Abstract

The Sundarbans is the name given to the heavily forested, tidally dominated region of the enormous Ganga-Brahmaputra-Meghna river delta shared by India and Bangladesh. Originally formed of alluvial deposits delivered directly by the Ganga, Brahmaputra and Meghna Rivers, sediments in the modern Sundarbans Delta are now mostly reworked from offshore by the large (2–4 m) semi-diurnal tides and seasonally enhanced by sea level set up from onshore monsoon winds, and by tropical cyclones that regularly form in the Bay of Bengal. Sediment annually deposited in the tidal delta continues to be dominated by river-borne sediments even though it has not been directly connected to the main river system for at least hundreds of years. Today, much of the Sundarbans tidal delta is covered by a vast mangrove ecosystem, which remains the world's largest despite significant historical reductions through deforestation. It provides a habitat for the Royal Bengal Tiger, as well as storm surge protection for the millions of people living outside of its boundaries.

Keywords

Sundarbans • Tidal delta • Ganga-Brahmaputra River • West Bengal • Bangladesh • Mangroves

1 Introduction

Deltas are low-lying dynamic landforms that are constructed by the deposition of sediment where a river meets the sea. Once formed, a delta's shape evolves to reflect the balance of river, wave and tidal processes operating at its front. In a delta where tides are the dominant geomorphologic agent, tidal energy is the principal mechanism for the delivery of sediment to the coast away from the main river mouth. The ebb and flow of strong tidal currents causes shoals and bars to form parallel to the flow direction in the river mouth. Once these bed features emerge above the water, they may coalesce to form elongated peninsulas separated by tidal creeks oriented perpendicular to shore. This accounts for the digitate coastline commonly observed in many tidal deltas. Diurnal flow into and out of tidal deltas maintains the relative position of tidal channels, but can also cause continuous erosion and aggradation of tidal islands. Some modern tidal deltas receive sediment directly from their rivers (e.g., Yangtze River Delta, China), while others are "abandoned" by riverine input when distributaries silt up or the river system migrates to another part of the delta. Nonetheless, strong tidal action continues to shape these deserted tidal deltas long after their rivers have migrated away.

K. G. Rogers (✉)
Institute for Arctic and Alpine Research, University of Colorado, UCB 450, Boulder, CO 80309, USA
e-mail: kgrogers@colorado.edu

S. L. Goodbred Jr
Department of Earth and Environmental Sciences, Vanderbilt University, Nashville, TN 37235, USA
e-mail: steven.goodbred@vanderbilt.edu

V. S. Kale (ed.), *Landscapes and Landforms of India*, World Geomorphological Landscapes,
DOI: 10.1007/978-94-017-8029-2_18,

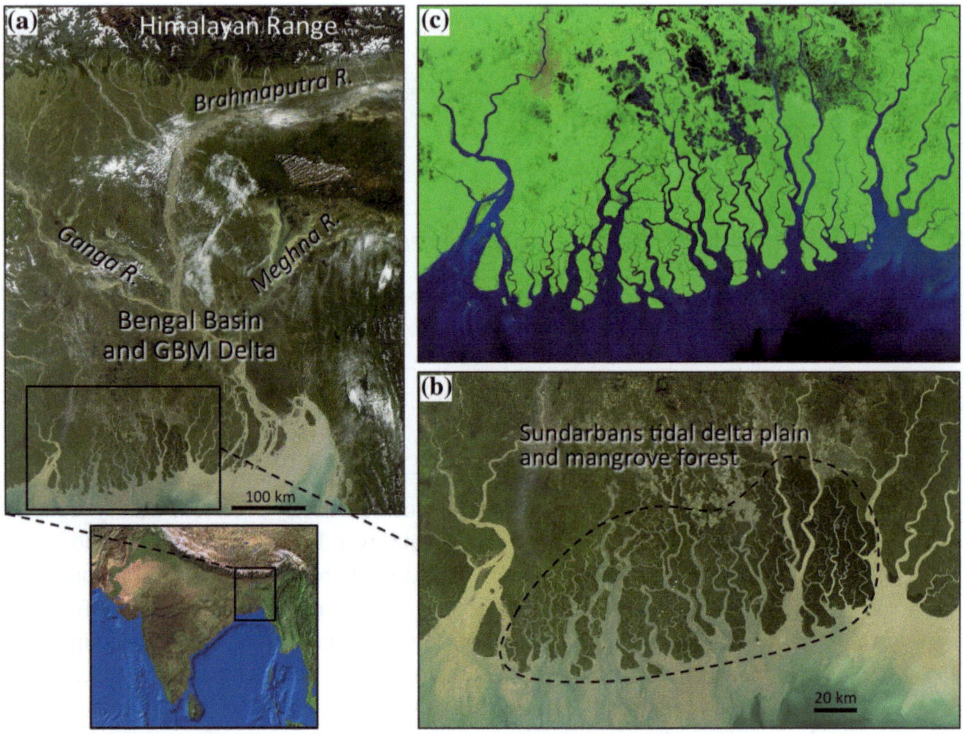

Fig. 1 Satellite images of the Bengal Basin, Ganga-Brahmaputra-Meghna (Bengal) River Delta and Sundarbans tidal delta with a regional inset map of India and South Asia. **a** True-color MODIS satellite image showing the Bengal delta and its proximity to the Himalayan front range. **b** Close-up of the Sundarbans delta plain (*dashed line*) showing the vast network of interconnected channels that twice daily confer tidal waters onto the landscape. The water in and around these channels are earth colored due to the high concentration of suspended sediment from the Ganga-Brahmaputra River that are transported by tides and currents and deposited on the Sundarbans landscape. **c** Infrared MODIS satellite image highlighting extensive areas of flooding landward of the Sundarbans, where much of the landscape has been cleared for rice cultivation but is now being converted to shrimp aquaculture. (images from NASA MODIS website—http://modis.gsfc.nasa.gov/)

Straddling the border between the Indian state of West Bengal and Bangladesh is the world's largest tidal delta: the Sundarbans (Fig. 1). The Sundarbans is the tide-dominated region of the enormous Bengal Delta, which was formed from sediments delivered to the Bengal Basin by three great rivers: the Ganga, Brahmaputra and Meghna. Over a billion tonnes of sediment is annually discharged by the rivers, which serve as conduits routing sediment and fresh water from the Himalaya to the Bay of Bengal. The highest rainfall and ~95 % of the sediment load is delivered to the Bengal coast during the strong summer southwest monsoon from May-September (Coleman 1969). Since the early Holocene this massive discharge has been filling the Bengal Basin and building the delta upward and outward into the Bay of Bengal. Since at least the 17th century, gradual decline in the discharge of muddy river water to the Sundarbans region of the delta (Allison et al. 2003) has led to the increasing influence of tidal processes in shaping and maintaining the Sundarbans morphology and landforms.

Today, the Sundarbans covers 10,000 km^2 of the lower Bengal subaerial delta plain. Meaning "beautiful forest" in the Bengali language, the Sundarbans tidal delta contains the most extensive and biologically productive halophytic mangrove forest on Earth and is home to several endangered plant and animal species (including the Royal Bengal Tiger; Fig. 2). The complex network of estuaries and islands is dissected by 10,000 km of navigable tidal creeks ranging in size from a couple of meters to several kilometers across. The thick mangrove forest also imparts protection from storm surges caused by frequent tropical cyclones in the Bay of Bengal. As a result, the Sundarbans tidal delta contributes critically important ecological and societal benefits to the region.

2 Geographical Setting

2.1 Tidal Plain

The Sundarbans is situated at the northern end of the Bay of Bengal between the Baleshwar River in southern Bangladesh, and the lower Hooghly River floodplain of India; the north-south international boundary follows the

Fig. 2 Images of Sundarbans tidal delta environments. *Clockwise from upper left* Typical mangrove forest boundary vegetated by Nipa palm, or Golpata (*Nypa fruticans*) and the early colonizing Keora tree (*Sonneratia apetala*). *Upper right* Typical sandy shoreline of the Sundarbans Delta fronting the Bay of Bengal. The exposed tree stumps reflect retreat of the shoreface in response to storms and shifting delta sedimentation patterns. Alternating bands in the sand are formed by waves separating the heavy and light mineral sands delivered by the Ganga-Brahmaputra-Meghna River from the Himalaya. *Lower right* Typical forest floor of the Sundarbans tidal delta showing shallow inundation by tidal flood waters and abundant rhizophores, or 'breathing roots', characteristic of several mangrove species, including the Sundri (*Heriteria fomes*) and Passur (*Xylocarpus mekongensis*) trees. *Lower left* Muddy forest floor showing fresh pugmarks of the Royal Bengal Tiger (*Panthera tigris*) and tracks of the Spotted deer (*Axis axis*), as well as crab (*Uca sp.*) burrows

Kalindi River of West Bengal. Reduced freshwater and sediment input has caused the Sundarbans to evolve from a network of river mouth distributaries into a system of interconnected tributive tidal channels with little or no connection to the upstream rivers. The largest of these tidal channels open into funnel-shaped estuaries near the coast. The northern boundary of the tide-dominated portion of the delta averages about 70 km inland of the coast and is marked by a sharp transition from dense mangrove forest to a highly cultivated, agriculture and aquaculture landscape. The area of the mangrove forest has decreased by 40 % over the past 300 years from harvesting of wood for timber and fuel (Islam et al. 1997). In 1987 (India) and 1997 (Bangladesh) the Sundarbans were classified as a protected UNESCO World Heritage Site, significantly reducing deforestation. The large tidal range (2–4 m) and funnel-shaped geometries of the larger estuarine channels amplify the tidal range from 2–3 m at the coast to 3–4 m 100 km inland, enabling the transport of sediment laden marine water beyond the northernmost reaches of the mangroves. North of the forest boundary, extensive embankments (or polders) along tidal channels prevent cultivated areas from being flooded with potentially damaging saline water. In turn, these embankments have reduced tide-related sediment deposition on the delta's surface, resulting in siltation within tidal creeks and sediment starvation inside of the embankments.

Fig. 3 Images showing the impact of Cyclone Sidr, a category-5 tropical storm that struck the Sundarbans coast on November 15, 2007. *Left* Strong winds led to widespread defoliation and mortality of trees along the coast, as well as deposition of sandy sediments eroded from the shoreface. *Right* Extensive scour by waves and the storm surge led to widespread erosion and local shoreline retreat

2.2 Climate

Climate in the Sundarbans region is controlled by both the tropical southwest monsoon and the Bay of Bengal. About 80 % of the rainfall occurs during the five months of the summer monsoon (June–October), with mean annual rainfall ranging from about 1,800 to 2,800 mm. Monsoon precipitation mixes with saline tidal waters in the Sundarbans, effectively flushing the soils and providing the brackish water conditions required to maintain the mangrove ecosystem. The mean annual relative humidity varies from 70–80 % and average annual temperatures range from 21 to 32 °C with humidity and heat being greatest in the pre-monsoon months (April–May). The Sundarbans' position north of the Bay of Bengal makes it vulnerable to large tropical cyclones that frequently form in the bay from October to December and April to May. The Bay of Bengal has been called a "breeding ground" for tropical cyclones due to the broad shallow shelf, warm sea surface temperatures and funnel shape of the bay and its numerous inlets (Murty et al. 1986). Storm surges have been documented up to 12 m high along the Bengal coast, and can be exacerbated by high winds, particularly if they coincide with high tide. Tropical cyclones, while historically devastating to people and property along the Bengal coast, play an important role in shaping the Sundarbans delta front by causing widespread erosion and scour within tidal channels and redistribution of sediment onto the land surface (Fig. 3).

2.3 Sediment Dispersal

Although the Sundarbans covers over half of the lower Bengal delta plain, it is separated from the modern Ganga-Brahmaputra-Meghna River mouth—where sediment is discharged to the Bay of Bengal—by over 100 km. Despite this great distance the Sundarbans continues to aggrade each year from the deposition of river-sourced sediment. From May to September, when river discharge is highest, sediment is transported along the coast westward from the river mouth and into tidal channels by twice daily tidal flooding and sea level set up from strong onshore-directed monsoonal winds (Rogers et al. 2013). When water carrying suspended sediment overtops channel banks, sediment is transported to the interior of tidal islands and deposited when the floodwaters recede. While the Sundarbans as a whole receives new sediment this way, gradual east-west gradients in elevation, flooding frequency, salinity and biodiversity exist, often corresponding with each other. For instance, mean elevations in the eastern Sundarbans (i.e., east of the Pussur River in Bangladesh) average 2 meters above sea level (m.a.s.l.), compared to 1.5 m.a.s.l. to the west near the international boundary (Ellison et al. 2000). These elevation differences allow year-round flooding of the lower lying western Sundarbans during normal high tides, while the interior of eastern tidal islands remain dry except during the highest tides and summer floods. Extensive flooding brings new sediment to the delta's surface, though the more frequent elevation-related flooding of the western Sundarbans has resulted in higher salinity and reduced mangrove diversity compared to the east.

2.4 Cyclones

Cyclones in the Bay of Bengal significantly affect the morphology and ecosystem of the Sundarbans. The winds and waves generated by a low-level storm can cause extensive defoliation of mangrove trees lining the delta front. A major cyclone can displace vast swaths of beach sand and dune fields, and render hundreds of acres of forest into mounds of organic rubble. Storm surges can travel several kilometers up estuarine channels and may cause deep subaqueous scour at the base of peninsular islands. Although the Sundarbans bears the brunt of frequent cyclones, the overall role such storms play in the maintenance and stability of the delta remains uncertain. Nevertheless, the forested delta plain provides an effective buffer that attenuates the impact of cyclones on the Bengal coast and its large human population.

3 Deltaic Landforms

Landforms in the Sundarbans reflect the influence of tides, as well as waves and storms formed in the Bay of Bengal. These landforms range from swamps and intertidal mudflats to sandy beaches and dunes.

3.1 Tidal Islands, Mudflats and Khals

Islands within the Sundarbans are dissected by an intricate network of interconnecting waterways that allow tides and the sediment they carry to penetrate deeply into the interior of the mangrove forest. These small drainages, locally known as "*khals*", allow new sediment to be brought in with tidal floodwaters and also efficiently drain water from the forest interior (Fig. 4). It is common for levees to form around the edges of tidal islands, creating low-lying interiors that remain waterlogged year-round. The continual input of new sediment causes tidal islands and smaller creeks to aggrade, resulting in the formation of new drainages that may eventually breach levees and drain inland swamps. Newly formed intertidal mudflats are often quickly bioturbated by fauna and mangrove seeds, developing a microtopography that affects the subsequent flow patterns of tidal currents and sediment accumulation. Erosion along channel banks is also common, particularly within larger channels. The banks of estuary channels are often fringed with gently sloping unvegetated intertidal mudflats reflective of depositional processes on one side, while the opposite bank has vertical walls similar to the cutbanks of river channels. This morphodynamic balance of deposition and erosion is ongoing throughout the interior of the Sundarbans tidal delta.

3.2 Beaches, Dunes and Chars

The cohesiveness of the fine grained sediment forming tidal islands permits them to resist erosion by all but the strongest tidal currents. Coarser-grained channel sands are more easily eroded and are found around the mouths of tidal estuaries where they build up to form sandy beaches and dunes on the seaward side of the muddy-floored mangrove forest. On the wind-protected lee of dunes, mudflats may form. Silty sediment of the mudflats is overlain by windblown sand sourced from the dunes, providing a favorable substrate for grasses. Grasses and sedges further stabilize these young landforms, which are eventually populated by early-succession mangrove vegetation that further trap sediment, leading to the creation of new islands. In the absence of a major storm event, the emergence of new islands, or "*chars*", along the Sundarbans delta front can occur within a single season; stabilization and colonization by mangroves can occur in less than a decade. The formation of new landforms along the Sundarbans delta front mostly compensates for the loss of islands from erosion and scour by storm waves, with local losses of about 2 km^2/year balanced by a net growth of 7 km^2/year over the entire delta (Allison et al. 2003).

4 Evolution and Geomorphic History of Bengal Delta

The Bengal Basin formed at the convergence of the Indian, Eurasian, and Sunda tectonic plates, where fluvio-deltaic sediments from the Himalaya have been accumulating since

Fig. 4 *Left* High tide and *Right* low tide images of a typical small-order tidal creek that dissects the vast tidal islands of the Sundarbans. These drainages, or *khals*, serve as conduits for sediment delivery to the interior of tidal islands, helping to keep their elevation above sea level and providing drainage to prevent water from stagnating on the forest floor

the late Eocene. The latest stage of construction occurred during the Holocene transgression of sea level, with delta growth beginning ~11,000 years ago and continuing to present. Sediment discharged by the Ganga, Brahmaputra and Meghna Rivers began filling the basin in the early Holocene, creating the enormous delta that now has a subaerial surface area of 110,000 km^2. The onset of early Holocene delta formation is stratigraphically marked in the subsurface by the widespread deposition of intertidal muds containing wood fragments and estuarine shells (Hait and Behling 2009). This suggests the young delta front was occupied by an extensive coastal mangrove system that persisted for 2,000–4,000 years, despite the rapid rise in global sea level that was occurring at that time (Goodbred and Kuehl 2000). In the mid-Holocene, tectonic shifts in the Bengal Basin was causing frequent shifts in the Brahmaputra River course and its sediment depocenters, while the Ganga continued to feed and build the western subaerial delta and its subaqueous shoals into the Bay of Bengal. About 3,000–5,000 years ago, however, the Ganga River began to migrate to the east, resulting in a pattern of abandonment and infilling of the distributary channels that originally connected the river system to the western lower delta plain (Sarkar et al. 2009). As the river system moved away from the western delta, tides and waves began to dominate and reshape the alluvial deposits originally laid down by the river system. The larger (>1 km wide) north-south oriented channels that dissect the modern Sundarbans tidal delta are relict distributary mouths of the ancestral Ganga River.

5 Concluding Remarks

The Sundarbans tidal delta has been described as moribund and sediment starved, though it remains a dynamic and actively aggrading coastal system. The immense monsoonal discharge of the Ganga, Brahmaputra and Meghna Rivers, though not directly connected to the Sundarbans, remains key to maintaining the elevation of the subaerial tidal delta over mean sea level. The nexus of river discharge and strong tidal currents along the Bengal coast allows river-borne sediment to reach the remote interior areas of tidal islands. The delivery of fine grained river sediment and the annual freshening of soils and channel waters have provided optimal conditions for mangrove proliferation for thousands of years. Though the Sundarbans tidal delta and mangrove forest appears to be morphologically stable, it is commonly cited as a deltaic ecosystem particularly vulnerable to anthropogenic and climate related environmental change (Giri et al. 2007). For example, deforestation in the Sundarbans has been considerably reduced since its classification as a preserved forest and tiger reserve, though illegal harvesting of mangroves for timber still occurs. Also, increasing salinity in the western Sundarbans has been attributed to water diversion projects along the Ganga River, but increasing groundwater extraction during the dry season may play a role as well. Other potential threats not yet well understood include changes in sea level, temperature and the frequency and magnitude of tropical storms. Any degradation of the Sundarbans mangroves resulting from these processes would be a significant loss and potentially threaten the storm surge buffer that the region currently provides to the people of Bangladesh and West Bengal. Over the long term, this "beautiful forest" has proven its resilience since the early Holocene by maintaining its relative position in the face of rapid sea level rise. While this geologic history may bode well for the future of the Sundarbans in response to rising 21st-century seas, the challenge is balancing protection of this valuable ecological and societal asset with continued development and well being of the people occupying the region.

References

Allison MA, Kahn SR, Goodbred SL, Kuehl SA (2003) Stratigraphic evolution of the late Holocene Ganges-Brahmaputra lower delta plain. Sed Geol 155:317–342

Coleman JM (1969) Brahmaputra River: channel processes and sedimentation. Sed Geol 3:129–239

Ellison AM, Mukherjee BB, Karim A (2000) Testing patterns of zonation in mangroves: scale dependence and environmental correlates in the Sundarbans of Bangladesh. J Ecol 88:813–824

Giri C, Pengra B, Zhu Z, Singh A, Tieszen LT (2007) Monitoring mangrove forest dynamics of the Sundarbans in Bangladesh and India using multi-temporal satellite data from 1973 to 2000. Estuar Coast Shelf Sci 73:91–100

Goodbred SL, Kuehl SA (2000) The significance of large sediment supply, active tectonism and eustasy on margin sequence development: late Quaternary stratigraphy and evolution of the Ganges-Brahmaputra delta. Sed Geol 133:227–248

Hait AK, Behling H (2009) Holocene mangrove and coastal environmental changes in the western Ganga-Brahmaputra Delta, India. Veg Hist Archaeobot 18:159–169

Islam MJ, Alam MS, Elahi KM (1997) Remote sensing for change detection in the Sundarbans, Bangladesh. Geocarto Int 12: 91–100

Murty TS, Flather RA, Henry RF (1986) The storm surge problem in the Bay of Bengal. Prog Oceanogr 16:195–233

Rogers KG, Goodbred SL, Mondal DR (2013) Monsoon sedimentation on the 'abandoned' tide-influenced Ganges-Brahmaputra Delta plain. Estuar Coast Shelf Sci 131:297–309

Sarkar A, Sengupta S, McArthur JM, Ravenscroft P, Bera MK, Bhushan R, Samanta A, Agrawal S (2009) Evolution of Ganges-Brahmaputra western delta plain: clues from sedimentology and carbon isotopes. Quatern Sci Rev 28:2564–2581

The Spectacular Belum and Borra Caves of Eastern India

A. C. Narayana, M. G. Yadava, Farooq A. Dar, and R. Ramesh

Abstract

The Belum and Bora Caves of Andhra Pradesh form exemplary karst features of great research importance to earth and archeological sciences, besides being attractive tourist destinations. These caves were formed by the dissolution of carbonate rocks (such as limestones and dolomites) through the action of groundwater. This process has contributed to the creation of a variety of surface and sub-surface features. The karstic features in and around these caves include karren, sinkholes, dolines, disappearing streams and karst springs. Sub-surface features include large caves with impressive speleothems (stalagmites and stalactites). These karstic features probably developed during more moist periods in the past. Subsequent change in climate to semi-aridity likely reduced the rate of dissolution of the carbonate to the currently observed rates.

Keywords

Karst topography • Belum Caves • Borra Caves • Speleothems • Stalagmites • Stalactites

1 Introduction

Karstification typically occurs in biogenic, biochemical and chemical sedimentary rocks, such as limestones and dolomites. Solution-enlarged fractures and channels (karren), closed depressions of differing dimensions (sinkholes, dolines, poljes), caves with speleothems (cave deposits that include stalactites, stalagmites and flowstones) characterize karst landscapes. In monsoonal tropics speleothems are valuable archives of past monsoonal changes (Yadava and Ramesh 2006; Sinha et al. 2007).

Although carbonate-rich rocks occur in many parts of India (Fig. 1), classic karst landscapes are conspicuously absent because one or more of the following key conditions are not favourable—(i) presence of soluble rocks (such as limestone and dolomite) close or near the surface (ii) high density of joints (iii) considerable relative relief (iv) moderate to heavy rainfall to cause solution of rocks, and (v) presence of vegetation. Nevertheless, caves and caverns occur in many parts of India. Some of the well known caves are—Guptadham Cave in Bihar, Sahastradhara Caves near Dehradun, Kutumsar and Kailash Caves near Jagdalpur, Mawsmai Caves near Cherrapunji, Gupteswar and Dandak Caves in Odisha, and Borra, Belum and Karnool Caves in Andhra Pradesh.

A. C. Narayana (✉) · F. A. Dar
Centre for Earth and Space Sciences, University of Hyderabad, Hyderabad, 500046, India
e-mail: a_c_narayana@yahoo.com

F. A. Dar
e-mail: farooq.dar1@gmail.com

M. G. Yadava · R. Ramesh
Physical Research Laboratory, Navrangpura, Ahmedabad 380009, India
e-mail: myadava@prl.res.in

R. Ramesh
e-mail: rramesh@prl.res.in

V. S. Kale (ed.), *Landscapes and Landforms of India*, World Geomorphological Landscapes,
DOI: 10.1007/978-94-017-8029-2_19, © Springer Science+Business Media Dordrecht 2014

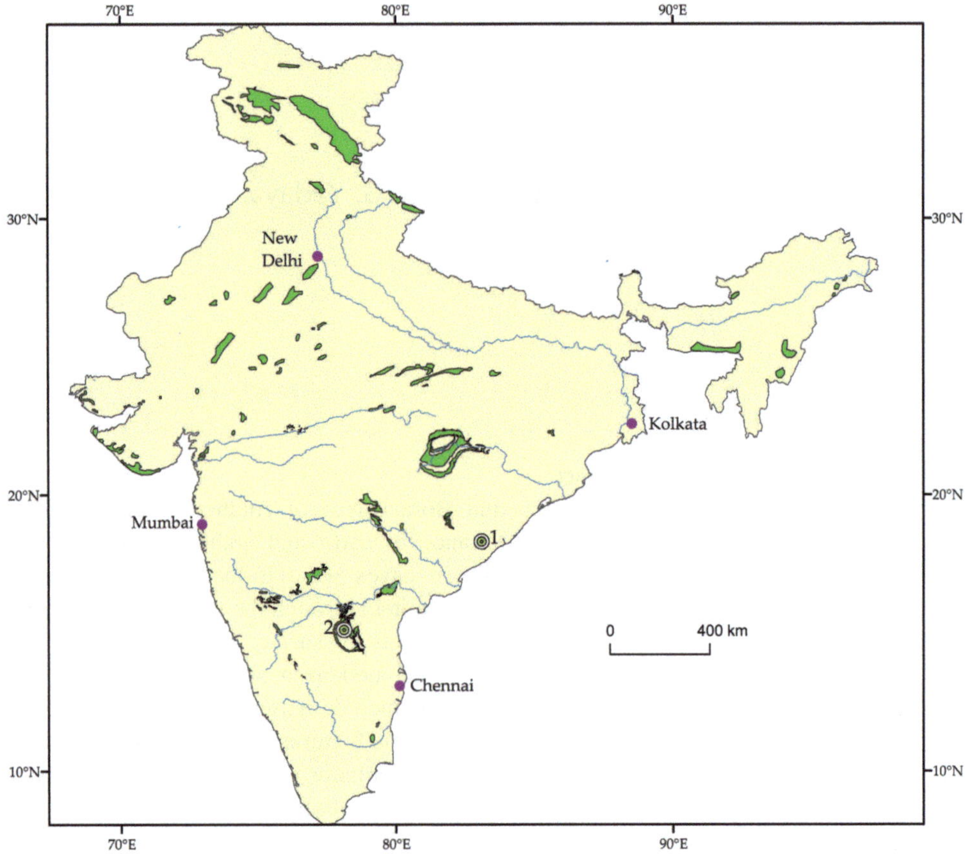

Fig. 1 Map showing the distribution of carbonate rocks in India and the locations of Borra *1* and Belum *2* Caves

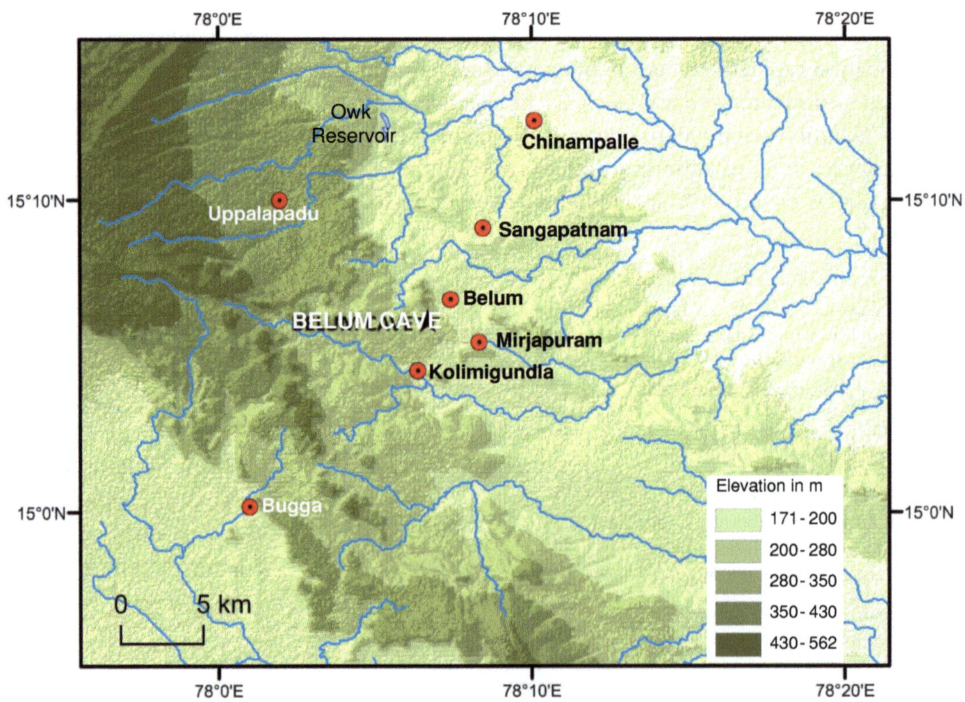

Fig. 2 Relief characteristics in the vicinity of Belum Cave

Here, the geomorphic features of the karst landscape and the architecture of speleothem formations from two well-known caves located in eastern Peninsular India, namely, the Belum and the Borra Caves are described.

2 Location of the Caves

The Belum Cave (Fig. 2) lies in southern Andhra Pradesh in the karstified Narji Limestone of the Cuddapah Basin of the Kurnool District. The *Bilam* (meaning a hole or burrow in *Sanskrit*) or Belum Cave is quite famous for tourism with about 0.2 million visitors annually. The cave is one of the longest caves in India, with a linear development of 3.2 km and a depth of ~29 m (Gebauer 1985). The region is characterized by a typical semi-arid climate with hot summers. The average annual rainfall is ~725 mm.

The most promising gallery is locally known as Parvatama Guha and is more than 300 m long, extending in NW–SE direction (Fig. 3). The gallery is very narrow (0.2–0.6 m height) and is very difficult to trek due to high humidity inside. The galleries are interconnected as rectangular blocks on the right side of the cave. The total length of galleries is about 3 km. Parallel to the main gallery, there is another narrow gallery on the eastern side of the cave extending for about 600 m in the NW–SE direction.

The Borra Cave (Fig. 4), discovered by William King George in 1807 CE, is located in the thickly forested Ananthagiri Hills of the Araku Valley in the Eastern Ghat, about 93 km from Vishakhapatnam. Borra in *Oriya* means "hole". The Borra Cave, having religious, historical, archaeological (e.g., Yadava et al. 2007) and economic importance, is an important tourist spot with about 5 million visitors annually. The average annual rainfall here is ~950 mm.

The total length of the Borra Cave runs is about 825 m, of which a distance of 200 m can be easily trekked. The entrance of the cave is at ~705 m a.s.l. The deepest point of the cave is ~80 m below the entrance level. The main entrance of the cave is 35 m wide and 40 m deep, but the width of the main chamber is ~15 m. A narrow gallery of 150 m long is connected to the main chamber. A parallel chamber, from the central part of the cave, of 50 m wide and 10 m deep runs for a distance of 75 m. It is observed that this chamber is connected to a stream at the deeper levels.

The rainfed Gosthani River (Fig. 5) originates from Ananthagiri Hills. The region is known for rich diversity of flora, fauna and mineral resources and is a habitat of diverse fauna, particularly bats and the golden gecko. Mosses and brown-to-green algae form the main flora of the cave. The native vegetation consists of semi-evergreen deciduous forests. Geologically, the Borra region is complex and the

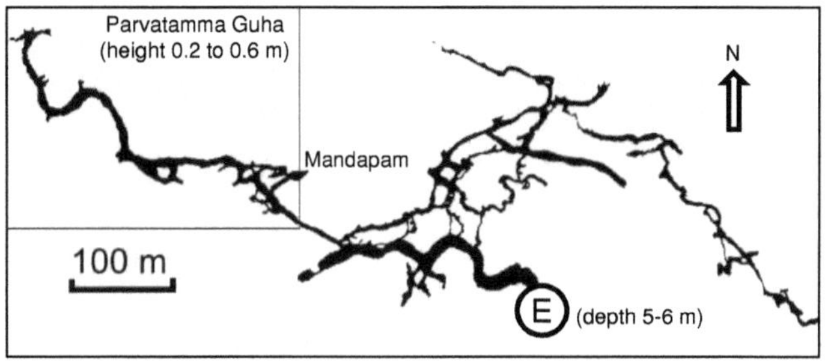

Fig. 3 A generalized ground plan of the Belum Cave. Main entrance is denoted by E

carbonate rocks are pure white, coarsely crystalline and form a 2 km^2 triangular complex in highly folded and metamorphosed Eastern Ghat Granulite Mobile Belt (EGGMB). The carbonate and silicate rocks are of roughly Proterozoic age (~1,000–550 Ma).

3 Landforms and Landform Diversity Around Caves

The landforms in Belum include well-developed surface and sub-surface karst features. The surface features include typical geomorphological features, such as disappearing streams, sinkholes, surface depressions and karren fields as well as mesas and buttes. A number of small, shallow natural depressions/sinkholes around Belum vary in dimensions (20–50 m diameter and 1–3 m in depth). These depressions guide the surface waters underground (Dar et al. 2011). The sinking streams disappear underground after a short distance and carry a large amount of soil with them during heavy rains that gets collected inside the cave. Sub-surface karst features, including solution modified features, vary in scale from enlarged bedding planes to large caves. Epikarst (near surface zone of weathered limestone) features are observed at a number of locations. The Belum Cave developed possibly by extensive dissolution and has evolved into many forms of conduits and chambers (Fig. 6). A stalagmite from the Belum Cave has been dated using Uranium-Thorium mass spectrometry to be at least 100 ka old (i.e., Marine isotopic stage 5b) (Narayana et al. 2011).

The Araku Region, where the Borra Cave is located, exhibits a tectonically developed unique landscape, composed of ridges, valleys and low-lying hills scattered all over (Figs. 4 and 5). The geomorphic features are mostly controlled by structure, with some colluvial/alluvial features. The elevation ranges from 800 to 1,300 m a.s.l. Generally, the hills are extended in NE-SW direction with the prominent Ananthagiri Range composed of igneous and metamorphic rocks. Weathered pediplains rise up to an elevation of 600–700 m a.s.l. Denudational hills have elevation ranging from 1,000 to >1,200 m a.s.l. A structurally

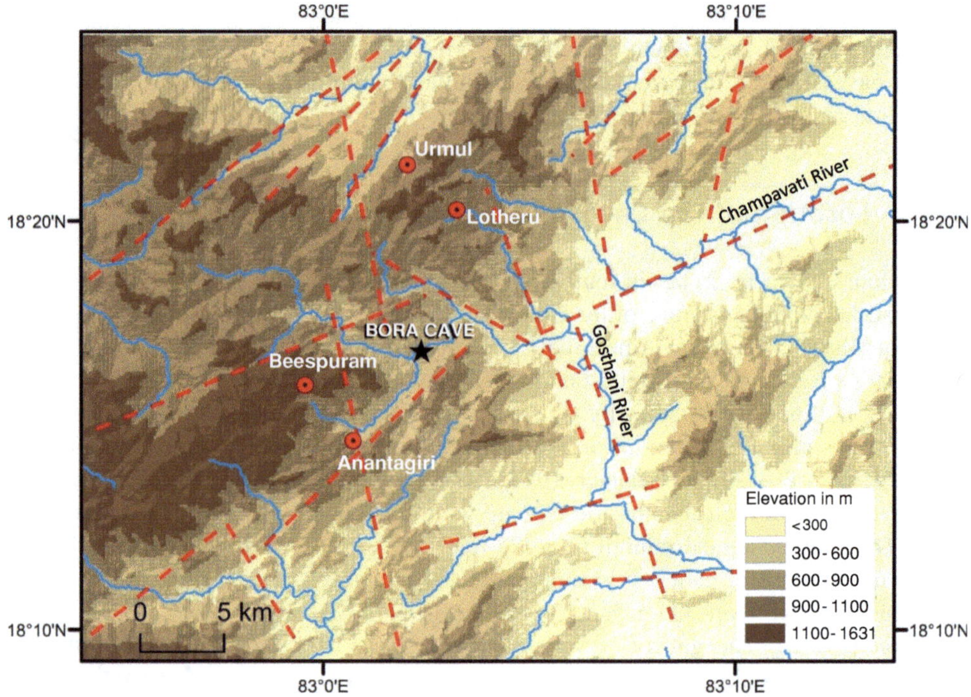

Fig. 4 Relief characteristics in the vicinity of Borra Cave. Broken lines and red dots represent lineaments and settlements, respectively

Fig. 5 The Araku Valley. **a** Rolling topography of Araku Valley near Borra Cave **b** Gosthani River flowing very close to the Borra Cave with dissolved $CaCO_3$ material appearing as white patches along the channel

Fig. 6 **a** A tunnel with solution channels and conduits within Belum Cave **b** stalactites and stalagmites with curvature tips and merger of stalactites with stalagmites

controlled dome on the southwest is the major feature which rises to an elevation of more than 1,600 m a.s.l. The valley floors have a general elevation of 60–400 m a.s.l. Drainage of the region has a typical structure that follows two major fractures/fault traces (Fig. 4). The Gosthani River (Fig. 5b) flows from west to east from its origin and then takes a sharp (90°) swerve and flows NNW-SSE towards the Bay of Bengal. The pattern of the streams is trellis to dendritic with the second and higher order streams joining orthogonally.

The Borra Cave opens at an elevation of 705 m a.s.l in the form of a 100-m wide window. The interior of the cave distinctly exhibits a variety of impressive speleothems of various sizes and irregularly shaped stalactites and stalagmites (Fig. 7). The stalactites range in length from ~0.1 to 3.5 m. The stalagmites are around 1.2 m long with columns of 6 m long and 0.75 m wide. The springs in the cave show abundant iron precipitating bacteria.

4 Environment of Deposition of Caves

Geomorphology of the area around Belum is highly karstified. It has been documented that the dissolution and release of calcium from carbonate minerals and its accumulation and precipitation in the form of secondary features occurs massively during humid conditions. It is clear that major part of the cave remains mostly inactive or only seasonally active. This indicates that the seepage water drains through largely porous bedrocks of the karst system. The major phase of carbonate dissolution (calcium release) and its subsequent precipitation may have occurred during humid periods that probably enhanced karstification (≥200 ka). Reprecipitation of secondary calcium as calcrete deposits around Belum can be linked to weathering cycle of calcrete profiles developed on silicate rocks (Durand et al. 2006). However, present hydrological observations in the Narji Limestone indicate active flow in conduits that feed many springs. This strongly suggests that the ongoing karst development and dissolution processes are restricted to deeper levels and at a slower rate. The Borra Cave might have also developed contemporaneously with the Belum Cave as they lie in the same topographic and climatic conditions. Primitive stromatolitic type microbial activity in the speleothems indicates active biological influence (Sushmitha et al. 2007) in the modern times.

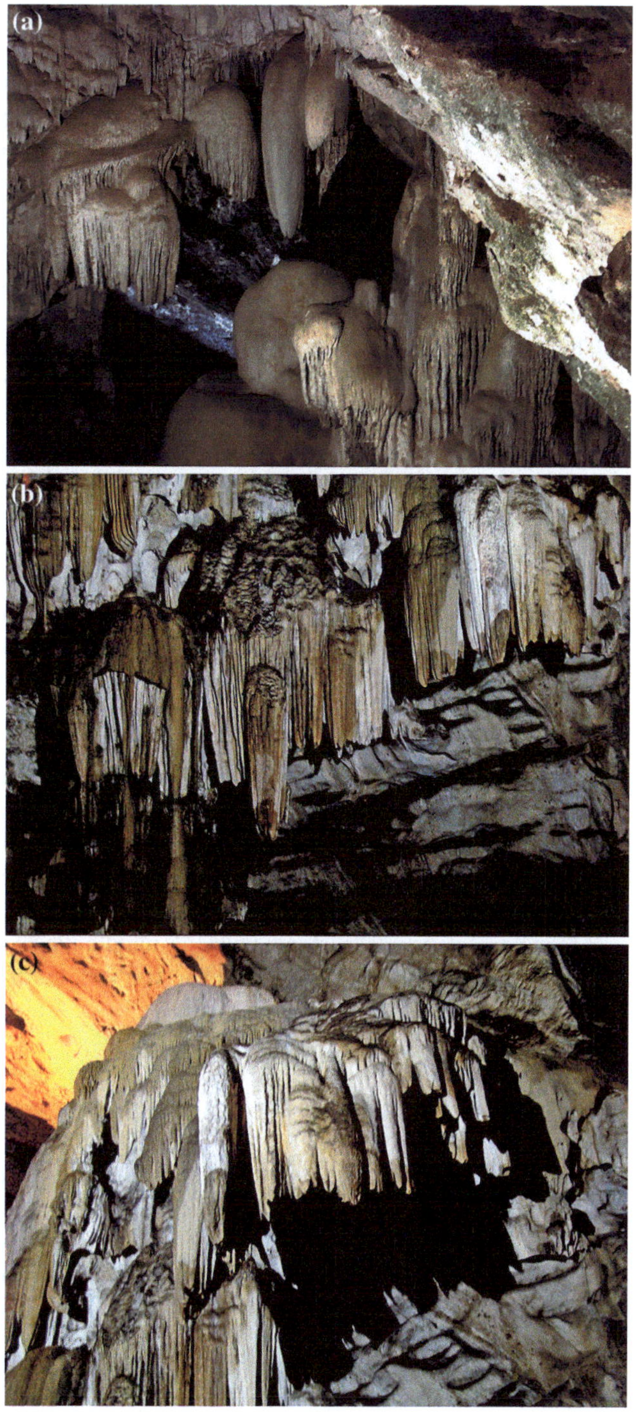

Fig. 7 Various architectural styles of stalagmites and stalactites inside the Borra Cave. **a** hanging stalactites, solution channels, corroded borders and merged stalactites and stalagmites **b** stalactites with needle pattern, dissolutional features and horizontal limestone beds with solution channels, and **c** hanging, sharp-tipped, broken stalactites and paired, sharp-edged stalagmites

5 Evolutionary History of the Caves

Based on geological, geomorphological and patterns of caves, the genetic history of karstification for Belum Cave can be explained. The carbonate rocks were likely deposited during the middle to upper Proterozoic under marine conditions in a discontinuous sequence. Though karstification might have started earlier, the evolution of the cave is believed to be Pre-Pliocene during more humid conditions, which coincides with the initiation of monsoon activity in

South Asia and the evolution of Himalaya at ~8–10 Ma. A eustatic fall in sea-level along with uplift of the Western Ghat could have caused structural fragmentation, outcropping of rocks, and change in the base level enhancing karstification. Later, less humid conditions might have prevailed. Thus, dissolution might have progressed towards water-table conditions promoting incision at deeper levels. This stage is evidenced with an active inaccessible conduit network at greater depths and a seasonally active vadose flow in the caves.

The evolutionary history of Borra Caves is quite complex. The carbonate rocks were deposited during Proterozoic in a heterogeneous EGGMB. The regional geology of EGGMB, where Borra Cave is located, is characterized by khondalite suite of rocks of Archean Age and red bed sediments, laterites, pediment fans, colluvium and alluvium of Quaternary age. The Gosthani River, which flows on karstic type formation, is the cause for the development of odd shapes of stromatolites in the caves. The tectonics and subsequent Gosthani River course associated with EGGMB might have influenced the evolution of the Borra complex as observed in the form of erosional dissection of the linear ridges forming denudational, scattered and isolated hillocks (Fig. 5). Denudational hills are the remnants of structural hills and linear ridges. On regional scale, they follow a definite pattern that matches with the original structural hills. The present-day floodplains are of narrow extension along the course of Gosthani River.

Archeological artifacts, stone tools of middle Palaeolithic culture, dating back to 30-50 ka, confirm human habitation for a long period in the vicinity of Borra Cave.

6 Concluding Remarks

The Belum Cave exhibits exemplary karst features and karstification is still in progress, with several characteristics of semi-arid karst. The karstic landscape is less developed than in a more humid setting, and is typified by a limited number of dolines, preponderance of runoff and evaporation over infiltration, and a lower density of caves. The information on the landscape evolution of Araku Valley is very scanty particularly in relation to the present-day geomorphological set up. The present day landforms are a remnant of erosional processes which evolved under structural and tectonic influence.

The Belum Cave is formed on a plain setting due to the development of sinkholes and surface depressions, whereas Borra Cave is formed on the flank of a hilly region because of a tectonic process followed by karstification. Both the caves attract a large number of tourists in most seasons. The caves might have evolved during Pre-Pliocene period. The Borra Cave provides archeological significance in the form of Palaeolithic tools. The speleothems of both the caves have been studied for paleoclimatic investigations by the authors among others.

References

Dar FA, Perrin J, Riotte J, Gebauer HD, Narayana AC, Ahmed S (2011) Karstification in the Cuddapah sedimentary basin, Southern India. Implications for groundwater resources. Acta Carsologica 40:457–472

Durand N, Ahmad SM, Hamelin B, Curmi P (2006) Origin of Ca in South Indian calcretes developed on metamorphic rocks. J Geochem Explor 88:275–278

Gebauer HD (1985) Kurnool 1984: report of the speleological expedition to the district of Kurnool, Andhra Pradesh, India. Abhandlungen zur Karst und Höhlenkunde (Essays on the karst and speleology), vol. 21, Munchen, Germany, pp. 77

Narayana AC, Tiwari M, Yadava MG, Ngugen DC, Shen CC, Sarang PB, Ramesh R (2011) Stable isotopic variations of a stalagmite from Belum Cave, India: potential for monsoon reconstruction. In: AGU Fall Meeting, San Francisco, pp 31C–1886

Sinha A, Cannariato KG, Stott LD, Cheng H, Edwards RL, Yadava MG, Ramesh R, Singh IB (2007) A 900 year (600–1500 A.D.) record of the Indian summer monsoon precipitation from the core monsoon zone of India. Geophys Res Lett 34:L16707. doi:10.1029/2007GL030431

Sushmitha B, Baskar R, Kaushik A (2007) Evidences for microbial involvement in the genesis of speleothem carbonates, Borra Caves, Visakhapatnam, India. Curr Sci 92:350–355

Yadava MG, Ramesh R (2006) Stable oxygen and carbon isotope variations in speleothems as monsoon proxies: a comparative study of four different locations in India. J Geol Soc Ind 68:461–475

Yadava MG, Saraswat KS, Singh IB, Ramesh R (2007) Evidences of early human occupation in the limestone caves of Bastar, Chhattisgarh. Curr Sci 92:820–823

Granite Landforms of the Indian Cratons

Yanni Gunnell

Abstract

The granite landforms of peninsular India, particularly in the semi-arid continental interior, form rugged clusters of oddly shaped bedrock monoliths and boulders corresponding to stripped weathering fronts. By no means unique to India, the array of landforms typical of most tropical landscapes and commonly described in geomorphology textbooks is present in various associations across the region. Some spectacularly tall inselbergs carved out of poorly jointed, potassium-rich late Archean or younger granitic rocks form mythical landmarks in the scenery. The most unique signature of Indian granitic landscapes, however, arises from the pervasive imprint of human endeavour to harness the amenities of granite landforms for worship (temples), warfare (historical hill forts), runoff agriculture, and dimension-stone quarrying—all of it on a scale rarely encountered among other cultures living amidst granitic landscapes on other continents.

Keywords

Granite landforms • Cratons • Inselbergs • Hill forts • Quarrying

1 Introduction

Granitic bedrock landforms have fascinated geomorphologists because they form distinctive landmarks, often with aesthetically suggestive shapes, and yet can be classified into categories based on form, scale and sometimes process (Twidale and Vidal-Romani 2005; Migon 2006). The cratons of the Indian shield, namely the Dharwar, Bastar, Singhbum, and Aravalli–Bundelkhand, form outcrops of Archean basement separated by platform rock sequences of Proterozoic to Phanerozoic age. All consist of Archean tonalite-trondhjemite gneiss (TTG), the dominant country rock also known as Peninsular Gneiss. The TTG was originally plutonic, sometimes forming mappable diapiric domes, but was metamorphosed to orthogneiss during shallow-angle (Archean-type) subduction between microplates during the Precambrian. The TTG is interspersed with younger granitic intrusions of variable size which play a prominent part in the topographic relief of the Deccan Plateau.

The ~400,000 km^2 Dharwar Craton is the main focus of this overview (Fig. 1) because it provides a spectacular and diverse assemblage of landforms ranging up in scale from corestones, tors, ruwares, castle koppies (kopjes), to bornhardts and inselbergs—all comparable to their counterparts in Australia, Africa and Brazil but nonetheless set in a unique cultural setting endemic to South Asia. Few of these Indian landforms are gazetted as unique from a scientific perspective but some may eventually become protected geomorphosites (Ranganathan and Jayaram 2006). The most extensive areas of granitic outcrop coincide with the Eastern Dharwar Craton (EDC) in the states of Karnataka

Y. Gunnell (✉)
Department of Geography, Université Lumière–Lyon 2, UMR 5600 CNRS, 5 avenue Pierre Mendès-France, 69676, Bron Cedex, France
e-mail: yanni.gunnell@univ-lyon2.fr

V. S. Kale (ed.), *Landscapes and Landforms of India*, World Geomorphological Landscapes,
DOI: 10.1007/978-94-017-8029-2_20, © Springer Science+Business Media Dordrecht 2014

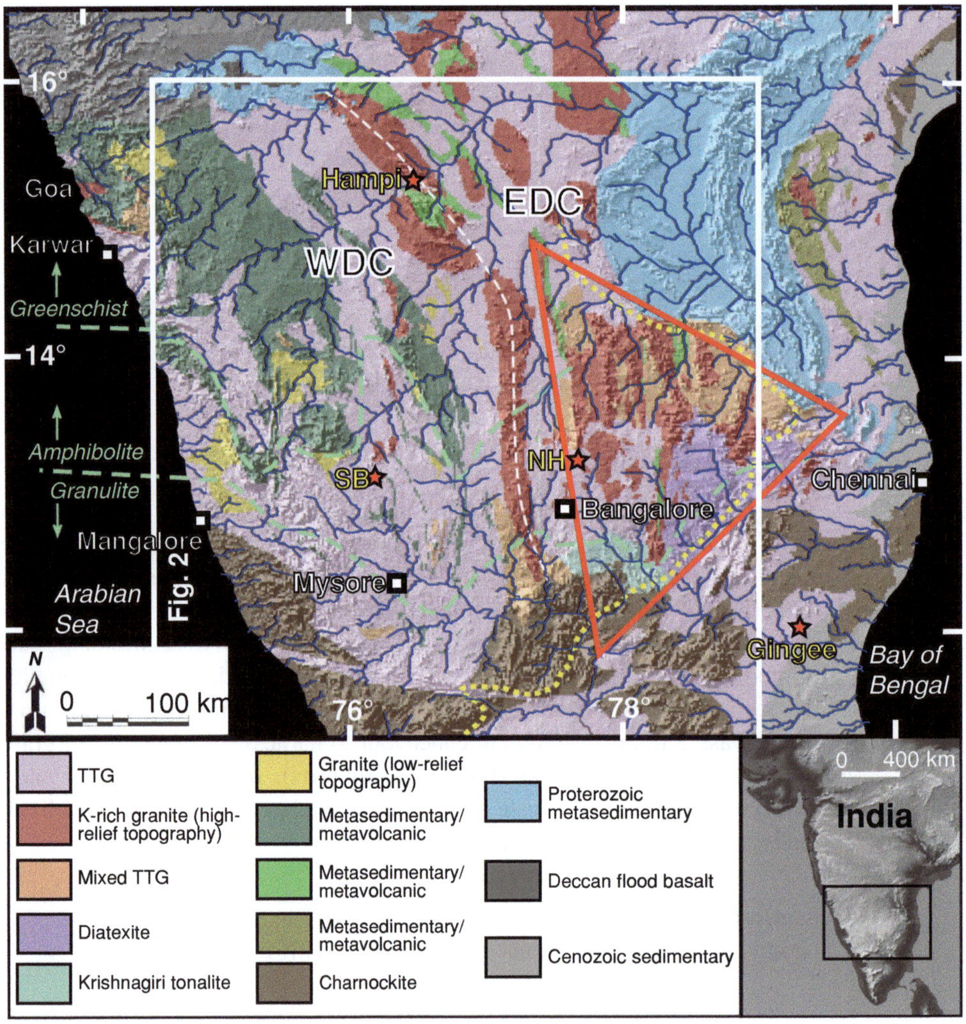

Fig. 1 Simplified lithostratigraphic map of the Dharwar Craton. *Red triangle* locates area of higher resolution geological mapping (Moyen et al. 2003b), revealing a mosaic of four units where previous maps only distinguished one or two outcrop categories. The Krishnagiri tonalites are transition rocks between the younger TTG and the southern charnockites; diatexite is highly heterogeneous migmatite; 'mixed TTG' is TTG, migmatite and granite involving exposure-scale variation. *Dashed green lines* approximate boundaries between metamorphic-grade rocks (greenschist, amphibolite and granulite facies). *Dashed yellow line* SE-Indian Fall Line. *SB* Sravanabelagola, *NH* Nandi Hills. *White dashed line axis* of Closepet Granite intrusion, displayed in Fig. 2

and Andhra Pradesh, in contrast to only 5 % of granite exposure in the Western Dharwar Craton (WDC, Fig. 1). Granitic outcrops also occur sporadically in Tamil Nadu and Kerala, on occasions corresponding to alignments of minor Proterozoic intrusions and ring complexes along deep crustal fractures (syenite, anorthosite, carbonatite, gabbro). Finally, the charnockites of South India deserve a mention because, by forming mountain blocks such as the Nilgiri, Palni and Anamalai Hills at 2.4–2.6 km a.s.l., they form the most elevated topography in South Asia south of the Himalaya (Gunnell and Radhakrishna 2001). Being classified as orthopyroxene granites, like the TTG these charnockites were originally igneous. However, they often exhibit a metamorphic (orthogneiss) fabric, suggesting a complex crustal history. Granite, therefore, is present in the lithological DNA of most crystalline rocks of the Indian cratons.

2 Lithology and Landforms: A Landscape-Scale Perspective

The extensive outcrops of TTG tend to underlie most of the extensive, low-relief erosion surfaces of the Indian cratons (Fig. 1), suggesting greater susceptibility to weathering of the gneiss among other available rock types. Despite a few exceptions (mostly in the WDC: see yellow polygons in Fig. 1), the dominant feature of the cratonic landscape is

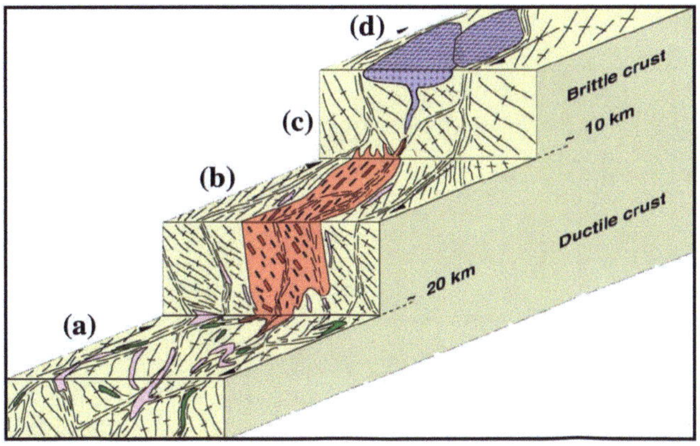

Fig. 2 Longitudinal cross-section of the Closepet batholith (after Moyen et al. (2003a) with permission from the author). At progressively shallower palaeo-depths, the model distinguishes from south to north a continuum made up of four segments: **a** a root zone, where magmas collected in active shear zones between the Eastern and Western cratons; **b** a transfer zone, featuring crystal–liquid partitioning during magma ascent; **c** a granite-free 'gap', which occurs where a phenocryst-rich lag deposit, too dense to rise, was segregated from more liquid phases able to rise to shallower crustal levels; and **d** an assemblage of shallow stocks, filled with this liquid and forming a petrographically diverse mosaic of granite and diorite intrusions exhibiting very weak fabrics and, on occasions, orbicular textures. The Hampi pluton is one example among this mosaic

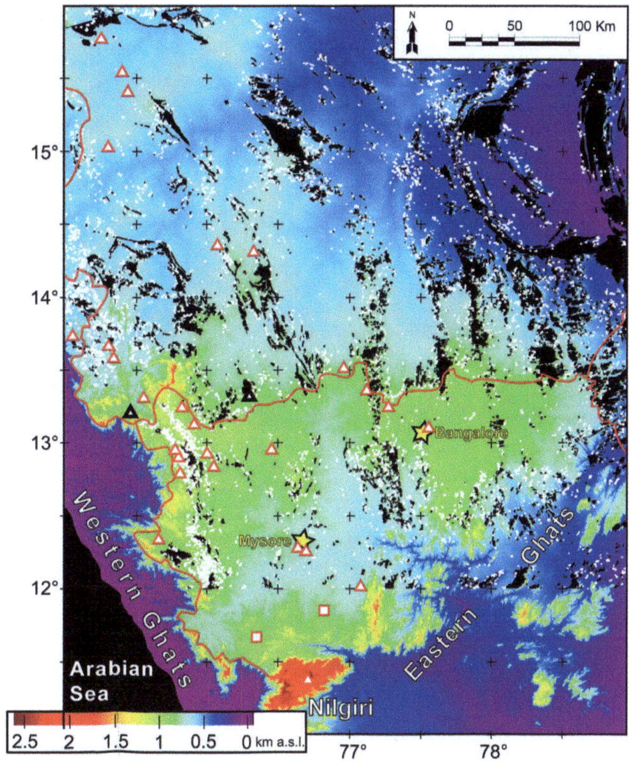

Fig. 3 Residual bedrock landforms, relief and denudation across the Dharwar Craton. *Black* and *white* overlays between 12 and 16°N: residual bedrock landforms with slopes >5°. *Black* landforms with relief >80 m above surrounding plain. *White* landforms with relief ≤80 m above surrounding plain. Low-relief terrain is underlain by TTG. The map reflects the geological footprint of resistant Precambrian supracrustal rocks and Younger Granites across the plain. Note the banded patterns related to the NNW–SSE Precambrian structures but the totally random distribution of relief on a more local scale (mixed clusters of *black* and *white*). *Red lines* major drainage divides. *Triangles* apatite fission-track samples. *Squares* cosmogenic radionuclide samples

that all Late Archean syntectonic or younger granitoids underpin areas of topographic relief (Fig. 1). The younger intrusions include Chamundi Hill near Mysore (at 800 Ma, the youngest in the WDC), and most granites of central–southern Tamil Nadu (~700–390 Ma) and of Kerala (7 mappable units, never exceeding 15–50 km^2: Ambalavayal, Munnar, Ezhimala, Kalpatta, Chengannur, Perilamala, Sholayur). These so-called Younger Granites are related to terrane accretion and crustal anatexis peripheral to the Dharwar Craton during the Panafrican orogeny. Contrary to the granites of the Archean crust, they originate from recycled (partially melted) older crust, and are thus more similar to modern granites than to their Archean counterparts.

In contrast to the WDC, where an older generation of TTG was emplaced ca. 3,300–3,100 Ma and where granite plutons form small, neatly circumscribed outcrops (e.g. Chitradurga, Arsikere, Shimoga: Fig. 1), granitic outcrops in the EDC are less distinctive. This arises because the EDC crust is dominated by a younger generation of TTG (2,900–2,700 Ma) that underwent extensive syn- or late-orogenic melting. As a result, it exhibits imbricated areas of TTG and migmatite among which areas where granitic melts separated out more neatly from the magma are difficult to map. Much of the EDC is thus a poorly differentiated mix of Late Archean TTG, migmatite and granite exhibiting rapid lateral variation at exposure level, with up to four varieties of granitoids identified (red triangle, Fig. 1): (i) Na-rich granites of TTG composition; (ii) K- and Mg-rich monzonites and granodiorites (i.e. sanukitoids, named after Mg-rich andesites occurring at Sanuki in Japan); (iii) K-rich and Mg-poor anatectic biotite granites; and (iv) a variant of the sanukitoids corresponding to the Closepet Granite (Moyen et al. 2003a). This typology is important because it applies to Late Archean granitoids on all continents, making the EDC a possible type area for Precambrian granitoid geodiversity. The geochemical and petrographic signatures of the EDC granitoids remain distinct from more modern subduction-related granitoids, implying a shift in modes of magma generation and an increase in subduction angle at convergent margins between the Archean and the present time.

The Closepet, a 400 km-long and 30 km-wide intrusion (Radhakrishna 1956; Moyen et al. 2003b) dated at 2,520 Ma (Fig. 1), is the flagship granitic feature of the Dharwar Craton. Its pink and grey granite outcrops form the

(a)
(b)
(c)
(d)
(e)
(f)

◀**Fig. 4** Aspects of the granitic scenery in South India. **a** Nandi Hills, 500 m of bedrock and joint-controlled relief above the TTG plain, with a typical vegetated apron of rockfall debris, **b** large granitoid bornhardt in humid northern Kerala (Wayanad), where deep weathering of the country rock in the foreground has attracted a mixture of upland agroforestry on the interfluves and paddy cultivation in low-order valleys floors (or dambos), **c** site of Gingee Fort, Tamil Nadu. Note the fortifications amidst the rockfall debris, **d** Top of a granitic inselberg near Tumkur, Karnataka, exhibiting preserved and collapsed components of an outer shell of rock, with shallow weathering pans (foreground) and rillenkarren (main rock face). A temple dedicated to the goddess Kali (the shakti of Shiva) is lodged in a cave concealed under the top shell of the dome, **e** agricultural flats during the monsoon season, lodged amidst chaotic granite boulder terrain near Hampi; fairly typical scenery among the porphyritic granite landscapes of the EDC (P. Dharani, free of rights), **f** asymmetrically stripped granite inselberg, ascribable to a more stable plateau hinterland to the *left* (note weathered and vegetated rock sheet terminating by a small scarp) and to undercutting by an incising river gorge (not visible) to the *right* (in response to this lowered local base-level, the upper rock sheet envelope has been stripped clean half way across the dome, exposing the unweathered bedrock surface beneath it). Note in foreground the red soils (chromic Luvisols) typical of semi-arid conditions on the crystalline rocks of the Dharwar Craton

boundary between the EDC and WDC (dashed white line, Fig. 1). Like its counterparts in the EDC it is bounded by N–S shear zones, suggestive of a major crustal discontinuity. It also provides a unique natural cross-section through the Archean crust, with greater pluton emplacement depths ($\sim$25 km, 7 kbar), today exposed by deeper continental denudation in the south, progressively shallowing northward ($\sim$5 km, 3 kbar). This N–S gradient (Fig. 2) matches the exposure of rocks of increasing metamorphic grade in the WDC, from greenschist, to amphibolite, to granulite facies (Fig. 1), again indicating much greater depths of long-term denudation in the south compared to the north.

3 A Few Clues to Landscape Evolution in the Granitic Cratons

The southward rise in mean elevation of the WDC land surface (Fig. 3) is partly ascribable to the greater resistance to weathering of the low-porosity charnockites in the south compared to the weakly metamorphosed metasedimentary rocks in the north (Gunnell and Radhakrishna 2001). Bedrock denudation rates measured on the summits of selected granite and charnockite domes by terrestrial beryllium-10 in Karnataka are extremely low, typically <2 m/Ma and not exceeding 8 m/Ma (Gunnell et al. 2007). In contrast, denudation rates in the surrounding TTG terrain based on apatite fission-track and beryllium-10 measurements (Fig. 3) have consistently provided values one order of magnitude greater. Assuming (given the low tectonic activity of the craton) that these respective rates have remained constant throughout the Cenozoic, it can be inferred that larger monoliths such as, for example, the Nandi Hills (500 m of relative relief, Fig. 1 and 4a) have been growing by differential erosion between the more resistant granite and the weaker TTG over the last 10–100 Ma. The characteristic growth time of such large inselbergs thus integrates many cycles of climatic change, in contrast to much smaller granitic tors that are typically <100 ka old (Gunnell et al. 2013). Relict outcrops of laterite capping the TTG around the Nandi Hills and Bangalore area, for example, testify to the fact that the landforms of this semi-arid continental interior also evolved under climates wetter than currently prevalent. The denudation has thus been driven by climate in eroding catchments where retreating knickzones along rivers have incised the topography, captured streams, shifted watershed boundaries by headward retreat, and steepened local topography as result. A succession of knickzones forming falls and rapids (dashed yellow line, Fig. 1) locate deep gorges in the crystalline basement (in effect: a fall line). Active tectonics having not played any detectable role thus far in enhancing the relief of Peninsular India, the granitic landscapes of the Indian cratons are essentially weathering-limited, hinging on differential erosion among the heterogeneous outcrops of Precambrian bedrock.

4 Granite, Culture and Conquest: Anthropogenic Geomorphology and Geoheritage

4.1 The Economics of Granite Landforms

Perhaps the most unique feature of South Indian granitic landforms is the economic and cultural imprint left on them by society. Building and ornamental stone is a huge economic growth area that has been conquering foreign markets since the 1930s (Radhakrishna 1996), largely explained by the low cost of labour. The major processing units are located in Bangalore, the historical hub of the industry. However, few quarries are fully mechanized. Instead, gangs of labourers can be seen peeling away exfoliated layers and splitting them with simple tools into slabs of required size, thus exploiting the physical properties of the rock (joint and crack density, strength, porosity, granularity, colour) and progressively revealing the onion-layer structure of the succumbing monoliths. Under the pressures of urban growth, some bornhardts of the EDC have been blasted or hand-chiselled out of existence, whether for export to the Far and Middle East, Europe and America via the ports of Mangalore and Karwar, or for the domestic market. Paving and cladding of prestigious modern buildings in Indian city centres, business parks and campuses is commonplace, with granite having surpassed marble for its durability and competitive price. Beams of granite and TTG up to 20 m long, hewn manually from such quarries, are commonly erected throughout the land as temple pillars, posts, and more recently vine trellis poles. Upstanding alignments of rough slabs of gneiss or granite driven into the soil commonly form ordinary palisades around urban properties and industrial compounds.

At least 2,600 quarrying concessions affecting ~ 3 % of the total land area occur in Karnataka. In semi-arid Tamil Nadu, resource estimates of dimension-stone granites exceed 24×10^6 m^3. The local diversity of rock facies and textures escalates into a dizzying catalogue of commercial names, with every possible crystalline lithotype of the Precambrian shield exploited as a dimension stone variety. Examples just among granites include raw silk, vanji pink, samantha pink, grey pearl, moon stone, chilli red, imperial red, ruby red, desert brown, honey dew, tippu white, rosa verde, green onyx. The dimension stone industry in the humid tracts of Kerala and western Karnataka remains comparatively undeveloped, largely because of the larger depths of weathering overburden and the steep terrain (Fig. 4b). In the drier interior, natural processes exacerbated by historical soil erosion have stripped weathering fronts and offer unimpeded access to corestones and rock sheets.

4.2 Granite Hills: Places of Warfare, Wealth and Worship

High topography throughout the world has often been put to advantage for military or religious purposes, and although relevant constructions occur on a variety of elevated bedrock outcrops at many places in the Deccan, the summits of granitic landforms have acquired their share of prestige through the construction of forts and temples of great fame. The summit of one large dome, the Nandi Hills (Fig. 4a), can be reached by road; elsewhere, Indian inselbergs are inaccessible, or else they require patient scaling up tortuous trails or steps cut manually into the bedrock. Among the more impressive forts of India, perched amidst a bouldery landscape of 2,250 Ma migmatite, is Gingee (Fig. 4c) in Tamil Nadu. Its labyrinth of walls and bastions dates from the Chola period (900–1,100 CE) and it was restored and expanded by the Vijayanagara kings. Its rainwater harvesting structures around the dome flanks were designed to live-out sieges by invaders. Guarding an important trading route between the coastal factories and the interior kingdoms, the fort was held for 11 years by the French, who eventually surrendered to the British in 1762 and definitively lost their initially stronger foothold on India. Likewise, Golconda Fort, 9 km west of Hyderabad in Andhra Pradesh, was described as a flourishing city by Marco Polo in 1292, renowned thereafter for its gemstone-cutting and -polishing craftsmen (a famous example is the locally mined Kooh-i-Noor diamond, received by Queen Victoria in 1877). Golconda was taken over by Muslim conquest in the 16th century and made a capital until it fell to Aurangzeb in 1687.

Granite landforms are also sites of worship. Whether in the semi-arid interior or along the more humid west coast, few inselbergs lack a small temple, sacred cavity at their summit (Fig. 4d) or sacred forest grove at their base, where trees persist naturally due to rainwater running off the bedrock. The 143 m-high inselberg at Sravanabelagola (Karnataka), for example, is with Mt Abu (another granitic inselberg in the Aravalli Craton, in Rajasthan) the most celebrated Jain pilgrimage site in South Asia. Today, it displays the largest monolithic granite sculpture in India: an 18 m-tall naked statue, over 1,000 years old, of the Jain saint Bahubali. The most illustrious geomorphosite endowed with cultural, spritual and historical value remains the UNESCO World Heritage site of Hampi (gazetted in 1999; Fig. 4e). Here, the ruins of the former capital of the Vijayanagara dynasty (1336–1565), situated in the northern portion of the Closepet Granite belt (Fig. 1) at the entrance to the Tungabhadra River gorge, extend over a core area of 30 km^2. Already a sacred site of pilgrimage for devotees of lord Shiva in the 10th century, this City of Victory became the single most powerful urban centre in the Deccan between the 14th and 16th centuries and one of the ten largest cities in the world. Reported by Renaissance Portuguese and Persian traders as an awesome achievement comparable to Rome, it stood as a bastion of traditional Hindu values dedicated to fighting back encroachments by Muslim sultans from the north, who were soon to be operating from this other granitic stronghold: Golconda. By being associated with key episodes of the Ramayana (planning of the campaign to Sri Lanka) it is also imbued with mythological significance. The giant granite boulders, caves and other geomorphic features have been suitably framed in the 1998 Franco-Indian movie *Hanuman*, which was filmed among the ruins of Hampi. The scenes capture the essence of the semi-arid granitic scenery prevalent elsewhere in the EDC.

5 Conclusions

A significant feature of the granitoid masses of the Indian cratons is that their geomorphological expression in the landscape does not always reflect in any obvious way the finer petrographic distinctions relevant to igneous petrologists.

In the Dharwar Craton, the regional topography essentially alternates between (i) N–S parallel ribs of rugged and untrafficable porphyritic granite (Fig. 1), which is fragmented by joint patterns into boulder-strewn monoliths and mostly covered by patchy dry deciduous woodland and scrub; and (ii) by more deeply weathered TTG corridors that benefit from groundwater recharge provided by runoff from the neighbouring granite hills. Human settlements and ingenious indigenous forms of water harvesting for agriculture known as tank systems (Gunnell and Anupama 2003) cluster in those areas. Landform patterns among the porphyritic, K-rich granites remain characteristically dominated by joint density. In conjunction with weathering-limited bedrock alteration controlled by climate, the stripping of weathering fronts is spatially controlled by contrasting depths of incision by drainage systems encroaching headward into the peninsula (Fig. 4f).

References

Gunnell Y, Radhakrishna BP (eds) (2001) Sahyadri, the Great Escarpment of the Indian subcontinent. Patterns of landscape development in the Western Ghats, 2 vols. Geological Society of India Memoir No. 47, Bangalore

Gunnell Y, Anupama K (2003) Past and present status of runoff harvesting systems in dryland peninsular India: a critical review. Ambio 32:320–324

Gunnell Y, Braucher R, Bourlès D, André G (2007) Quantitative and qualitative insights into bedrock landform erosion on the South Indian craton using cosmogenic nuclides and apatite fission tracks. Geol Soc Am Bull 117:576–586

Gunnell Y, Jarman D, Braucher R, Calvet M, Delmas M, Léanni L, ASTER Team (2013) The granite tors of Dartmoor, Southwest England: rapid and recent emergence revealed by Late Pleistocene cosmogenic apparent exposure ages. Quat Sci Rev 61:62–76

Migon P (2006) Granite landscapes of the world. Oxford University Press, Oxford

Moyen JF, Martin H, Jayananda M, Auvray B (2003a) Late Archean granites: a typology based on the Dharwar Craton (India). Precambrian Res 127:103–123

Moyen JF, Nédélec A, Martin H, Jayananda M (2003b) Syntectonic granite emplacement at different structural levels: the Closepet granite, South India. J Struct Geol 25:611–631

Radhakrishna BP (1956) The Closepet granites of Mysore State, India. Mysore Geologists' Association Special Publication 3, Bangalore

Radhakrishna BP (1996) Mineral resources of Karnataka. Geological Society of India, Bangalore

Ranganathan N, Jayaram S (2006) Geomorphology of Karnataka (South India). Karnataka Geologists' Association, Bangalore

Twidale CR, Vidal Romaní JR (2005) Landforms and geology of granite terrains. Taylor & Francis, London

The Andaman Archipelago

Jyotiranjan S. Ray

Abstract

The Andaman Archipelago of the north-eastern Indian Ocean is made up of more than 570 islands and islets, formed along the convergent margin of the Indian and Burmese Plates. These islands are either located on the accretionary prism or on the volcanic arc of the subduction zone. Began forming more than 20 Ma ago the archipelago is home to some of the spectacular and scientifically significant tectonic, volcanic, coastal and erosional landforms. This chapter describes some of these landforms and discusses the current views on their origin and evolution.

Keywords

Andaman and Nicobar Islands • Convergent margin • Subduction zone • Accretionary prism • Volcanic arc • Mud Volcano • Corals reef • Cliff

1 Introduction

When two tectonic plates converge, the plate with higher density sinks into the mantle beneath the lighter plate forming a subduction zone. The former is usually an oceanic plate whereas the latter is either oceanic or continental. Many tectonic landforms are generated along such plate margins as a result of subduction zone processes. Transfer of oceanic sediments and crustal rocks from the underthrusting plates onto the overriding plates produces accretionary prisms, which develop next to deep ocean trenches that mark the boundaries of converging plates. Because subduction zones are seismically active regions, a whole host of earthquake related landforms also get generated on the accretionary prisms. Due to incorporation of soft sediments at high rates of deformation, accretionary prisms produce some of the world's most complexly deformed rocks and their subsequent growths result in the formation of series of hill ranges and valleys. Sediments and altered oceanic crust of the underthrusting plate undergo defluidization during their journey into the mantle. Part of this fluid mixed with fine grained sediments gets detached from the plates and rises through the fault networks of the accretionary prisms and form mud volcanoes on surface. Deeper subduction of these materials aids melting process in the adjacent mantles to produce arc of volcanoes (volcanic/magmatic arc) on overriding plates. Worldwide, there exist only a handful of subduction zones where all the above typical landforms of a convergent margin are present and exposed. The Indonesian-Andaman convergent margin is one of such zones.

The Andaman and Nicobar Islands of India is an archipelago that is located on the outer arc of the Indonesian-Andaman subduction zone at which the Indian Plate subducts obliquely beneath the Burmese Micro-plate along the Andaman Trench (Fig. 1). The current convergence rate at the trench off Sumatra is ~5 cm/year. Andaman Sea is an active extensional basin that encompasses backarc and forearc basins of the Andaman subduction zone separated by an

J. S. Ray (✉)
Physical Research Laboratory, Navrangpura,
Ahmedabad, 380009, India
e-mail: jsray@prl.res.in

V. S. Kale (ed.), *Landscapes and Landforms of India*, World Geomorphological Landscapes,
DOI: 10.1007/978-94-017-8029-2_21,

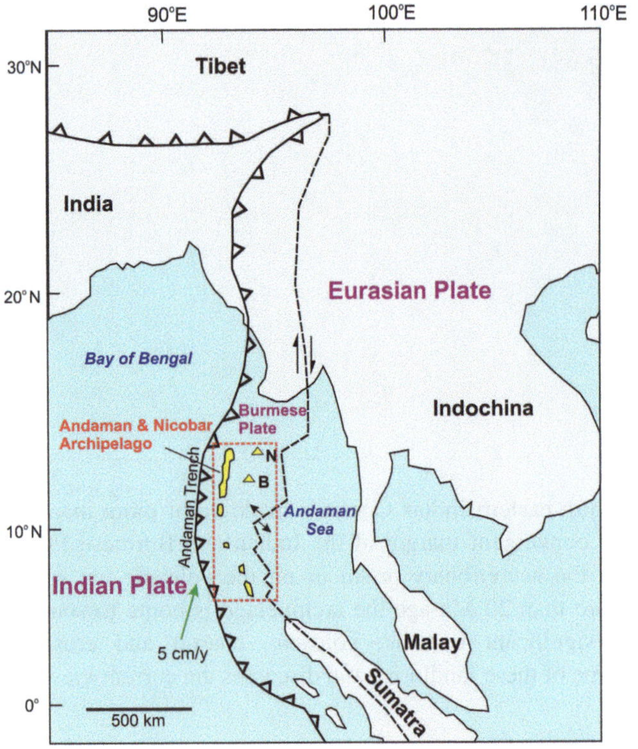

Fig. 1 Map of north-eastern Indian Ocean and adjoining landmasses showing the location of the Andaman Archipelago (*rectangle*) and major tectonic features of the region. The *triangles* are the locations of the volcanic islands of Narcondam (N) and Barren (B). 5 cm/year vector on the map is the direction (and rate) of motion of the Indian Plate

island arc ridge (or volcanic arc) containing two subaerial volcanoes: the active Barren Island and the extinct Narcondam (Figs. 1 and 2). The Andaman–Nicobar Ridge, the main component of the Andaman forearc, is an accretionary prism (Fig. 2). The ridge is an imbricate stack of thrust slices (Fig. 2) consisting of segments of oceanic lithospheric sequences (ophiolites), pelagic sediments, and deep and shallow water sediments. This ~800 km long (6.7–13.7° N), arcuate shaped ridge consists of numerous islands spreading over a breadth of ~170 km (92.2–94.3° E). The archipelago, though largely an Indian Territory, is located much closer to Myanmar and Indonesia and forms a boundary between the Bay of Bengal and the Andaman Sea (Fig. 1).

2 Landform Diversity

The Andaman and Nicobar Islands possess a diverse group of landforms that include coastal and marine, tectonic, erosional and volcanic morphologies. Development of these landforms is a highly dynamic process because of high frequency of earthquakes and resulting deformation. Changing climatic conditions, short-term extreme events and urbanization also contribute to their constant modification.

2.1 Coastal and Marine Landforms

(a) Islands and Islets

The Andaman archipelago is made up of more than 574 islands and islets forming a roughly north-south chain between the Irrawaddy Delta of Myanmar and the Ache Province of Indonesia (Fig. 1). The northern group of islands (552) are known as the Andamans, whereas the southern group (22) is called the Nicobars. The Andamans and the Nicobars are separated by a 140 km wide seaway, called the Ten Degree Channel. Most of the islands in the Andamans are separated by narrow tidal creeks, whereas distances between individual islands are much larger in the Nicobars. The longest island is the South Andaman, whereas there are countless smaller islands and islets of variable dimensions (Fig. 3a). All the islands that lie close to the central axis of the chain have ophiolitic or basaltic cores, whereas those that lie off the axis, particularly in the east (e.g., the Ritchie's Archipelago or the Havelock group of islands) are made entirely of pelagic limestones and shales. Coral reefs surround most of the islands. About 86 % area of the archipelago is covered by dense forest (Fig. 3) and has a tropical climate receiving rain from both the southwest and northeast Indian monsoons. The inland forests are predominantly wet-evergreen with their canopy at 30–40 m above the ground, whereas the coastlines are forested with mangroves (Fig. 3b).

(b) Coral Reefs

The coral reefs of Andaman and Nicobar Islands are the largest and have the richest fauna amongst all the major coral reef sites of India. The reefs of the archipelago are generally of *fringing* type (Fig. 3a), where the reefs grow near the shore and extend out into the sea like a submerged platform. An ~320 km long barrier reef with a 4 m deep lagoon has also been reported on the west coast of Middle and South Andaman Islands, separated by a 20–30 km wide and 80 m deep channel. The United Nation Development Programme's survey of 2001 had identified 197 species of coral in the region. These reefs are vulnerable to earthquakes, rise in sea surface temperature, high rate of sedimentation and human interference. Deformation of shorelines by past earthquakes have permanently damaged many reefs, the most recent being the destruction caused by the December 26, 2004 earthquake and associated tsunami (e.g. Awasthi et al. 2013).

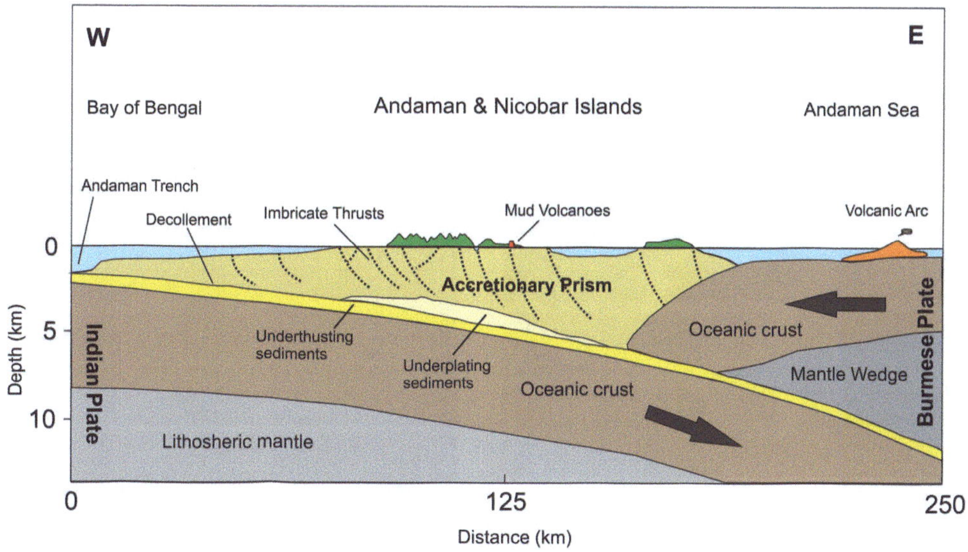

Fig. 2 A schematic cross sectional view across the Andaman subduction zone showing position of various tectonic features at surface and at depth. Andaman and Nicobar Islands represent the subaerial portion of the Andaman accretionary prism. The arrows show the direction of motion of the Indian and Burmese Plates

(c) Rocky and Sandy Coasts

The coastlines of the Andaman and Nicobar Islands are highly diverse. Most of the large islands, which lie on the main axis of the chain, have rocky coasts. These coasts are either built of rocks that make up the basement of the islands (Fig. 4a) or are dead coral reefs (Fig. 5a). Such coasts are observed predominantly along the eastern shoreline of the main three islands of the Andamans (North, Middle and South), Rutland, Interview, Narcondam and Barren. Sandy coasts are common along coral rich shores and shores dominated by sedimentary rocks. Such coasts are common all along the western shore of the main Andaman group islands, Interview Island and around the Ritchie's Archipelago (Fig. 4b), North Reef Island, North and South Sentinel Islands, Little Andaman and most of the islands of the Nicobar group. Many of these islands possess some of the most beautiful sandy beaches in the world; however, access to general public is limited.

(d) Marine Terraces

Marine terraces are observed on the coasts of the island chain. Earthquakes along the faults cause uplift and/or subsidence creating these terraces. Raised beaches, thrusted basement rocks (Fig. 4a), dead coral reefs (Fig. 5) and submerged coasts in the Nicobars are evidence of such events. Studies based on ages of exposed corals on some of these terraces (e.g., Fig. 5) reveal that there have been at least 14 major landscape-changing seismic events during the last 40 ka and that a major earthquake, associated with a tsunami occurred ~600 years ago (Awasthi et al. 2013).

Fig. 3 **a** An aerial photograph of the western coast of the South Andaman Island showing islands, islets and coral reefs. **b** A photograph of the Saddle Peak, the tallest hill of the Andaman Archipelago, taken from the Turtle Beach at Kalipur, North Andaman Island. It also shows wet evergreen forest cover and mangrove

Fig. 4 **a** A rocky coast with a cliff along the eastern coastline of South Andaman Island. The rocks here are predominantly ophiolites. **b** A carbonate sand beach at Kalapathar, Havelock Island. Below the turquoise water lie the coral reefs

Fig. 5 **a** Dead coral reefs forming a terrace at Kalipur. The number in the bracket represents the radiocarbon age of the corals. **b** Uplifted basement rocks and dead coral reefs at Radhanagar Beach, Havelock Island. The dead reef was determined to be alive (was below the sea) ~7770 years ago

2.2 Tectonic Landforms

(a) Oceanic Trench

The Andaman Trench lies about 100 km west of the main axis of the Andaman-Nicobar Islands (Figs. 1 and 2). This ~950 km long segment of the Indonesian-Andaman oceanic trench forms an arc and is much shallower than its southern extension near Sumatra and Java, which is the result of extremely high rate of sedimentation linked to the Bengal Fan. The bathymetry of the Andaman region east of the trench is shallower by more than 1000 m because of sudden rise of the accretionary prism (Fig. 2).

(b) Hill Ranges and Valleys

The islands of the Andaman Archipelago, except the two volcanic islands, are located on the accretionary prism (Fig. 2). The western slope of the prism is very steep, descending to depths of 1500–2000 m within a distance of 30 km and the seafloor in the east is smoother with gentler slopes (Cochran 2010). The morphology of the prism is structurally controlled by several eastward dipping thrusts, strike-slip faults and large scale folds that run parallel to the trench (Fig. 2), particularly in the Andamans, creating long N-S trending hill ranges/ridges and valleys. Most of these ridges are of moderate elevation (100–300 m a.s.l.). The tallest hill of the Andamans is the Saddle Peak (~732 m a.s.l.) in North Andaman (Fig. 3b), whereas that of the Nicobars is Mt. Thullier (~642 m a.s.l.) in Great Nicobar. Many fault bounded ridges are also present below the sea surface on the eastern margin of the prism.

(c) Mud Volcanoes

Mud volcanoes are subaerial/submarine sedimentary structures whose surface morphology resembles that of a real volcano, but on a much smaller scale (Fig. 6a). They form as a result of emission of depressurized pore water, gases and argillaceous material from deep seated sources and occur either on top of surface-piercing shale diapirs or along faults or fractures. Mud volcanoes of the Andaman Archipelago belong to the second type

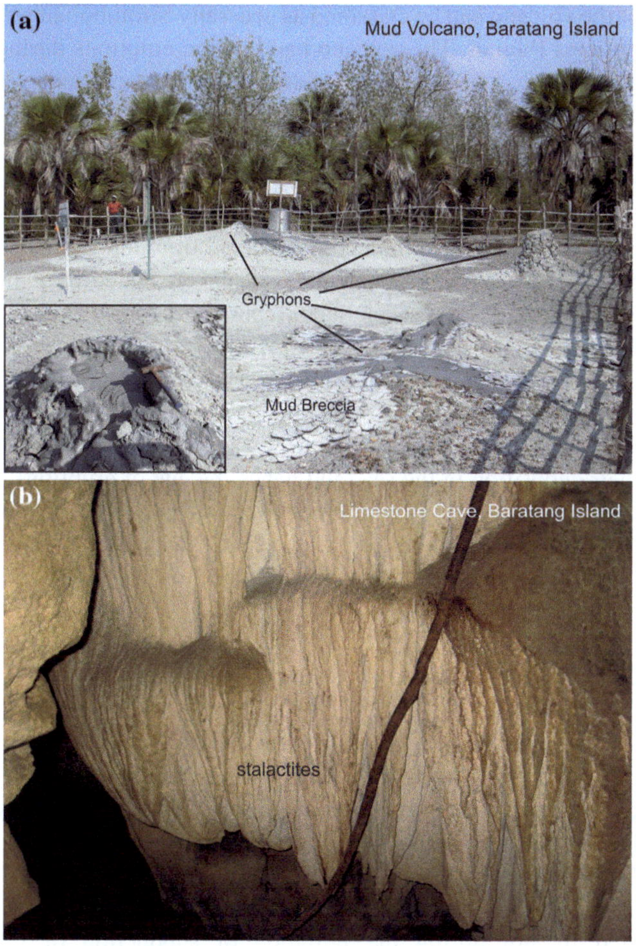

Fig. 6 **a** The main mud volcano of Jarawa Creek, Baratang Island showing ghryphons through which mud breccias is ejected. The inset photo shows interior of a gryphon. **b** Stalactites in Nayadera limestone cave in Baratang

Fig. 7 **a** The Barren Island Volcano showing active tephra emission from the central cinder cone on December 24, 2010. Also shown are the a′a′ lava field and a portion of the caldera wall. **b** The Narcondam Island as seen from a boat ~500 m away from its eastern coast

(Fig. 2). These have been reported only from the Middle (Baratang) and North Andaman Islands. The famous ones are located in the Jarawa Creek region of Baratang (Fig. 6a) and Hati Level of Shyamnagar in Diglipur. A typical Andaman mud volcano is 10–20 m high and contains 8–10 gryphons through which gases and mud breccia (comprised of water, clay rich mud matrix and rock clasts from underlying formations) are ejected. They emit thermogenic hydrocarbon gases, water with much lower chlorinity than seawater and young (<40 ka) smectite-illite-kaolinite-chlorite dominated argillaceous sediments, which are derived from the marine sediments and altered oceanic crust of the Indian slab at shallow depths (2–6 km) of the Andaman Subduction Zone and carried to the surface from the decollement through the fault networks of the accretionary prism (Ray et al. 2013a).

2.3 Erosional Landforms

(a) Cliffs and Arches

Rock cliffs are the most common wave dominated erosional landforms in the Andaman Archipelago. Tectonic forces and sea level changes also contribute to their formation. Most of these cliffs are found along the eastern margins of the South and Middle Andaman Islands, where resistant ophiolitic/basaltic rocks form the coastline (e.g., Fig. 4a). Such cliffs are also observed on the volcanic islands of Narcondam and Barren. Cliffs made up of sandstone, limestone and shale are common on islands of the Ritchie's Archipelago. On Narcondam, wave-cut arches and caves are observed along its north-eastern coastline. Such features are the result of preferential erosion of less resistant pyroclastic material against more resistant basaltic lava flows. Arches of limestones are generally found on the Ritchie's Archipelago.

(b) Limestone Caves
The karst features of Baratang and Middle Andaman are prominent landforms of the archipelago. The sandy limestone formations of Baratang have been dissolved by rainwater to produce limestone pavements on surface, sink holes and small cave networks, some of which have rich speleothem deposits. The limestone caves at Nayadera, Wrafters Creek, Baratang Island (Fig. 6b) are a major tourist attraction. Although the cave is not as spectacular as those on the mainland India, it contains all major types of speleothems, such as the dripstones (straws, stalactites, stalagmites and columns/pillars), flowstones, pore deposits (helictites) and pool deposits (lily pads/shelfstone).

2.4 Volcanic Landforms

The volcanic arc of the Andaman subduction zone has two subaerial volcanoes, east of Andamans and one submarine volcano east of Nicobars (Kamesh Raju et al. 2012). Existence of more submarine volcanoes in this arc cannot be ruled out. Amongst the two subaerial volcanoes, Barren Island (Fig. 7a) is located in the south (12.3° N, 93.9° E), ~108 km from the Middle Andaman, whereas Narcondam (Fig. 7b) is located in the north (13.4° N, 94.3° E), ~135 km from the North Andaman and both rise from the seafloor of the Andaman Sea from depths of ~2,000 and ~1,000 m, respectively. Both are stratovolcanoes having layered appearance with alternating lava flows, airfall tephra, pyroclastic flows, volcanic mudflows (lahars), and debris flows—built around a central cone or dome.

The older and extinct Narcondam forms an elliptical island with long and short axes of 3.8 and 2.4 km, respectively. The highest peak on the island is ~710 m a.s.l., which is a collapsed lava dome. Most of the volcanological aspects of the island remain largely unknown because of limited accessibility, thick vegetation and difficult terrain to work on. The estimated submarine and subaerial volumes of the volcano are 95 and 4 km^3, respectively (Streck et al. 2011). While the age of the youngest activity of the volcano remains unknown, two of the subaerial lava flows have been dated to 550 and 700 ka, indicating a much older age for the emergence of the volcano (Streck et al. 2011).

The younger and active Barren Island is roughly circular with a diameter of ~3.2 km and has an average height of 300 m a.s.l. A polygenetic, 230 m tall cinder cone is located roughly at the centre of a circular caldera (Fig. 7a), which has a diameter of ~1.6 km. The cone has an elliptical summit crater (200 m × 100 m). The caldera wall has a breach on the north-western side, through which most of the historic and recent lavas have flowed into the sea. These lava flows are a′a′ type and form an ~1 km lava field. The eruptive style of the volcano can be classified as generally Strombolian and at times Plinian. The volcano had historic eruptions during 1787–1832. After 159 years of dormancy it became active in 1991 and has remained active since then with almost continuous tephra eruptions and produced at least four major lava eruptions in 1991, 1995–1996, 2005–2006 and 2009 (Sheth et al. 2009). Radiocarbon dating of marine sediments interlayered with ash in a core collected from the Andaman Sea, ~32 km southeast of the volcano, revealed that the volcano had at least seven major tephra eruptions during the last 70 ka (Awasthi et al. 2010). Ray et al. (2013b) dated plagioclase separates from the ~69 ka ash layer from the same core by ^{40}Ar-^{39}Ar method and reported an age of 1.8 Ma. As this age of plagioclases was older than the age of their deposition in the sea (and hence the age of eruption) it was inferred that the grains came from older rocks present in the plumbing system of the volcano—which essentially means that the volcano is at least 1.8 million years old.

3 Evolutionary History

As mentioned above, the islands of the Andaman Archipelago, except the two volcanic islands, are the subaerial expression of the Andaman accretionary prism, currently located at the boundary of the converging Indian and Burmese plates (Fig. 2). According to some workers the ophiolitic basement of the accretionary prism was generated in situ during the initiation of subduction of the Indian Plate under the Eurasian Plate along the eastern boundary ~95 Ma ago (Sarma et al. 2010) and is equivalent of the Tethyan ophiolites of the northern boundary. Bulk of the materials present in the prism got deposited prior to the opening of the Andaman Sea. It is postulated that the ophiolites and trench sediments of late Cretaceous to Eocene age were uplifted by a series of N-S trending East-dipping thrusts. Continued subduction resulted in the formation of a forearc basin in which Oligocene siliciclastic turbidites and Mio-Pliocene carbonate turbidites and volcanic ash were deposited, which subsequently got uplifted along the thrust planes and got folded to form oceanic ridges, hill ranges and valleys. The islands along the main axis probably started emerging out of the sea surface sometime during 23–15 Ma and attained the present configuration during the past 4 Ma (Curray 2005). The evolution of the Andaman volcanic arc is very poorly understood. Based on the current location of the arc and available age data it appears that the present arc may not be older than the time of initiation of the latest phase of opening of the Andaman Sea (4 Ma). Formations of coral reefs, marine terraces, erosional features and volcanic landforms are continuous processes and must have begun after the emergence of the islands.

4 Concluding Remarks

The Andaman Archipelago situated in the north-eastern Indian Ocean is one of the youngest landmasses of the Indian subcontinent and is made up of more than 570 islands and numerous islets. Located along a convergent margin, the islands are home to numerous tectonic and volcanic landforms, which owe their origin to the subduction of the Indian Plate underneath the Burmese Plate. The islands are critical to our understanding of the evolution of a subduction zone. Most of these islands still possess their pristine natural beauty in terms of unpolluted environment and unaffected biodiversity, including underwater life, except for the rapid decline of ancient indigenous human population.

References

Awasthi N, Ray JS, Laskar AH, Kumar A, Sudhakar M, Bhutani R, Sheth HC, Yadava MG (2010) Major ash eruptions of Barren Island volcano (Andaman Sea) during the past 72 kyrs: clues from a sediment core record. Bull Volcanol 72:1131–1136

Awasthi N, Ray JS, Laskar AH, Yadava MG (2013) Chronology of major terrace forming events in the Andaman Islands during the last 40 kyrs. J Geol Soc India 82:59–66

Cochran JR (2010) Morphology and tectonics of the Andaman Forearc, northeastern Indian Ocean. Geophys J Int 82:631–651

Curray JR (2005) Tectonics and history of the Andaman Sea region. J Asian Earth Sci 25:187–232

Kamesh Raju KA, Ray D, Mudholkar A, Murty GPS, Gahalaut VK, Samudrala K, Paropkari AL, Ramachandran R, Surya Prakash L (2012) Tectonic and volcanic implications of a cratered seamount off Nicobar Island, Andaman Sea. J Asian Earth Sci 56:42–53

Ray JS, Kumar A, Sudheer AK, Deshpande RD, Rao DK, Patil DJ, Awasthi N, Bhutani R, Bhushan R, Dayal AM (2013a) Origin of gases and water in mud volcanoes of Andaman accretionary prism: implications for fluid migration in forearcs. Chem Geol 347:102–113

Ray JS, Pande K, Awasthi N (2013b) A minimum age for the active Barren Island volcano, Andaman Sea. Curr Sci 104:934–939

Sarma SD, Jafri SH, Fletcher IR, McNaughton NJ (2010) Constraints on the Tectonic setting of the Andaman ophiolites, Bay of Bengal, India, from SHRIMP U-Pb Zircon geochronology of plagiogranite. J Geol 118:691–697

Sheth HC, Ray JS, Bhutani R, Kumar A, Smitha RS (2009) Volcanology and eruptive styles of Barren Island: an active mafic stratovolcano in the Andaman Sea, NE Indian Ocean. Bull Volcan 71:1021–1039

Streck MJ, Ramos F, Gillam A, Haldar D, Duncan RA (2011) The Intra-oceanic Barren Island and Narcondam Arc Volcanoes, Andaman Sea: implications for subduction inputs and crustal overprint of a depleted mantle source. In: Ray J et al (ed) Topics in igneous petrology. Springer, The Netherlands, pp 241–273

Teri Red Sands, Tamil Nadu

R. Jayangondaperumal

Abstract

A conspicuous feature that occurs on the southeast coastal Tamil Nadu, in southeast India covering vast areas is the red sand, known as teri red sands. On the basis of geomorphic setting and optical ages, the teri sands can be broadly classified into three main types—(i) the inland fluvial teri sands, (ii) the coastal teri sands, and (iii) the near-shoreline teri sand dunes. The inland teri sediments have the highest percentage of clay and silty-sand components, indicating that these were brought and deposited by fluvial process during stronger winter monsoon >15 ka. Luminescence dating of the coastal teri dunes reveals their deposition was prior to ~11 ka, and the near-shoreline dunes were laid down at around 5–6 ka. These coastal dunes were formed during a period of lower sea level and the near-shoreline dunes were formed during a period of comparatively higher sea level. Red coating of the sand grains was post-depositional and occurred after 11 ka for the coastal teri dunes and after 5–6 ka (mid-Holocene) for the near-shoreline teri dunes.

Keywords

Coastal dunes • Red sands • Sand dunes • Northeast monsoon • Sea level • Microlithic industry

1 Introduction

The east coast of India is dominated by coastal depositional features such as long beaches, spits, tidal flats and sand dunes. Sand dunes in coastal areas are natural mounds of wind-blown sand, occasionally topped with vegetation. A stable sea shore has two main lines of sand dunes. The first one runs along the sea shore and the second one is also parallel but at a distance of about 100–500 m from the sea shore. The area between these two types of dunes is normally covered by sandy plain with minor dunes and coconut palms. The occurrence of dunes is of special significance to human habitats because they form a natural line of defence against the sea. They protect the hinterland from attack by waves, cyclones, storm surges, blowing wind and by deflecting strong winds in upward direction. Over a long term, the coastal dunes represent a fragile ecosystem reflecting changes in relative sea level. Their formation is related to dry conditions supported by strong winds. Wind plays a dominant role as only a speed greater than 16 km/hr can lift and transport the fine sand particles. Orientation of dunes and their age help in unravelling the past wind directions and reconstruction of past environments.

In some coastal areas, such as near Visakhapattnam, Puducherry, Tirunelveli, Tuticorin and parts of Sri Lanka, the sands are characterized by vivid red colour. In southeast Tamil Nadu such sands are described as teri red sands (teri = sandy waste in local parlance). The white dunes are found along the modern shoreline whereas the red dunes occur inland, stratigraphically above the marine calcareous

R. Jayangondaperumal (✉)
Wadia Institute of Himalayan Geology, Dehradun, 248001, India
e-mail: ramperu.jayan@gmail.com

V. S. Kale (ed.), *Landscapes and Landforms of India*, World Geomorphological Landscapes,
DOI: 10.1007/978-94-017-8029-2_22,

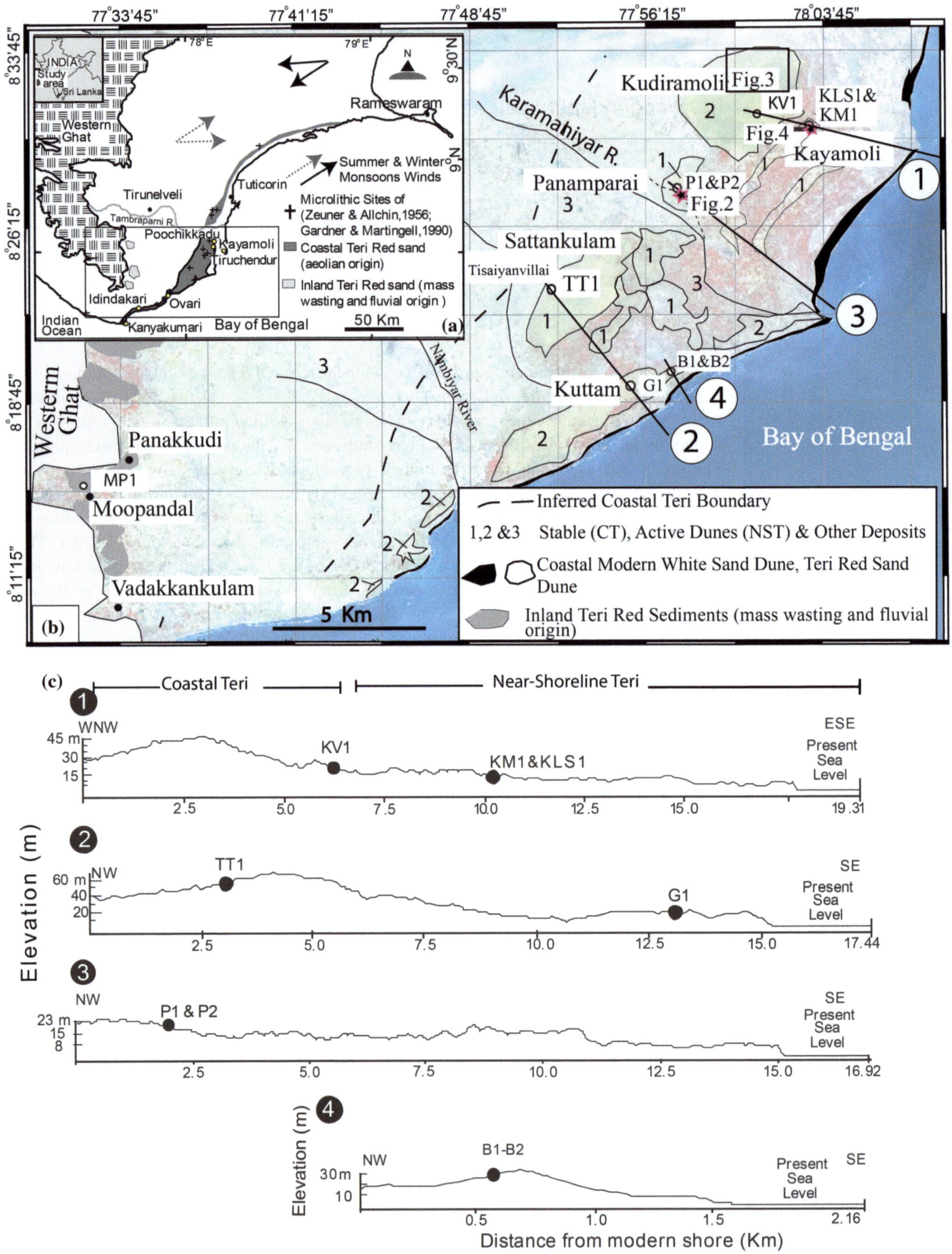

77°33'45"
77°41'15"
77°48'45"
77°56'15"
78°03'45"
8°33'45"
8°26'15"
8°18'45"
8°11'15"
INDIA
Study area
Sri Lanka
78°E
79°E
9°30'N
9°N
N
Western Ghat
Rameswaram
Summer & Winter Monsoons Winds
Tirunelveli
Tuticorin
Tambrapani R.
Microlithic Sites of (Zeuner & Allchin,1956; Gardner & Martingell,1990)
Coastal Teri Red sand (aeolian origin)
Inland Teri Red sand (mass wasting and fluvial origin)
Poochikkadu
Kayamoli
Tiruchendur
Idindakari
Ovari
Indian Ocean
Kanyakumari
Bay of Bengal
50 Km
(a)
Kudiramoli
Fig.3
KV1
KLS1& KM1
Fig.4
Kayamoli
Karamahiyar R.
Panamparai
P1&P2
Fig.2
Sattankulam
Tisaiyanvillai
TT1
B1&B2
Kuttam
G1
Nambiyar River
Western Ghat
Panakkudi
MP1
Moopandal
Vadakkankulam
Bay of Bengal
5 Km
(b)
Inferred Coastal Teri Boundary
1,2 &3 Stable (CT), Active Dunes (NST) & Other Deposits
Coastal Modern White Sand Dune, Teri Red Sand Dune
Inland Teri Red Sediments (mass wasting and fluvial origin)
(c)
Coastal Teri
Near-Shoreline Teri
WNW
ESE
45 m
30
15
KV1
KM1&KLS1
Present Sea Level
19.31
NW
SE
60 m
40
20
TT1
G1
17.44
23 m
15
8
P1 & P2
16.92
Elevation (m)
B1-B2
30m
10
2.16
Distance from modern shore (Km)

◀ **Fig. 1** **a** *Inset* map showing the southeast Tamil Nadu coastal region and the distribution of red teri sediments and the locations of major microlithic sites (Jayangondaperumal et al. 2012). **b** IRS-1D Satellite image draped over SRTM-DEM showing the distributions of teri red sand dunes together with location of OSL sampling sites as *open circles*. CT coastal teri, NCT Near-shoreline teri. **c** Topographic profiles across the dune systems as solid thick lines in **b** are shown as *solid circles* with numbers 1–4, *1* Kudiramoli, *2* Sattankulam, *3* NE of Panampari, *4* NE of Kuttam

grit of Pleistocene age. The main objective of this chapter is to describe the geomorphic setting of the teri sand dunes and their probable mode of origin.

2 Location and Geological Setting

The area under review is located along the southeast coast of Tamil Nadu between Tirunelveli (8° 43′ 58.8″ N; 77° 42′ 0.0″ E) and Kanyakumari (8° 5′ 5.69″ N; 77° 32′ 30.47″ E), extending ~15 km inland from the coast (Fig. 1). The teri deposits are continuous in the north and discontinuous in the south covering an area of ~500 km^2 and their thickness reaches up to 12 m. Six types of marine-coastal deposits occur in the area (Jayangondaperumal et al. 2012). Of these coastal deposits, the oldest deposits of coarse marine sandstones, corals, siltstones and shell fragments belong to the Ovari Series (OS) that lies unconformably over the basement of igneous and metamorphic rocks. The OS is overlain by Idindakarai Series that consists of terrigenous grains of gravel, sand, and marine shells. The U-series ages of marine shells (~112 to 124 ka, Bruckner, 1989) indicate that these deposits were laid down during a high-stand sea level in the last interglacial period. The Kanyakumari Series overlies the Idindakarai Series. It consists of fossil coastal dunes cemented by calcium carbonate (aeolianites). The overlying Poochikkadu Series comprises of calcrete and some marine sediments. The upper part of the calcrete has layers of land snail shells (*Helix vittata*) that gave radiometric ages between ~26 and 31 ka (Gardner 1986). These ages on the land snail shells provide a minimum estimate for the host sediments. The Teri Series (teri sand) overlies Poochikkadu series that consists of older marine aeolianite deposits (>31 ka). The teri sands were deposited between 11 and 15 ka (Jayangondaperumal et al. 2012). The young marine and lagoonal deposits that overlie the teri sands are known as Mandapam Series that belong to late Holocene (~2.9 ka). The modern coastal sands drape the Mandapam Series. The teri sands rest either on the crystalline basement, Ovari marine sandstone, aeolianite of Kanyakumari or the Poochikkadu Series. Presence of heavy minerals, such as ilmenite, rutile, zircon, garnet, monazite and sillimanite suggest the provenance of teri from Precambrian khondalite, charnockite and granite gneisses.

The region is in the rainshadow of the Western Ghat. The present climate is semi-arid with an average annual rainfall of ~700 mm. Nearly 60 % of the annual precipitation occurs during the northeast (winter) monsoon and the remaining falls during the summer monsoon. The wind direction varies considerably during the two monsoons. Strong westerly and south-westerly winds bring the summer monsoon whereas north-easterly and easterly winds are characteristic of the winter monsoon with a composite mean speed of ~4–5 m/s (Fig. 1a).

3 Classification and Depositional Environment

The teri can be classified into the following three types depending on the origin and distance from the modern shoreline (Fig. 1).

3.1 Inland Fluvial Teri Sands

These deposits were formed by mass wasting and fluvial processes. These are confined to the foothills of the Western Ghat, between 110 and 135 m a.s.l. and ~25 km inland from the modern shoreline (Fig. 1a). These are sand sheets, medium to coarse grained, with silt and clay. The percentage of clay is high. The sediments are reddish-yellow and the top surface exhibits soil development. Luminescence dates suggest that these sands were deposits between 23 and 15 ka during strong NE winter monsoon and close to the LGM or Last Glacial Maximum (Jayangondaperumal et al. 2012).

3.2 Coastal Teri Dunes

The coastal zone teri dunes are bounded by the foothills of Western Ghat (in the west) and by the near-shoreline teri plain (in the east). The dark red (Munsell colour 10 R 3/6) coastal teri sands show 2–12 m thick deposits that occur between 30 and 60 m a.s.l., and extend up to 15 km inland. The deposits show hummocky topography and at places occur as isolated mounds that are erosional remnants of former extensive ridges suggesting that these sands were formed by aeolian process. The teri sands are presently stabilized due to vegetation cover. The deposit shows microlithic (flakes and core pieces) randomly distributed and in concentrated sites in the upper layer of the red units, suggesting human occupation between 11 and 5.6 ka (Jayangondaperumal et al. 2012). The top portion of the coastal teri provides OSL (Optically Stimulated Luminescence) age of 11 ka. Toward inland ~1 km NE of Karamaniyar River, the coastal teri

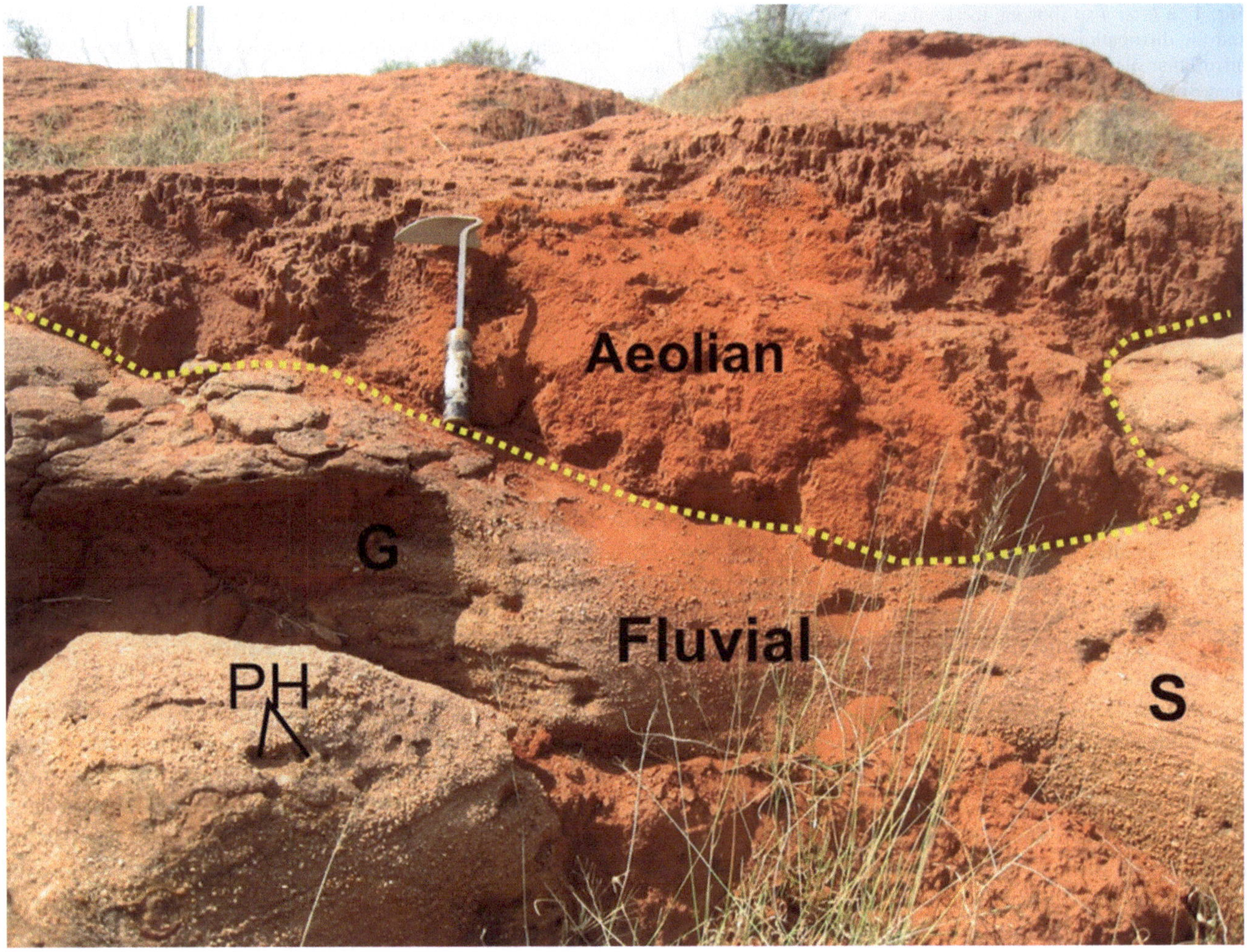

Fig. 2 Striations (*S*), groves (*G*) and Potholes (*PH*), underneath the unconformity (*yellow broken line*) between aeolian and fluvial surface (see location in Fig. 1b as a star)

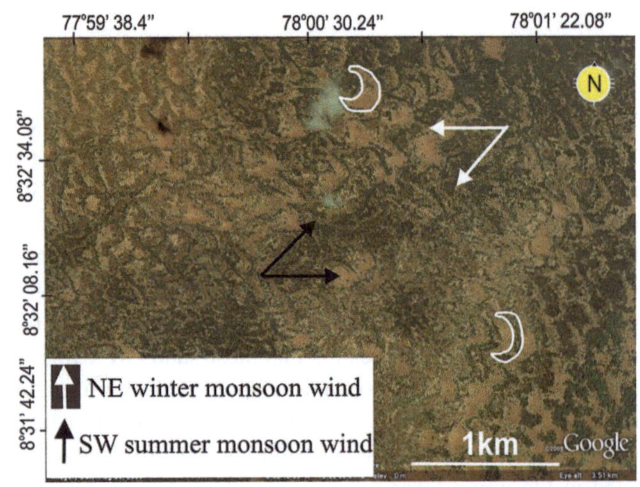

Fig. 3 Barachan dunes (outlines in *white*) seen at NW of Kayamoli in the near coastal teri dunes on Google Earth image. Arrows show wind direction. Location marked as a box in Fig. 1b

dunes show evidence of reworking, both by fluvial and aeolian activity (Fig. 2), suggesting that fluvio-aeolian source bordering dunes formed between 4.1 and 2.3 ka (Jayangondaperumal et al. 2012).

3.3 Near-Shoreline Teri Dunes

These dunes occur as discontinuous patches parallel to the present shoreline (Fig. 1b). These are yellowish red, 4–6 m thick and occur from 10 to 25 m a.s.l. The furthest occurrence is ~5 km from the modern coast. These teri dunes occur as sand-sheet deposits except around NW of Kayamoli village, where they occur as barchans (Fig. 3). The deposits at Kayamoli are ~2.5 m thick sand sheets of near-shoreline teri sands and are underlain by calcretes with land snail shells atop (Fig. 4). Analogous stratigraphic relations are observed in the red sediments from Bavanapadu–Ichchapuram, Andhra

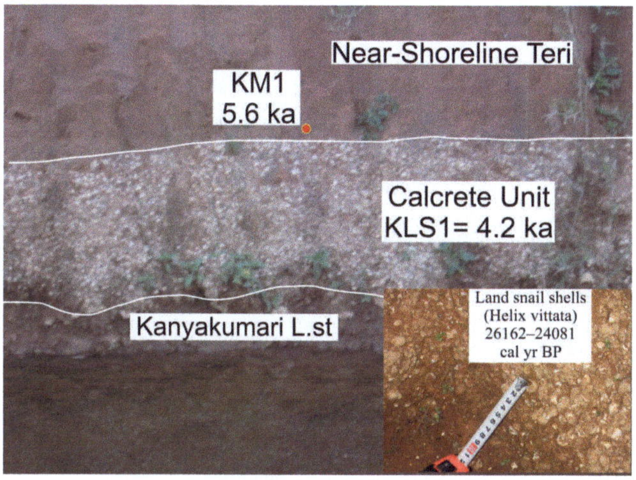

Fig. 4 Section showing the base of the near-shoreline teri stratigraphy as seen in an irrigation canal exposure at Kayamoli village. The near-shoreline teri rests on a calcrete unit of Poochikkadu Series (location is marked as a star in Fig. 1b). *Inset* Land snail shells found in the calcrete unit. Sample name followed by numbers with ka (kilo years) correspond to OSL age (Jayangondaperumal et al. 2012) and land snail shells age corresponds to ^{14}C age (Gardner 1986). *White lines* represent unconformity

Pradesh. Reworked coastal teri sands were then re-deposited as near-shoreline teri after ~5.6 ka (Fig. 4). Further, these dunes are loose and unconsolidated and were further reworked during ~0.13 and ~0.22 ka. The reworking of these dunes can be ascribed either to local fluvial reworking, a common feature during the monsoon, or to intense wind action during the Little Ice Age.

4 Duration of Reddening and Human Occupation on Teri Sands

Previously, the degree of reddening was linked to age (Walker 1967), but recent studies reveal that the colour and age are not related (Roskin et al. 2012) and the reddening can occur even over time scales of centuries (Jayangondaperumal et al. 2012). In the study area, the dune reddening is post-depositional and most likely occurred after ~11.4 ka for the coastal teri dunes and after ~5.6 ka for the near-shoreline teri dunes. The microlithic artifacts were found in scattered manner and these were not within the teri deposits suggesting human occupation after accretion of the teri sands (post 11 ka). Presence of microlithic sites associated with the coastal dunes suggests that the cultures existed in the region during ~11.4 and 5.6 ka.

5 Evolution of Red Teri Sands

Two views exist on the origin of the teri sands. The earlier view (Foote 1883) consists of detrital origin of the teri sands that was derived from red earths of the eastern foothills of Western Ghat. It was later transported by westerly and south-westerly winds during the summer monsoon (Fig. 1b). The more recent view (Gardner 1986) considers its origin as a result of easterly and north-easterly winds during the winter monsoon and thus the teri sand is the product of aeolian process. It was later reddened in situ as a result of weathering of feldspar, garnet, and some opaque minerals. Hence the deep red matrix is enriched in kaolinite, hematite and illite in the clay fraction.

Based on geomorphic setting and optical dating integrated with previous research, a model for the evolution and reddening of teri deposits is shown in Fig. 5 (Jayangondaperumal et al. 2012). The age of the oldest marine sediments of Ovari Series and Kanyakumari aeolianite are unknown. Idindakarai Series was deposited during the last interglacial highstand position at ~124 ka. The calcrete in the Poochikkadu Series was deposited between ~26 and 24 ka (Fig. 5a). Subsequently, the deposition of coastal aeolian sand or coastal teri took place between ~24 and ~11.4 ka, a period corresponding to the LGM when the sea lowered to a depth of −120 from −80 m. The regression continued until ~11 ka. The lower sea level facilitated a higher sediment supply (Fig. 5b). This was concomitant with stronger landward northeasterly and easterly wind associated with the NE monsoon. Transportation of sediments from the exposed shelf occurred at this time leading to the formation of coastal teri dunes. Since the NE monsoon was strong at ~15.4 ka, the precipitation eroded the uplands of the Western Ghat and deposited the inland fluvial sands along its flanks (Fig. 5b).

The exact timing of initiation of coastal teri deposition is unknown, because the base is not exposed. The teri sands lie on the leeward side of the Western Ghat. Weak summer monsoon during the LGM, and stronger NE monsoon as compared to the present lead to speculation that the coastal teri sand dunes deposition perhaps began post ~24 ka and ceased at 11.4 ka. Following this, the coastal teri sediments were weathered in situ and reddening took place after ~15.4 ka and 11.4 ka, respectively (Fig. 5c). The microlithic sites were occupied during weathering of coastal teri after 11.4 ka. The higher sea level between ~7.3 ka and 5.2–4.2 ka allowed the submerged coastal teri sediments to be reworked and deposited as near-shoreline teri. These sediments were reworked in situ and reddening took place

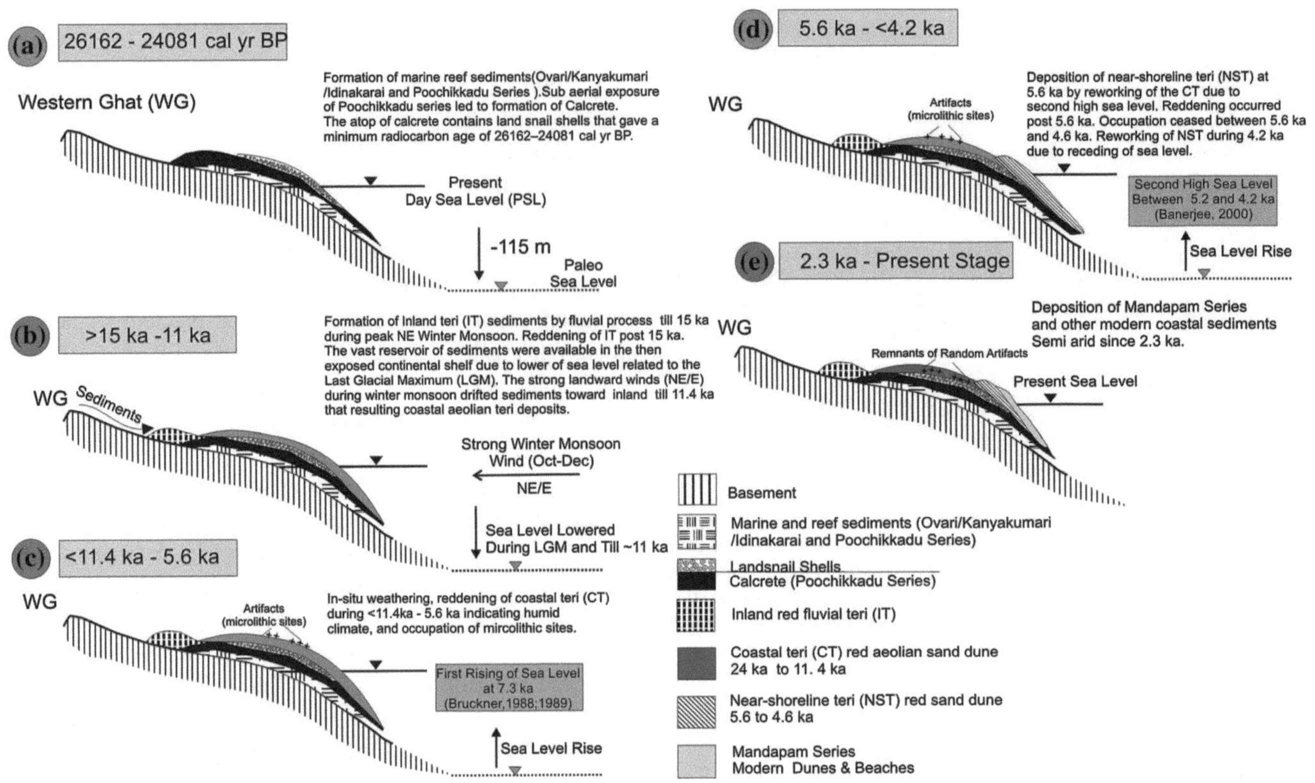

Fig. 5 A schematic illustration showing successive stages of coastal teri red sand dune development. The model is based on optical ages together with the known stratigraphy of the region (after Gardner 1986; Jayangondaperumal et al. 2012)

after 5.6 ka. The microlithic occupation ceased prior to or after deposition of the near-shoreline teri (Fig. 5d). Fluctuating sea level between 4.2 and 2.3 ka facilitated reworking of the near-shoreline teri by both aeolian and fluvial activities. The region then became persistently semi-arid with stable sea level since ~2.3 ka (Fig. 5e).

6 Conclusions

The inland teri sediments were deposited during strong NE winter monsoon before ~15.4 ka; the precipitation eroded the uplands of the Western Ghat, and deposited the inland fluvial sands along its flanks. The coastal teri sands were formed by aeolian process between ~24 and ~11 ka, and this period corresponded with the LGM, when the sea level was lower by 120 m. High sediment supply was facilitated by lower sea level (exposing the vast area of shelf sediments) and a stronger landward north-easterly and easterly wind associated with NE monsoon.

The coastal teri sediments were weathered in situ and reddening took place after ~15.4 ka and 11.4 ka, for fluvial teri and coastal teri, respectively. The microlithic sites were occupied during the weathering of coastal teri after 11.4 ka. The microlithic occupation ceased prior to or after deposition of the near-shoreline teri. Between 4.2 and 2.3 ka, fluctuating sea level facilitated reworking of the near-shoreline teri by both aeolian and fluvial activities, and then the region became persistently semi-arid with stable sea level since ~2.3 ka. Presently, the teri sands are stable wherever there is good vegetation cover.

References

Bruckner H (1989) Late Quaternary shorelines of India. In: Scott DB (ed) Late Quaternary sea level correlation and applications. Kluwer Academic, New York, pp 169–194

Foote RB (1883) On the geology of Madura and Tinnevelly districts. Mem Geol Surv India 20:1–103

Gardner RAM (1986) Quaternary coastal sediments and stratigraphy of southeast India. Man Environ 10:51–72

Jayangondaperumal R, Murari MK, Sivasubramanian P, Chandrasekar N, Singhvi AK (2012) Luminescence dating of fluvial and coastal red sediments in the SE coast, India, and implications for paleoenvironmental changes and dune reddening. Quatern Res 77:468–481. doi:10.1016/j.yqres.2012.01.010

Roskin J, Blumberg DG, Porat N, Tsoar H, Rozenstein O (2012) Do dune sands redden with age? The case of the northwestern Negev dunefield, Israel. Aeolian Res 5:63–75. doi: 10.1016/j.aeolia.2011.11.004

Walker T (1967) Formation of red beds in modern and ancient deserts. Geol Soc Am Bull 78:353–368

The Laterite-Capped Panchgani Tableland, Deccan Traps

Vishwas S. Kale

Abstract

Panchgani is well-known for its impressive tableland and string of mesas, and is a popular getaway for tourists from Mumbai and Pune. The flat-topped hills are the result of differential erosion. The mesas and the tableland are built of flat-lying caprocks and are scarp-bounded on all sides. Ferricrete duricrusts (or laterites) act as the caprock. The mesa tops are featureless. There is evidence of pseudo-karstic activity and mechanical disintegration of the crust rim. The broadly accordant heights of the mesas suggest that they were formed due to the breaching and fragmentation of an extensive lateritized surface and subsequent back-wearing of the cliffs. The formation of mesas appears to be a three stage process—formation of lateritized surface, breaching and fragmentation of the lateritized surface by stream incision, and slope retreat and circumdenudation of isolated patches.

Keywords

Tableland • Mesa • Duricrust • Laterite • Deccan traps • Panchgani • Mahabaleshwar

1 Introduction

One of the most popular tourist destinations in western India is the twin hill resorts of Panchgani and Mahabaleshwar (Fig. 1). Because of cool salubrious climate and breathtaking scenery, thousands of tourists from Mumbai and Pune and other parts of the country, flock to these two places to escape the summer heat and to relax. During the British Colonial period, these hill settlements were popular health resorts. Set in the midst of a spectacular landscape, comprising verdant hill ranges, laterite-capped plateaux, deep valleys and awe inspiring waterfalls, Mahabaleshwar is perched on the Western Ghat (*Sahyadri*). Panchgani, located about 19 km east of Mahabaleshwar, is one of the most visited tourist spots. This settlement was founded by the British in mid-19th century. The hill resort is also famous for its numerous residential schools. Apart from the picturesque valley of the Krishna River, the main attraction in Panchgani is the Tableland (Fig. 2).

V. S. Kale (✉)
Department of Geography, University of Pune,
Pune, 411007, India
e-mail: vskale.unipune@gmail.com

2 Tableland and Mesas

The tabular hills of Panchgani are classic examples of tablelands and mesas (Fig. 3). A mesa (Spanish for table) is an elevated area with a flat top and steep sides. Here the term 'tableland' *sensu lato* refers to a flat-topped hill, larger than a mesa but smaller than a plateau. There are five such flat-topped hills (therefore the name *Panchgani*), but further south more flat-topped hills can be seen. A laterite-capped plateau is present further south of Mahabaleshwar-Panchgani. This elongated plateau is commonly referred as the Kas Plateau (also Bamnoli Range) (Fig. 1).

The special identity of the Panchgani Tableland is largely due to its flatness, vastness and simplicity (Fig. 4). The expansive tableland and the adjoining mesas are built of

V. S. Kale (ed.), *Landscapes and Landforms of India*, World Geomorphological Landscapes,
DOI: 10.1007/978-94-017-8029-2_23,

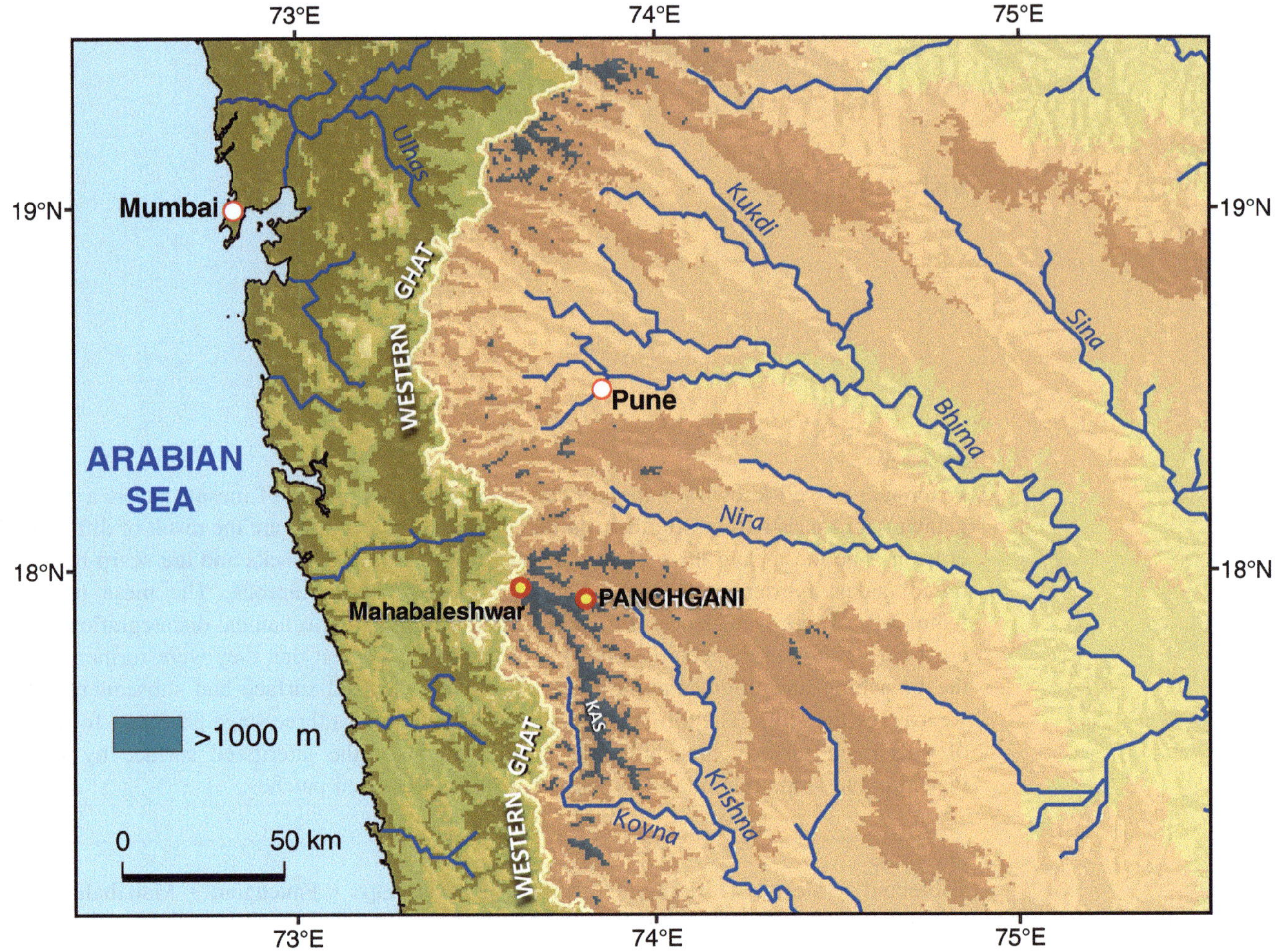

Fig. 1 Map of western Deccan Traps region showing the location of Panchgani and Mahabaleshwar. KAS Kas Plateau (also called Bamnoli Range)

flat-lying caprocks and are scarp-bounded on all sides. Thick laterite crust (ferricrete duricrust) acts as the caprock and protects the underlying Deccan Traps basalt flows from surface water erosion. The altitude of the mesa tops varies between ~1,300 and 1,350 m a.s.l. The edge of the mesas/tableland is delimited by steep scarps. The scarp slopes reveal a cliff-like morphology. The height of the laterite cliffs ranges between ~2 and 18 m (Fig. 5). The local relief in the area is generally less than 550 m. The average annual rainfall at Panchgani is about ~1,700 mm, which is significantly lower than Mahabaleshwar (>6,200 mm). This is because; Panchgani is located in the rainshadow zone of the Western Ghat.

The mesa-tops have a general slope towards the south and southeast. The tabular hills are connected by low, smooth ridges or separated by the valleys of the Bavdhan Stream and Kudali River. Sometimes cols have formed between two mesas by tributary streams on opposite sides of a ridge cutting back towards one another. As a result, the tableland/mesa perimeter is characterized by indentations and promontories. Evidence of mechanical disintegration of the indurated crust is provided by the presence of deep fissures and clefts, and partially to completely detached blocks of laterites at the margins of the crust (Fig. 6), particularly on the southern rim of the tableland (marked B in Fig. 3). Rockfalls constitute a major hazard to the houses and buildings located at the base of the Tableland.

Although not clearly visible in the field in one section, the laterite profile in the area generally reveals a thick ferricrete duricrust overlying a mottled, clay-rich saprolitic horizon (rock weathered *in situ*). The saprolite then passes downward into unweathered basalt. The thick ironstone cap is very resistant to erosion by surface runoff or by channelized flows, but the saprolite is easily erodible when uncovered. Consequently, as soon as the laterite carapace is stripped, the saprolitic horizon is rapidly eroded, giving rise to smoother and gentler slopes. The landscape seen at Panchgani, thus, is the result of differential erosion.

Fig. 2 The Panchgani Tableland. Note the remarkable flatness of the tableland. The Panchgani town is seen on the *right*

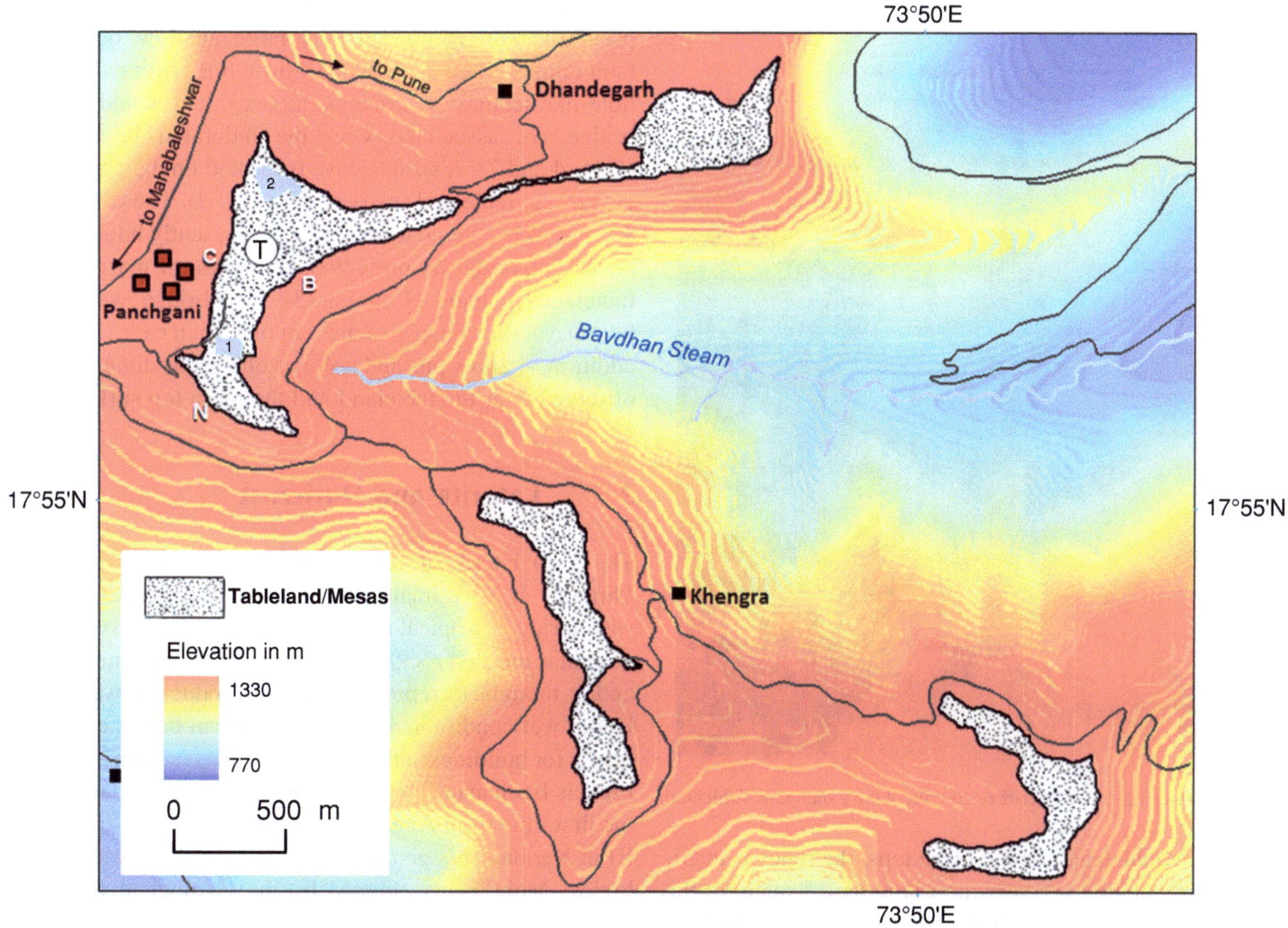

Fig. 3 Map showing the Panchgani Tableland *T* and mesas. *C* cave with café, *N* natural bridge, *B* site of cracking and separation of laterite blocks along the crust margin, 1 and 2 tanks. Major roads are shown by *thin black line*

Fig. 4 View of the featureless tableland top. Note the shallow depressions *D* filled with fine reddish sediments and major cracks *C* in the curst

Fig. 6 Evidence of mechanical disintegration of the indurated crust. Note the presence of clefts *C* and partially to completely detached blocks of laterites *DB* at the margins of the crust. Location indicated by *B* in Fig. 3

Fig. 5 A towering cliff in laterite on the rim of the Panchgani Tableland

Over the tableland and the mesa tops, the relief is negligible and vegetation is conspicuously absent. Except for the two man-made tanks (Fig. 3), the top surface of the tableland (and the mesas) is featureless (Fig. 4). This is mainly on account of the lack of surface water erosion on the resistant caprock, implying near-absence of the mechanism of denudational downwearing. However, this does not imply total absence of fluvial activity in the area. In fact, a number of pseudo-karstic forms such as large shallow depressions (sometimes filled with wind-blown fine sediments), sinkhole-like features, overhangs (alcoves), caves and a natural bridge are associated with the indurated laterite crust (Figs. 4 and 7). A sizable cave is located on the north facing cliff of the tableland (marked C in Fig. 3). The cave, which appears to be altered to some extent, presently houses a café. A natural bridge is present on the western margin of the tableland (marked N in Fig. 3). The bridge was formed because of the collapse of the roof of a small cave (Fig. 7). In addition, shallow pits and hollows of various dimensions are observed over the tableland and the mesa top surfaces.

3 Laterite and Duricrust

The term laterite is derived from *latericius* in Latin, for "brick". It is primarily the result of intense chemical weathering in tropical areas. Laterite is a hard cavernous or vesicular mottled rock, rich in iron and/or aluminum oxides and is thought to represent the end product of weathering. Under wet conditions it is softer, and can be cut easily into bricks for buildings and houses. It hardens on exposure to air. Francis Buchanan (1807) reported and described laterites for the first time from Angadipuram (10° 58′ 44″ N, 76° 12′ 12″ E) in Kerala. This geosite is now one of the National Geological Monuments enlisted by the Geological Survey of India (GSI).

Fig. 7 A cave *C* formed within the laterite crust. The roof of this cave has collapsed giving rise to a natural bridge *N*. See Fig. 3 for location of the natural bridge

According to Schellmann (1981) "*Laterites are products of intense subaerial rock weathering whose Fe and/or Al content is higher and Si content is lower than in merely kaolinized parent rock*". Duricrust (or cuirasse) is a general term for hard crust formed near the ground surface. Depending upon the dominant primary mineral the duricrusts are classified into four main types—ferricrete (iron rich), calcrete (calcium rich), silcrete (silica rich), and alcrete (alumium rich) duricrusts. At Panchgani the duricrust is iron rich i.e. ferricrete. In India, laterite duricrusts have formed on a variety of rocks, such as basalts, granites, gneisses, khondalites, charnockites, etc.

Because laterite hardens on exposure, it usually forms a hard surface capping following a drop in the base level (due to uplift) and aeration. Such a tropical landscape displays a string of mesas and cliffs below which are gentle slopes with a cover of laterite debris. Many workers believe that duricrusts play a role in relief inversion. Lateritized valley-bottoms may become ridges or chain of mesas bounded by younger valleys (Ollier and Sheth 2008).

The duricrusts in some environments are considered as indicators of palaeoclimates. Although controversial, it is generally agreed that ferricretes form under humid tropical conditions. Therefore, their occurrence in semi-arid landscapes (for example at Bidar in Karnataka) is interpreted in terms of climate change.

4 Age of Panchgani Laterites and Mesas

The Deccan volcanic activity ceased about 65 million years ago. Geochemical studies indicate that the high-elevation laterites occurring on the Western Ghat crest (for instance at Mahabaleshwar, Panchgani and Kas) have developed from the topmost (i.e. youngest) Deccan Traps flows (Widdowson 1997). Further, palaeomagnetic considerations suggest the late Cretaceous-early Paleogene as the most likely age of both the laterite (Schmidt et al. 1983) and the surface on which it is developed. It is plausible to suppose that the weathering of the youngest lava flows and laterite formation took place without a long time gap. This in turn suggests that the laterite surface represents the post-Traps lava surface. As the mesas (and the tableland) were produced by the dissection and fragmentation of the lateritized surface, they are younger on the geological timescale.

5 Formation of Tableland and Mesas

Duricrust-capped tablelands/mesas form due to dissection of a large plateau or lateritized surface and differential erosion. In the Panchgani area, the laterite-capped mesas are found on interfluves, and occur on both sides of the river valleys almost at the same elevation. The present-day laterite distribution and the broadly accordant heights of the hills suggest that the mesas are an erosional remnant of a former lateritized surface which used to extend over a large part of the Deccan Traps—from Mahabaleshwar (located at a higher level in the west) in all directions for several tens of km (Fig. 8). Subsequently, the lateritized surface was breached, first by headward erosion and then by vertical incision by streams and rivers, followed by lateral extension of the valleys (Fig. 8). Large patches of laterites on the divides/intefluves were further breached by tributary streams on opposite sides of the divide cutting back towards one another. This created an uneven surface, where laterite-capped outliers stood higher above the surrounding areas as mesas and tablelands, connected by low, smooth ridges or separated by younger valleys at a lower level (Fig. 8).

Once isolated, the area of the caprock outliers is consumed rapidly by scarp retreat by circumdenudation (denudation from all sides). Wasting of the caprock is favored by undercutting at the base of the indurated layer by percolating rainwater and caving. The development of pseudo-karstic forms further accelerates the process of the removal of the caprock, paving the way to reach the lower, more erodible saprolitic horizon.

The development of pseudo-karstic forms is favoured by the occurrence of clay-rich saprolitic horizon below the porous and vesicular laterite crust. During the monsoon season, rain water percolates through cracks, cavities and vesicles in the thick crust and reappears in springs and seepage zones at the caprock-saprolite interface. Away from the rim, pseudo-karstic activity creates sub-surface cavities and surface subsidence. This results in the formation of shallow depressions and sinkhole-like features on the top

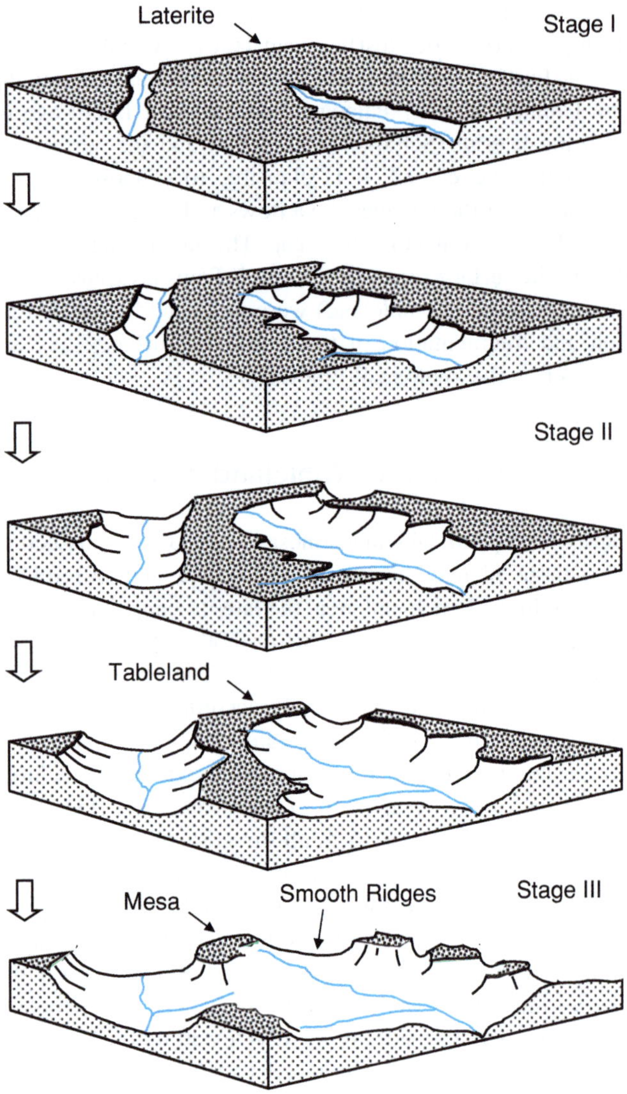

Fig. 8 Stages in the formation of mesas and tableland. *Stage I* lateritized surface is formed by deep weathering of the basement rocks and subsequent induration following fall in base level (uplift) and aeration. *Stage II* rejuvenation of drainage leads to erosion and breaching of the lateritized surface. *Stage III* back-wearing or slope retreat creates laterite-capped isolated hills or mesas connected by smooth ridges and separated by valleys

surface (Fig. 4). More often than not, these features are guided by cracks and fissures in the crust (Fig. 4). Along the crust margins, sapping and slope undercutting promote cliff failure and mechanical disintegration of the laterite crust (Fig. 6). The mesas steadily decrease in size through slope retreat and the steep slopes are gradually replaced by smooth ridges and gently sloping surfaces.

In a nutshell, the formation of the laterite-capped tablelands and mesas is a three stage process (Fig. 8). In stage I, a lateritized surface is formed by deep weathering of the basement rocks and subsequent induration following fall in base level (due to uplift) and aeration. In stage II, rejuvenation of drainage leads to erosion and breaching of the lateritized surface. Subsequent back-wearing or slope retreat (stage III) creates laterite-capped isolated hills or mesas connected by smooth ridges and separated by valleys.

6 Conclusions

The Panchgani (or five hills) resort, founded by the British in mid-1850s, has grown into a popular tourist destination. The Tableland is the special feature that draws large number of tourists every year. This outstanding landscape of flat-topped hills or mesas, bounded by high cliffs and picturesque valleys is the product of differential erosion. The main tableland and the mesas are crowned by thick, resistant laterite duricrust. Although the mesa tops are featureless, there is evidence of pseudo-karstic activity and mechanical disintegration of the crust margins. The expansive tableland and the isolated mesas are dwindling remnants of an ancient lateritized surface that draped much of the area during late Cretaceous-early Cenozoic period. Subsequent fragmentation of the expansive lateritized surface and back-wearing of the laterite cliffs of the outliers has given rise to these mesas. The flat-topped hills of Panchgani are, thus, relict landforms. The tabular hills have been preserved only due to the unusual resistance of the laterite duricrust. Once the carapace is stripped, the hills will be subdued by subaerial processes within short time. Panchgani-like tablelands, mesas and pseudo-karstic features are observed at a number of localities in the Western and Eastern Ghats, the west coast of India as well as in central India.

References

Buchanan F (1807) A journey from Madras through the countries of Mysore, Canara and Malabar. Asian Educational Services, New Delhi, Madras

Ollier CD, Sheth HC (2008) The high Deccan duricrusts of India and their significance for the laterite issue. J Earth Syst Sci 117:537–551

Schellmann W (1981) Considerations on the definition and classification of laterites. In: Proceedings of the first international seminar on lateritiztion processes, Trivandrum, Oxford and IBH Publishing Co., New Delhi, pp 1–10

Schmidt PW, Prasad V, Ramam PK (1983) Magnetic ages of some Indian laterites. Palaeogeogr Palaeoclimatol Palaeoecol 44:185–202

Widdowson M (1997) Tertiary palaeosurfaces of the SW Deccan, Western India: implications for passive margin uplift. In: Widdowson M (ed) Palaeosurfaces: recognition, reconstruction and palaeoenvironment interpretation, vol 120. Geological Society London Special Publications, London, pp 221–248

The Lonar Crater: The Best Preserved Impact Crater in the Basaltic Terrain

Makarand S. Bodas and Bibhas Sen

Abstract

The Lonar Crater, created by an impact of a bolide or a meteor about 570 thousand years ago in central part of the Deccan Volcanic Province, is one of the very few hypervelocity impact craters in the world carved out from the basaltic target rocks and the only crater in lava flow sequence of a Continental Flood Basalt Province. With its average diameter of 1.8 km, the crater is a simple, bowl-shaped and remarkably circular depression within a flat country dotted with sporadic conical hills. It has nearly 150 m depth and a rim that is raised nearly 20 m above the surrounding country. The evidence of it being of the impact origin is available in the form of its relatively unaltered morphology, identification of the subsurface breccias beneath the sediments in the crater, presence of shocked minerals, glasses and ejected melt breccias. Although the crater is located on the drainage divide of two moderate-sized rivers, namely Purna and Penganga; there is no evidence to suggest that the impact has resulted in the disruption, truncation or reorganization of the drainage network locally or regionally.

Keywords

Lonar • Impact crater • Deccan traps • Breccia • Ejecta blanket • Drainage

1 Introduction

Lonar Crater (Fig. 1), located in Lonar, a small town in Mehekar sub-division of Buldhana District of Maharashtra, is one of the very few hypervelocity impact craters in the world carved out from the basaltic target rocks and the only crater in lava flow sequence of a Continental Flood Basalt Province. To the earth scientists, the crater centered at 19° 58′ N and 76° 31′ E potentially offers an analogue for impact craters on terrestrial planets and planet-like bodies (Osae et al. 2005) with basaltic crust (e.g. Mars) whereas, for the ecologists and microbiologists, the crater lake with its saline and alkaline water gives a unique opportunity to study "extremophiles" (organisms that grow in extreme environments) as well as alkaliphilic and halotolerant bacteria (Kanekar et al. 2008).

M. S. Bodas (✉)
2, Meghsagar, Swastishri Society, Karvenagar, Pune, 411052, India
e-mail: makarandbodas@gmail.com

B. Sen
IV/ 16, GSI Complex, Alandi Road, Pune, 411006, India
e-mail: bibhas.sen@gmail.com

2 Regional Geological and Geomorphic Setting

Lonar is located in the central part of the Deccan Traps or Deccan Volcanic Province ($\sim$65 Ma) and the area exposes a sequence of seven basaltic flows. Among them, the lowest two flows have pahoehoe and the upper five flows have a'a morphology (Sen and Sable 2005). The lava flows range in

V. S. Kale (ed.), *Landscapes and Landforms of India*, World Geomorphological Landscapes,
DOI: 10.1007/978-94-017-8029-2_24, © Springer Science+Business Media Dordrecht 2014

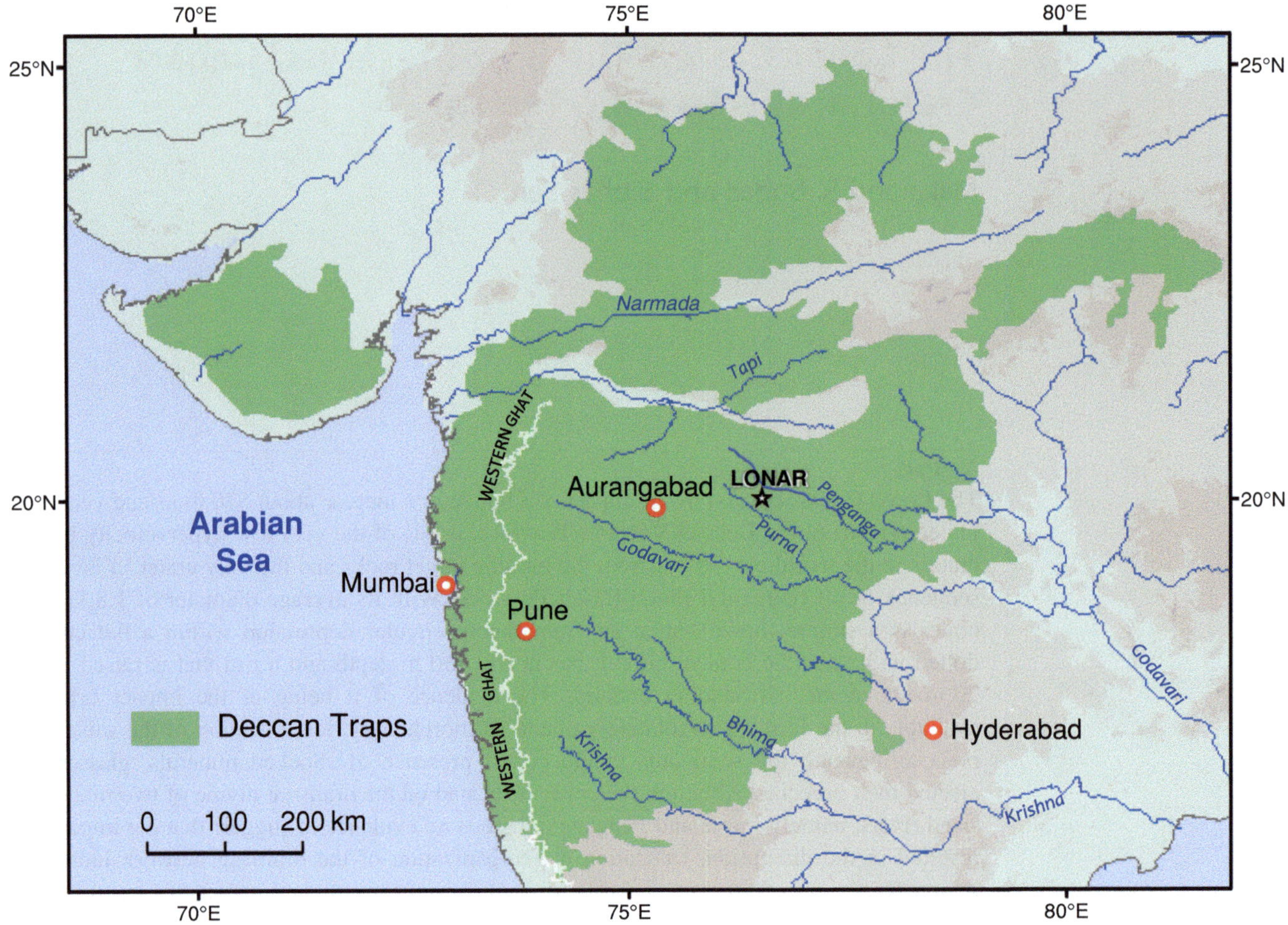

Fig. 1 Map of western India showing the distribution of Deccan Traps and the location of Lonar crater. Some major cities are also shown

thickness from 8 to 40 m and many of them can be easily separated from one another by the presence of red interflow horizons. The regional gradient of the flow sequence is southeasterly and the amount is 1:250. The lava sequence is overlain by 0.05–0.9 m thick, dark, clayey soil developed due to the alteration of the uppermost lava flow prior to the impact cratering and it can be best observed along several gullies and in well sections beneath the ejecta blanket produced as a result of the impact.

The Lonar Crater is located on the drainage divide between Purna and Penganga Rivers (Fig. 1), the southeasterly flowing tributaries of the Godavari River. Whereas the Penganga River drains the areas located due north of the crater, the Purna River flows on the southern side of the crater. Lower order streams originating in the vicinity of the crater flow either due south or roughly due north and meet the Purna and Penganga Rivers, respectively. The divide reveals undulating topography dotted with low hills reaching a maximum elevation of 616 m a.s.l observed due east-southeast of Lonar Crater. The climate over the area is semi-arid (average monsoon rainfall ~680 mm).

3 Morphology of Lonar Crater

The Lonar Crater with its average diameter of 1,830 m is a bowl-shaped and remarkably circular depression (Figs. 2–3) within a flat country dotted with sporadic conical hills. It has nearly 150 m depth and a rim that is raised nearly 20 m above the surrounding country (Fig. 5). The highest point (~610 m a.s.l.) on the rim is located in the southwestern part of the crater. Using gravity measurements, Fudali et al. (1980) have, however, suggested that the original diameter of the crater could have been 1,710 m, depth 230–245 m and the original average rim height ~40 m.

The inner slope of the crater is covered unevenly with dense vegetation and it can be divided into three slope categories: moderate slope (30–40°) observed in the uppermost part of the crater between 600 and 575 m a.s.l., followed downwards by steep slope (≥40°) and lastly the gentle slope in the toe region of the crater wall (Sen and Sable 2005). The crater interior shows signs of degradation in the form of development of gully erosion and debris flows. The almost flat crater floor is ~1,200 m in diameter

Fig. 2 Google Earth image showing the Lonar crater, the Ambar lake and the Lonar town. Note the notch and a fan-like feature in the northeast of the crater produced by the headward erosion of the crater rim by a major southwesterly flowing stream

Fig. 3 Panoramic view of the Lonar crater

and is occupied by a saline lake with nearly 5–7 m depth (Nandy and Deo 1961). The lake level is influenced by surface runoff during monsoon and groundwater input effective during monsoon and dry seasons (Komatsu et al. 2013).

The drainage within the crater is centripetal (radial inward) and is almost exclusively formed by the first order seasonal streams originating from the crater rim and debouching into the crater lake. In the northeastern part of the crater, there exists a prominent notch probably produced by the headward erosion of the crater rim by a major southwesterly flowing stream (Fig. 2) along which at least three perennial springs emerge. This Dhar Valley incision has led to the formation of a fan delta in the front of the valley (Komatsu et al. 2013).

In addition to the main Lonar Crater, there is yet another smaller crater (called *Chhota* Lonar/Ambar Lake) with nearly 300 m diameter and $\sim$20 m high rim located less than 1 km to the north of the main crater (Figs. 2 and 4). Although it is suggested that the *Chhota* (meaning small) Lonar too formed together with the Lonar Crater as a result of the impact of a fragment of the same impactor (Fredriksson et al. 1973), there is no confirmation as yet.

4 Geology of the Lonar Crater

The geological succession of Lonar Crater essentially consists of three components (Figs. 4 and 5)—the basaltic lava flows (target rocks), rocks formed as a result of the impact, and the unconsolidated lake sediments. Of these, the uppermost layer of the unconsolidated clayey to silty sediments with salt encrustations can be seen on the crater floor and the basaltic lava flows (7 in number) and rock debris giving rise to fall-back fragmentary breccia can be best observed along the foot tracts that join the crater floor with

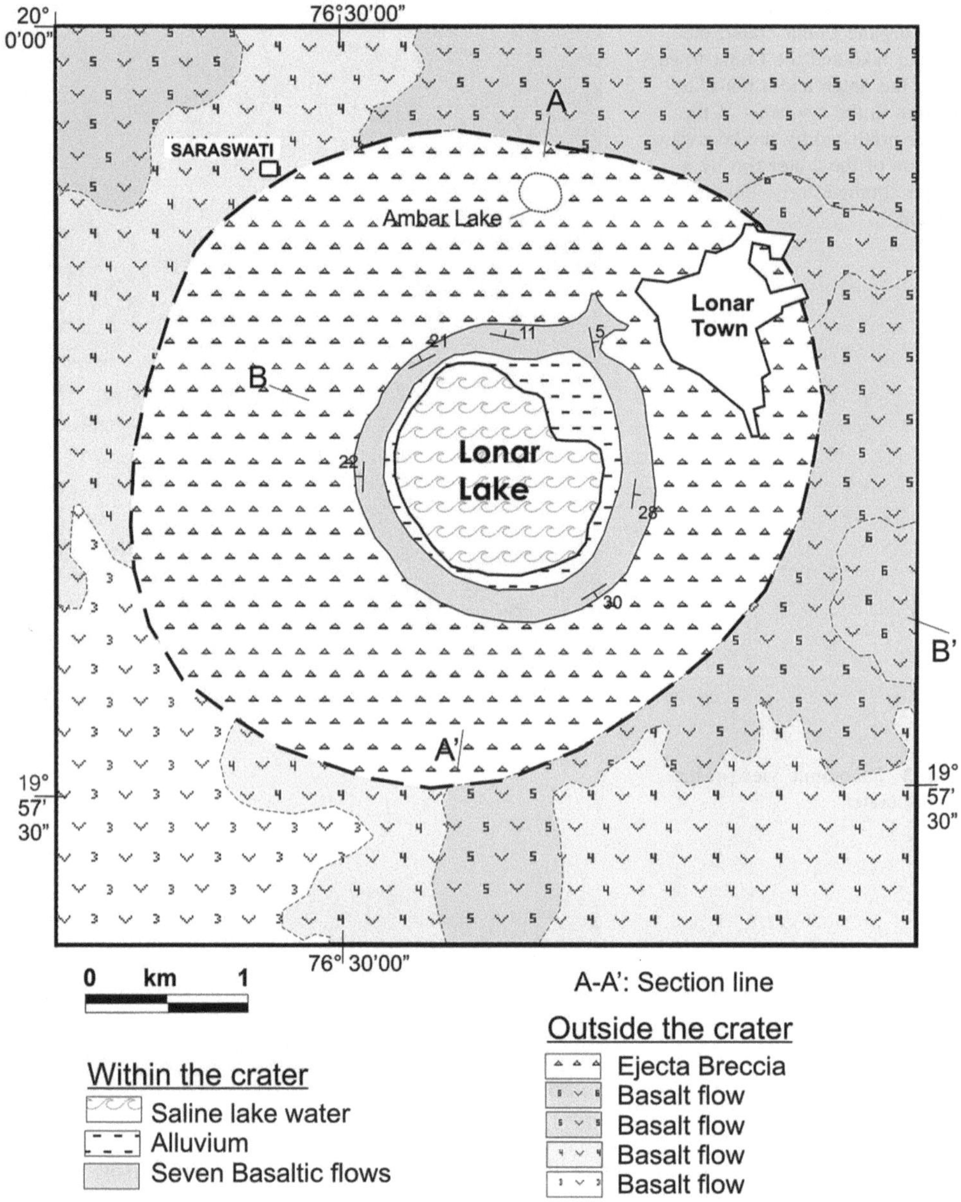

Fig. 4 Geological sketch map of the Lonar crater and its surrounding area. Cross-sections along *A–A'* and *B–B'* are presented in Fig. 5

the rim. The ejecta blanket with fragmental breccia and suevitic breccia formed as a result of impact is well exposed in the rim area of the crater and radially outside the crater for a distance of >2 km. In fact, the Lonar town is located on the ejecta blanket (Fig. 4).

The target rocks, especially the upper two flows exposed along the inner slopes of the crater, bear the signatures of deformation as a result of impact (Fudali et al. 1980). They are mainly seen in the form of upturning of flows in the crater walls and even their overturning in the rim portions (Osae et al. 2005). In addition, the development of three sets of fracture planes that are millimeter to meter apart is also seen in the basaltic rock-mass. The most conspicuous amongst them is a fracture that always strikes tangentially to the crater outline and dips quaquaversally (sloping in all directions) by 5–30°. The second set of fractures is nearly sub-vertical and it radiates out from the centre of the crater. The third set of fracture too strikes tangentially to the crater outline but dips inward (towards the centre of the crater) by 30–45° (Sen and Sable 2005). Cementation of large basaltic blocks (fraction of a meter to even >5 m across) by fine grained material of similar composition in the rim portion of the crater (Fig. 6) provides yet another evidence of deformation due to impact.

Few centimeters to nearly eight meter thick ejecta blanket resting over the palaeosol formed by the weathering of basaltic lava flows prior to the impact is spread around the Lonar Crater. It slopes gently (2–6°) away from the crater

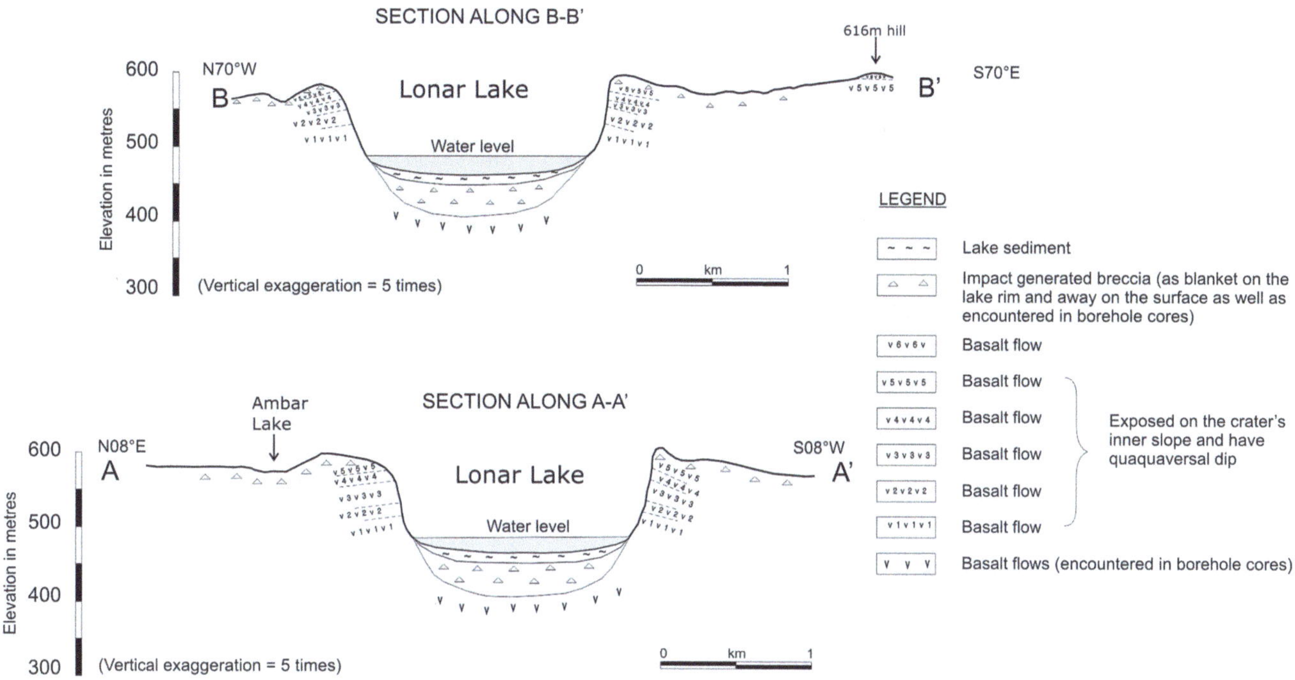

Fig. 5 Sections across the Lonar crater. For location of the cross-sections refer to Fig. 4

Fig. 6 Fragmentary clast supported breccia with large basaltic blocks as clasts and fine grained material of similar composition as matrix exposed near Gomukh temple, Lonar crater

with an undulating surface that represents the original post-impact surface that was not much modified by post-cratering erosion (Osae et al. 2005). The ejecta occurs in two different forms viz. crudely stratified, fine grained ejecta (Fig. 7) with clasts of a mixture of bed rock units that are either unshocked or have a range of shock effects (up to the formation of planar deformation features in minerals plagioclase and pyroxene and formation of myskelynite in the target rocks) and ejecta containing impact derived melt rock fragments and glasses (Fig. 8) that can be classified as the basaltic equivalents of suevitic breccias. The field characters suggest that the most likely mechanism for the transportation and deposition of the ejecta may have been "debris surge" similar to fluidized craters of Mars (Osae et al. 2005).

Fig. 7 A section through crudely stratified ejecta blanket (seen in the lower part of the photograph) exposed in the courtyard of Daityasudan temple, Lonar

Fig. 8 "Suevitic breccia" with a dark glassy fragments seen along foot track due southwest of MTDC guest house, Lonar

The lake sediments deposited on the floor of the crater have a total thickness of 60–100 m and they are underlain by brecciated rocks that pass downwards into a succession of basaltic lava flows (Nandy and Deo 1961). The lake sediments are valuable records of the past climatic and environmental conditions. On the basis of a study of a 10 m long core from the lake bed, Anoop et al. (2013) have been able to identify two major periods of weaker monsoons in this area between 4.6 and 3.9 ka and between 2.0 and 0.56 ka during the Holocene.

5 Age of the Lonar Crater

Although there is hardly any doubt that the Lonar Crater is very young as compared to the age of the target rocks, the Deccan Traps basalts; the age data on the possible impact dates generated in the later part of the past century varied a great deal. Till recently, the age of the Lonar Crater was thought to be constrained between 15 and 50 ka, an age that is similar to the Meteor Crater in Arizona, USA.

However, recent dating of the 'impact melt rock' samples from Lonar by $^{40}Ar/^{39}Ar$ technique has given an age of 570 ± 47 ka. This age is ten times older than the oldest date given by non-isotopic methods (Jourdan et al. 2011). The considerably younger ages by fission tracks and thermo-luminescence methods indicate that the Lonar impact rocks have been perturbed by the post-impact processes like alteration and wild fires that have partially erased and reset the fission tracks or would have affected electron charged traps used in thermo-luminescence dating (Jourdan et al. 2011).

6 Origin of Lonar Crater

The origin of Lonar Crater had been a matter of controversy since 19th century. A number of hypotheses ranging from it being the only available evidence of a volcanic outburst or a great gaseous extrication to a crypto-volcanic origin with deep seated carbonatite have been put forth to explain its origin. However, there is now abundant evidence that together suggests the Lonar Crater to be a result of an impact event. The impact was caused by a hypervelocity bolide or a meteor. The most important line of evidence includes the relatively unaltered morphology of the crater on one hand and the identification of the subsurface breccias beneath the sediments in the crater, presence of shocked minerals, glasses and ejected melt breccias on the other. However, till date, fragments of the impactor (meteor) have not been retrieved from the site and the geochemical evidence too has not yet been successful in unequivocally characterizing the nature of the impactor. Although the characterization of the impactor is still elusive, based on the circularity of the crater, it is likely that the angle of incidence of the impactor was >30° from the horizon and the greater extension of the ejecta in the west probably indicates that the impact was from the east (Misra et al. 2010).

7 Concluding Remarks

In spite of the fact that the meteor impact was a catastrophic event, there is very little geomorphic evidence in the topographical maps or satellite images suggestive of any kind of noteworthy disruption, truncation or re-organization in the drainage network of the Purna or the Penganga River systems or the drainage divide morphology, locally or regionally. This is very intriguing and has probably to do with the smaller size of this simple impact crater and the impact site being on the water divide. Further, given the age (~570 ka) of the impact event, the relatively unaltered morphology of the crater suggests very low overall denudation rates in this semi-arid part of the Deccan Traps region.

Although the Lonar Crater is undoubtedly the best preserved impact crater, it is certainly not the only example of impact crater available in the Indian subcontinent. Dhala Crater (25° 17′ 59.7″ N, 78° 08′ 03.1″ E) in Shivpuri District, Madhya Pradesh is the second confirmed impact structure in India. Unlike Lonar being a simple impact crater carved out of the basaltic lava flow sequence, the Dhala structure is a remnant of a large complex impact structure (estimated diameter ~25 km) occurring on the Archean crystalline basement (Pati and Pati 2013). In addition, Ramgarh structure (25° 20′ 08.9″ N, 76° 36′ 23.7″ E), a crater-like feature in Rajasthan and Shiva structure in the Arabian Sea too have been suggested to be of impact origin. However, they are yet to yield any unequivocal evidence in favour of the impact cratering.

Lonar is one of the National Geological Monuments enlisted by the Geological Survey of India (GSI). This geosite is also one of the most popular tourist destinations in Maharashtra. Although the main attraction is the crater, tourists also visit Lonar to see the ruins of historical temples as well as resident and migratory birds.

References

Anoop A, Prasad S, Plessen B, Basavaiah N, Gaye B, Naumann R, Menzel P, Weise S, Brauer A (2013) Palaeoenvironmental implications of evaporative Gaylussite crystals from Lonar lake, central India. J Quat Sci 28:349–359

Fredriksson K, Dube A, Milton DJ, Balsundaram MS (1973) Lonar lake, India: an impact crater in basalt. Science 180:862–864

Fudali RF, Milton DJ, Fredriksson K, Dube A (1980) Morphology of Lonar crater, India: comparisons and implications. Moon Planet 23:493–515

Jourdan F, Moynier F, Koebrl C, Eroglu S (2011) ^{40} Ar/ ^{39} Ar age of the Lonar crater and consequence for the geochronology of the planetary impacts. Geology 39:671–674

Kanekar PP, Joshi AA, Kelkar AS, Borgave SB, Sarnaik SS (2008) Alkaline Lonar lake, India-a treasure of alkaliphilic and halophilic bacteria. In: Proceedings of Taal 2007: the 12th world lake conference. 12:1765-1774

Komatsu G, Senthil Kumar P, Goto K, Sekine Y, Giri C, Matsui T (2013) Drainage systems of Lonar crater, India: contributions to Lonar lake hydrology and crater degradation. Planet Space Sci http://dx.doi.org/10.1016/j.pss.2013.05.011

Misra S, Arif M, Basavaiah N, Srivastava PK, Dube A (2010) Structural and anisotropy of magnetic susceptibility (AMS) evidence for oblique impact on terrestrial basalt flows: Lonar crater, India. Geol Soc Am Bull 122:563–574

Nandy NC, Deo VB (1961) Origin of Lonar lake and its alkalinity. TISCO 8:1–12

Osae S, Misra S, Koebrel C, Sengupta D, Ghosh S (2005) Target rocks, impact glasses and melt rocks from the Lonar impact crater, India: petrography and geochemistry. Meteor Planet Sci 40:1473–1492

Pati JK, Pati P (2013) Impact cratering from an Indian perspective. In: Sinha R, Ravindra R (eds) Earth system processes and disaster management, society of earth scientists. Springer, Berlin, Heidelberg, pp 213-229

Sen B, Sable AB (2005) A study of physical volcanological features and REE signatures of Lonar crater area, Buldhna district, Maharashtra. Geological Survey of India unpublished open file report (accession No. 22419) 53P www.portal.gsi.gov.in

The Great Rann of Kachchh: The Largest Saline Marshland in India

Navin Juyal

Abstract

The Great Rann of Kachchh in western India is a unique landscape in the Indian subcontinent. Barely rising above the sea level, it is a vast expanse of salt encrusted terrain which can neither be called a land or sea. Apparently looks topographically monotonous however, a close scrutiny indicates significant geomorphic variability. This is ascribed to a combination of earth surface processes and tectonics. In the absence of longer sedimentary records, the evolutionary history of the Great Rann is limited to the surface exposures of the deposits whose ages go back to about six thousand years. Landforms have a strong finger printing of marine process with subordinate fluvial contribution in landform evolution. The terrain witnessed two major earthquakes, one between 2.2 and 1.4 ka and another in 1819 CE. These earthquakes appear to have significantly influenced the terrain morphology and landform evolution. Although speculative, it seems that frequent earthquakes could have been one of the major reasons for the desertion of human occupation.

Keywords

Great Rann • Allah Bund • Mid-Holocene • Tidal flat • Earthquakes • Sea level changes

1 Introduction

Climate, tectonics and sea level changes are the major contributing factors responsible for the landscape evolution during the Quaternary. This is particularly true for areas proximal to the continental margins; where slight change in land-sea configuration brings about amplified changes in the style of landform development. The Rann of Kachchh (*rann* = ephemeral playa lake) is one such area and is analogous to the delta-flank depressions commonly found on the margins of deltaic areas (Glennie and Evans 1976). Divisible into the Great and Little Rann of Kachchh, the Great Rann of Kachchh, which occupies an area of ~16,000 km^2, is bounded by the Arabian Sea and the Indus Delta in the south and west, the Thar Desert in the north, the Aravalli Ranges in the east and the Kachchh upland in the south (Fig. 1). Compared to this, the Little Rann of Kachchh (area ~4,100 km^2) is located in the southeast of the Great Rann and is connected with the Great Rann by a narrow saline stretch in the north. It is bounded by the mainland from three sides and opens into the Gulf of Kachchh in the southwest. Structurally, the northern boundary of the Great Rann is demarcated by the Nagar Parker Fault (NPF), whereas the southern boundary is delineated by the Kachchh Mainland Fault (KMF). In addition, the Island Belt and Allah Bund Faults are the major geological structures within the Great Rann (Fig. 2). The present configuration of the Kachchh Basin is ascribed to the differential movement along these faults which resulted in the development of series of horsts e.g. the hills

N. Juyal (✉)
Geosciences Division, Physical Research Laboratory, Navrangpura, Ahmedabad, 380058, India
e-mail: navin@prl.res.in

V. S. Kale (ed.), *Landscapes and Landforms of India*, World Geomorphological Landscapes,
DOI: 10.1007/978-94-017-8029-2_25, © Springer Science+Business Media Dordrecht 2014

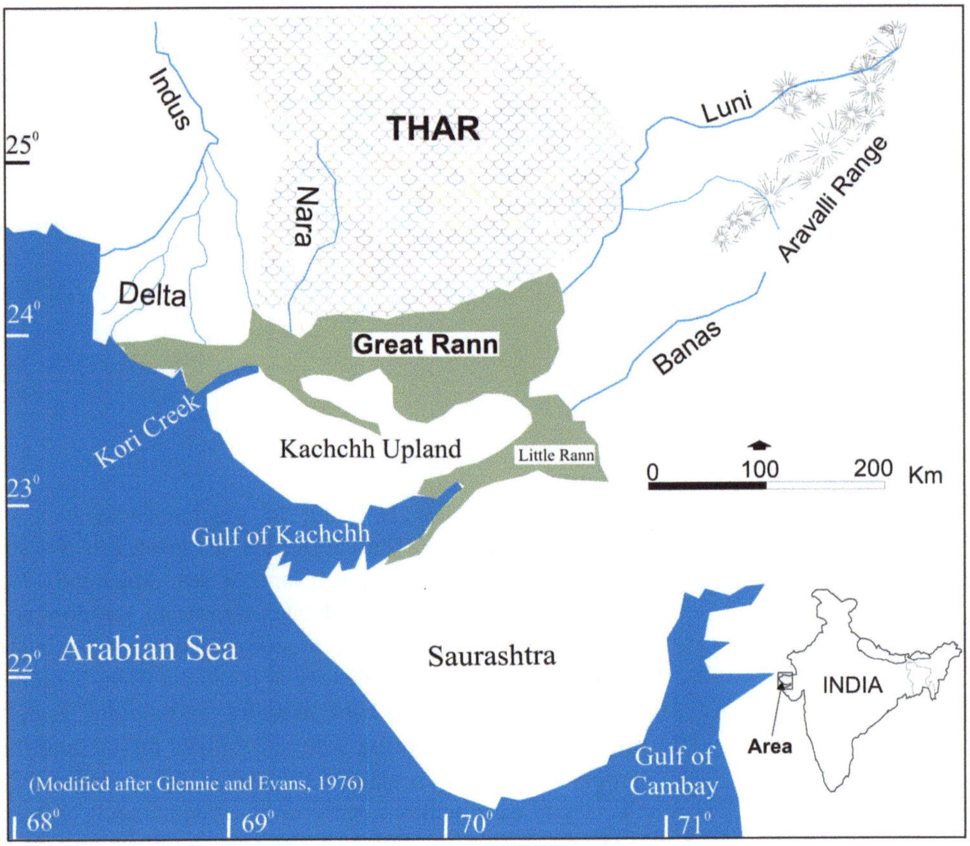

Fig. 1 Map showing the location of Great Rann and Little Rann of Kachchh. The shaded area to the north of the Great Rann is the sand dune field of the Thar Desert

of Kachchh and Island Belt and the graben viz. the Banni Plain and the Great Rann (Biswas 1987). Presently, the Kachchh Basin is passing through the stress reversal phase with the uplift of large fault bounded blocks in response to the compressive stresses active on the Indian Plate. The salt encrusted terrain, known as Ranns of Kachchh, are recent depositional basins formed within the structural depression resembling half graben formed during later phase of uplift during the late Tertiary. The Ranns existed as a shallow embayment and inlets of the sea from Paleogene–Neogene until recent (Biswas 1987). During the last 2000 years, silting up of the Rann, probably accompanied by episodic tectonically induced land level changes has converted the marine embayment into dry salt covered mud flats. Presently, the Great Rann lies marginally above the tidal range of the Arabian Sea and is flooded annually by the sea water through the Kori Creek in the south and the Luni River in the east (Fig. 1; Glennie and Evans 1976). The terrain receives ~200 mm of annual rainfall, majority of which falls during the months of July and September (southwest monsoon). Because of low rainfall and high average temperature, the rate of potential evaporation is very high.

2 Landforms

The major landforms of the Great Rann can be grouped under two broad topographic units viz. (i) the uplifted rocky islands, and the (ii) low lying salt encrusted Rann surface.

2.1 The Uplifted Rocky Islands

These are four east–west trending islands, also known as the Island belt. From west to east, these are: the Pachham (618 km^2, maximum elevation 463 m a.s.l.), Khadir (346 km^2, maximum elevation 283 m a.s.l.), Bela (542 km^2, maximum elevation 246 m a.s.l.), and Chorar (area 617 km^2 maximum elevation 87 m a.s.l.). Dominated by south dipping Mesozoic and Neogene marine rocks (Biswas 1987), the islands are characterized by a steep northern scarp and a gentle southern slope which eventually merges with the Great Rann. The scarp is a geomorphic expression of the tectonically active east-west trending Island Belt Fault (IBF). Island belt which preferentially occur along the east-west transect of the Great Rann, are suggested to be the result of tectonic

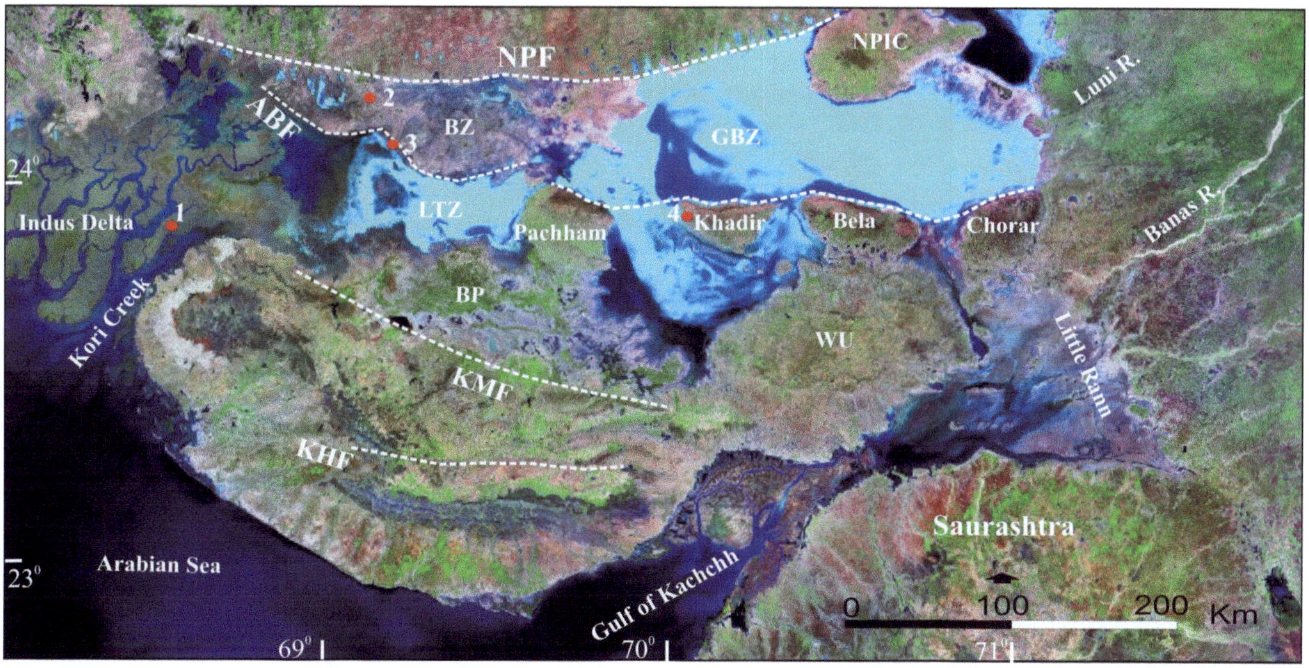

Fig. 2 Satellite image showing the geomorphology of the Great Rann. *NPF* Nagar Parkar Fault, *ABF* Allah Bund Fault, *KMF* Kachchh Mainland Fault, *KHF* Katrol Hill Fault, *WU* Wagad Uplift, *NPIC* Nagar Parkar Igneous Complex, *BZ* Bet Zone, *LTZ* Linear Trench Zone, *GBZ* Great Barren Zone, *BP* Benni Plain. *1* Basta Bunder Fort, *2* Vigakot, *3* Karim Shahi, *4* Dholavira. *Source of satellite imagery* http://www.mmnt.net/db/0/0/ftp.glcf.umiacs.umd

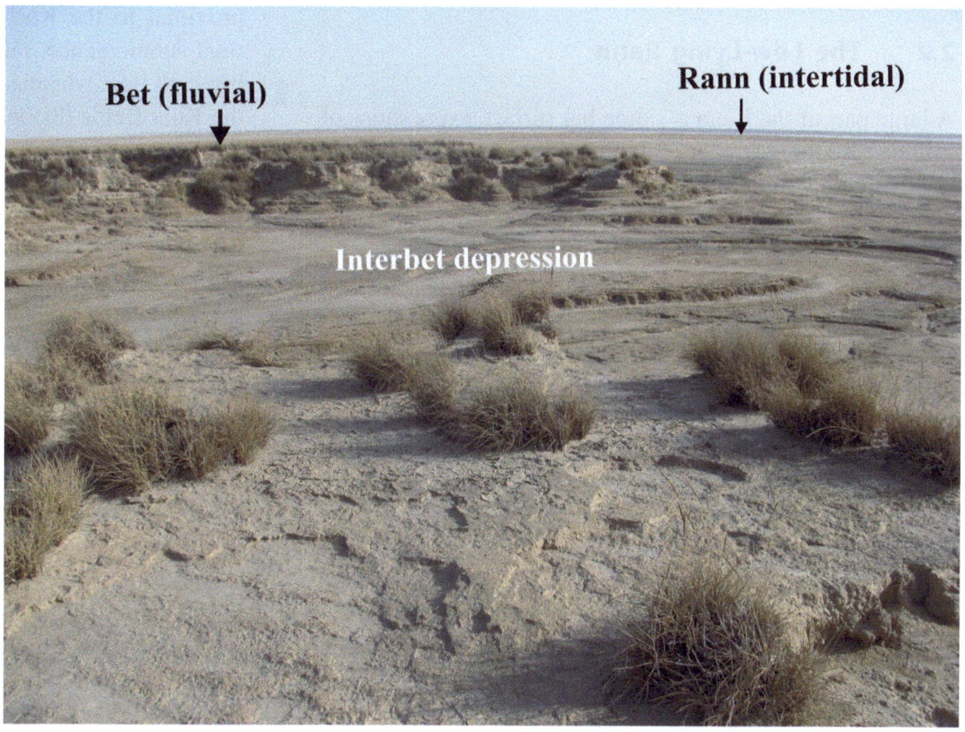

Fig. 3 Fluvial dominated sandy Bets surface and the intertidal Rann surface

Fig. 4 Dissection of the Allah Bund Fault Scarp (ABFS) by gullying, close to the mouth of the Nara River. *Inset* is the picture of a NW–SE trending beheaded stream at ABFS. The *arrow* indicates the direction of flow

activity along the east-west trending IBF during the rift reversal phase (Biswas 1987). During the monsoon, when the Great Rann is submerged under flood waters, the island belt stands out as imposing geomorphic feature in the region.

2.2 The Low-Lying Rann

A large part of the Rann is saline but surface expressions of salt encrustations are concentrated more in areas which are relatively low lying. Within the low lying Great Rann, considerable variability in the landforms and geomorphic processes can be observed. Based on the surface assemblages and flooding pattern, geomorphologically the low lying Rann is divided into four east-west trending zones (Roy and Merh 1982)—(a) Bet Zone (b) Linear Trench Zone (c) Banni Plain and (d) Great Barren Zone (Fig. 2).

(a) **Bet Zone**: These are slightly elevated and discrete patches of medium to fine micaceous sand-dominated incised tablelands which are above the tidal influence and are separated from each other by interbet depressions (Figs. 2, 3). High concentration of Bets (*bet* = island) is found south of the parabolic dune field of the Thar Desert. On the satellite imagery they appear as linear to curvilinear features covered with vegetation and terminate abruptly in the vicinity of 1819 Allah Bund Scarp in the south (Fig. 2).

(b) **Linear Trench Zone**: This is a narrow east-west trending low lying tract extending from the Kori Creek in the west to Kuar Bet in the east. Flanked by the Banni Plain in the southeast and the 1819 Allah Bund Scarp in the northwest, the zone gets regularly inundated by tidal waters of the Arabian Sea through the Kori Creek (Fig. 2). The magnitude and extent of inundation varies in accordance with the tidal amplitude during the monsoon. For example, the western part, proximal to the Kori Creek, is perennially under the tidal submergence, the central part gets inundated during high tide, whereas the northeastern part proximal to the Kuar Bet is flooded during exceptionally high tides, which on evaporation gives rise to a salt encrusted plain (Fig. 2).

(c) **Banni Plain**: Lies between the Kachchh Mainland in the south, the Pachham, Wagad and Bela uplifted areas in the west and the east. The Banni Plain can be considered as tectonically uplifted mudflats along the southern margin of the Great Rann (Fig. 2; Kar 2011). It is a vast salt-affected plain with sparsely distributed halophytic grasses and bushes. Although the relief is <10 m, the change in slope across the surface is barely perceptible. The sediments of the Banni are distinctly fine textured and are composed of stratified deposits of silt and clay, along with fine sand. The steeper slope is towards the north and the gentler one is to the south, while the presence of gullies cutting into the Quaternary alluvium is ascribed to recent tectonic activity in the region and resultant erosion (Kar 2011).

(d) **Great Barren Zone**: This is a shallow saucer like depression towards the eastern part of the Great Rann and abuts against the dunes of Thar Desert in the northwest, alluvial plains in the east and rocky islands of Pachham, Khadir, Bela and Chorar in the south

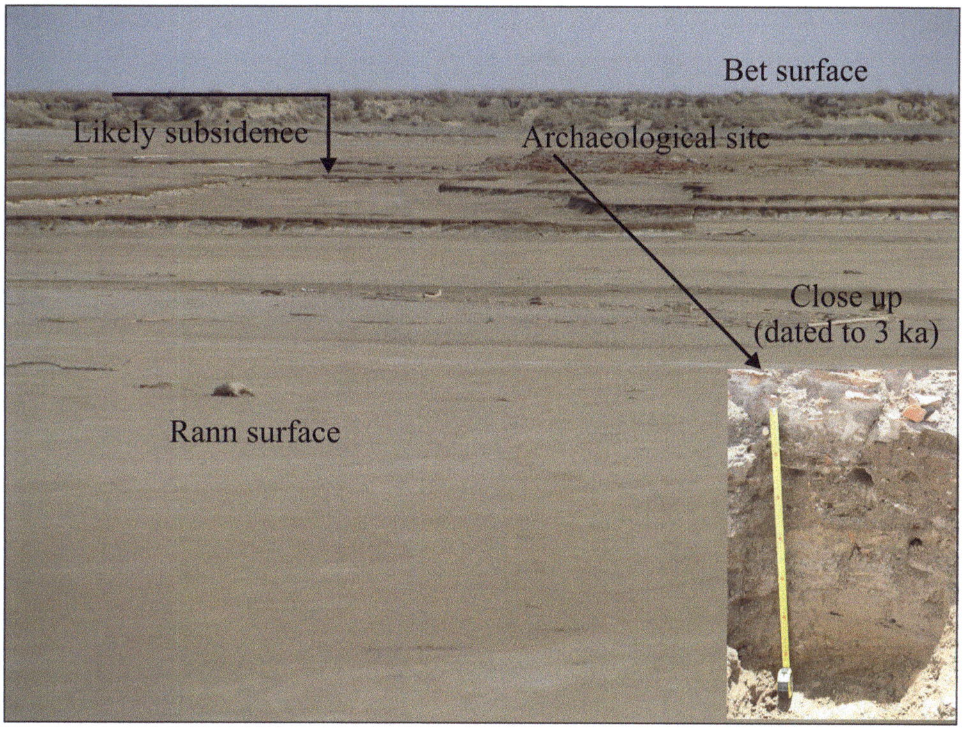

Fig. 5 The archaeological site at Karim Shahi on the tidal affected Rann surface. The *inset* shows the exposed section

(Fig. 2). It is demarcated from the western Great Rann by a narrow uplifted area around Kuar Bet which prevents the eastward migration of tidal surges. Flooding of the eastern Great Rann occurs during the monsoon by the rivers draining from the east, such as the Luni River and other smaller streams emanating from northwest and rocky island belt in the south (Figs. 1, 2).

3 Evolutionary History

Sedimentation: Historical record suggests that the Great Rann was a marine gulf when Alexander the Great visited India in the late 4th century BCE. This is further supported by recent studies, which indicate that the terrain emerged from marine transgression after 2 ka. With the recession of the sea, an estuarine condition prevailed and numerous rivers and streams from the upland region, particularly from the east and north, would have drained into the Great Rann. Glennie and Evans (1976) were of the opinion that considerable part of sediment was probably derived from the combined Indus and Nara Rivers, which extended southwards towards the present Indus Delta with some of its waters escaping into the Arabian Sea via Kori Creek (Fig. 1). There are references to suggest the existence of a mighty river, called the Saraswati, receiving water and sediment from the Himalaya and which was flowing through the eastern Nara-Hakra into the Great Rann of Kachchh (Oldham 1926).

Therefore, spatial and temporal changes in Great Rann sedimentation—an integral part of its evolutionary history—can be ascribed to a combination of fluvial and marine processes. This is reflected adequately in the sedimentary architecture and textural attributes e.g. presence of clay and silt with sandy lenticles suggesting deposition under the tidal flat low energy protected environment, whereas the occurrence of fluvial sand rich Bets suggests their contribution through fluvial regime (Roy and Merh 1982). Based on the sediment characteristics, micropaleontology and optical dating of alluvium abutting the Khadir Island, in the eastern Great Rann, Khonde et al. (2011) invoked the existence of a high sea level until around 500 years. Tyagi et al. (2012) postulated a marginally high sea level in the Great Rann between 5.5 and 2 ka. Majority of the western Great Rann was dominated by tidal flat environment and the continental contribution to the Rann sediment was limited to the northern fringes.

Seismicity: The Great Rann is tectonically active terrain and has experienced earthquakes of varying magnitude in the past. For example, the earthquakes of 893 CE and 1668 CE which were located in the north and northwest of the Great Rann were severe. The 1819 Allah Bund earthquake (magnitude 7.5) not only inflicted severe damage to the life and property of the people in Kachchh, but also brought in significant geomorphic changes. One of the major geomorphic expressions of the 1819 earthquake is the development of 4–5 m high, northward tilted and ~90 km long east-west trending linear scarp (Rajendran and Rajendran

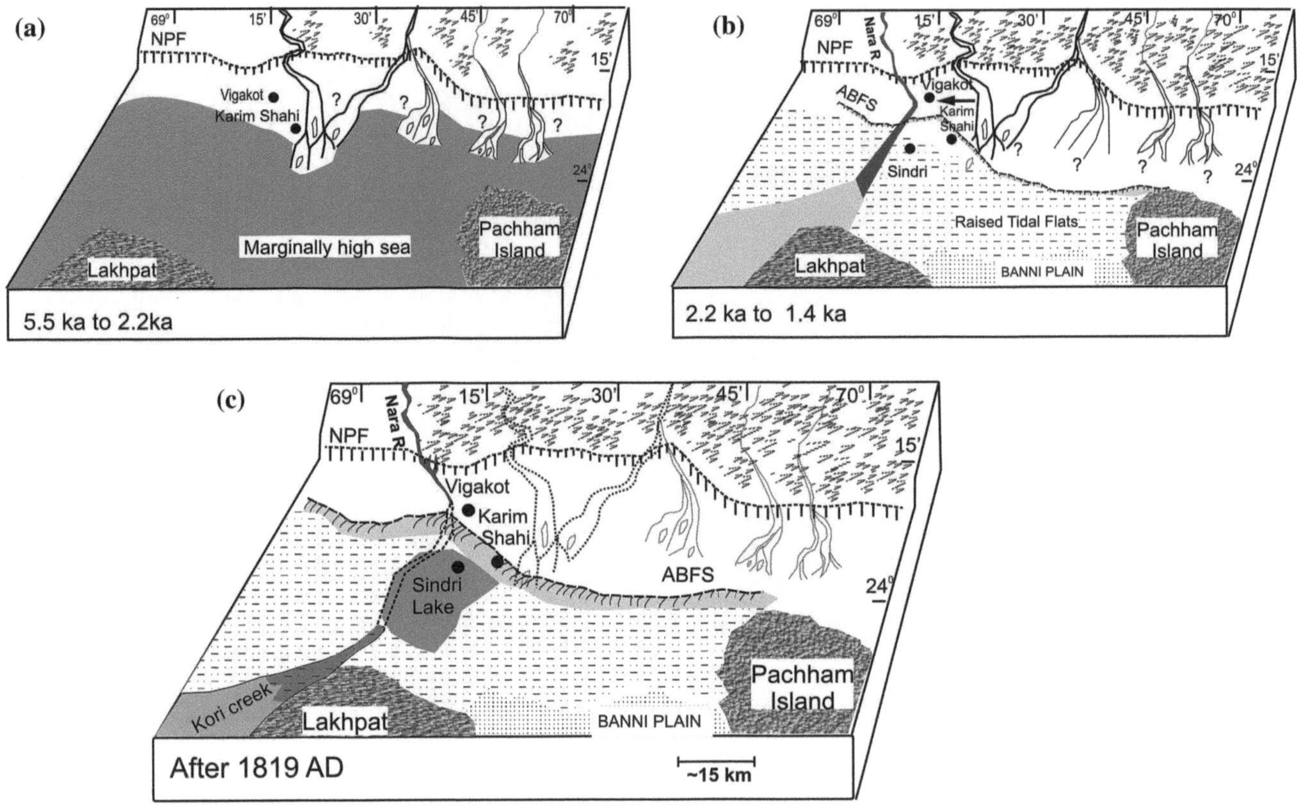

Fig. 6 Block diagrams depicting the evolutionary stages of the western Great Rann since the last 5.5 ka. **a** Tidal flat sedimentation dominated during 5.5 and 2 ka. **b** A major transformation in the landform and earth surface processes took place after 2.2 ka, a major earthquake between 2.2 and 1.4 ka brought the land surface above the intertidal influence along with the westward shift of the Nara River. Probably the surface expression of the Allah Bund Fault Scarp (ABFS) appeared during this earthquake. **c** The present land configuration was achieved after 1819 Allah Bund earthquake. Modified after Tyagi et al. (2012)

2001). The dominant northern tilt of the scarp surface is also manifested by the tilt of the laminated Rann sediments (Fig. 4). Two prominent channels can be seen on satellite imagery. The one which is trending north–south is the relict Nara River channel, whereas an oblique beheaded channel (Fig. 4 inset) seems to be an older river course that became defunct probably due to an older earthquake (prior to the 1819). High density of incipient streams (flowing north or west) in the vicinity of raised channels is found. The northward flow direction accords well with the regional tilt of the Allah Bund Scarp, however the westward flowing young streams are attributed to the obstruction caused by the structural impounding (by the Allah Bund Scarp) of Nara River from 1819 to 1826 CE (Thakkar et al. 2012 and references therein).

Based on the east–west extension of the 1819 Fault Scarp, the spatial distribution of beheaded streams, abandoned channels, defunct scroll bars, headward erosion of incipient streams and shifting of channel courses, including obstruction caused to the Nara River, the east-west and north-south extent of landform deformation would have been ~90 km and ~15 km, respectively (Rajendran and Rajendran 2001 and references therein). Thakkar et al. (2012) based on the discovery of a submerged Basta Bunder Fort (Fig. 2) between the Kori and Sir Creek, inferred that the magnitude of landform deformation was much larger compared to what so far has been estimated.

Rajendran et al. (1998) believe that the 1819 Allah Bund is a cumulative scarp formed by multiple earthquakes. The terrain seems to have been visited by similar magnitude earthquake sometimes between 800 and 1000 years ago.

Towards the areas around the Nara River channel in the vicinity of the Allah Bund Scarp, the relics of human civilization in the Great Rann were investigated to ascertain the impact of pre-1819 earthquake (Tyagi et al. 2012). To the north of the Allah Bund, the Nara River has cut ~2 m deep channel through the inter-tidal Rann sediment containing fossil gastropod shells preserved in their living position suggesting prevalence of inter-tidal environment before the Nara River occupied the present course between 2.2 and 1.4 ka.

An abrupt withdrawal of tidal flat environment to an earthquake-induced land level change (river bed rise) between 2.2 and 1.4 ka led to the incision of the tidal flat sediments (Tyagi et al. 2012). Therefore, the westward shift of the Nara River (present position) and the first surface expression of the Allah Bund would have appeared after 2.2 and before 1.4 ka earthquake, thus supporting the suggestion that Allah Bund is a cumulative scarp.

The Great Rann was once occupied by the Harappan people as evidenced by the presence of sites such as Dholavira, Vigakot and Karim Shahi (Fig. 2). The location of these sites was selected meticulously considering the proximity of the fresh water and safety from the tidal inundation. For example, Vigakot was located on the bank of Nara River, whereas a northwest-southeast trending stream flowed proximal to the Karim Shahi (Fig. 5). The deposits of this stream have been dated between 5 and 3 ka. Radiocarbon and optical dating of Karim Shahi archaeological site indicates that people inhabited the area until around 3 ka. Currently, Karim Shahi is close to the line of present-day tidal inundation which seems to be an unfavorable location. This can be possible only if the land has subsided. At this stage it is difficult to determine whether the subsidence occurred during the pre- or post-1819 earthquake.

4 Concluding Remarks

Apparently looks monotonous, the Great Rann of Kachchh contains diversified landscape which owes its genesis to a combination of land, sea and tectonic interaction during the Quaternary. Due to the inaccessibility, the knowledge of the history of landform evolution is limited to the last few thousand years. Existing data indicate that the sedimentation during the mid-Holocene was influenced by a combination of continental and marine processes. It appears that marine sedimentation dominated particularly the western Great Rann through the Kori Creek (Fig. 6a). The terrain witnessed two major earthquakes, the older earthquake (between 2.2 and 1.4 ka) was responsible for the westward shift of the Nara River (present course) along with the withdrawal of marginal high sea from the major part of the Great Rann (Fig. 6b). The 1819 Allah Bund earthquake not only caused the complete withdrawal of the sea from the region but also obstructed the Nara River course along with the creation of Sindri depression (Wynne 1872) and subsidence of Basta Bunder Fort in Kori Creek. It can be speculated that the present day landform configuration particularly in the western Great Rann was achieved after the 1819 Allah Bund earthquake (Fig. 6c). Although preliminary in nature, the evidence indicates that earthquake induced land-level changes could have been one of the major factors responsible for the present configuration of the Rann and the disruption of the human occupation in the Great Rann.

References

Biswas SK (1987) Regional tectonics framework, structure and evolution of the western marginal basins of India. Tectonophysics 135:307–327

Glennie KW, Evans G (1976) A reconnaissance of the recent sediments of the Ranns of Kutch, India. Sedimentology 23:625–647

Kar A (2011) Geomorphology of the Arid Lands of Kachchh and its importance in land resources planning. In: Bandyopadhyay S et al (eds) Landforms Processes & Environment Management. ACB Publications, Kolkota, pp 388–414

Khonde N, Maurya DM, Singh AD, Chowksey V, Chamyal LS (2011) Environmental significance of raised rann sediments along the margins of Khadir, Bhanjada and Kuar Bet Islands in Great Rann of Kachchh, Western India. Curr Sci 101:1429–1434

Oldham RD (1926) The Cutch (Kachh) earthquake of 16th June 1819 with a revision of the great earthquake of 12th June 1897. Mem Geol Surv India 46:71–147

Rajendran CP, Rajendran K (2001) Characteristics of deformation and past seismicity associated with the 1819 Kutch earthquake, Northwestern India. Bull Seismol Soc Am 91:407–426

Rajendran CP, Rajendran K, John B (1998) Surface deformation related to the 1819 Kachchh earthquake: Evidence for recurrent activity. Curr Sci 75:623–626

Roy B, Merh SS (1982) The Great Rann of Kutch: an intriguing Quaternary terrain. Recent Res Geol 29:519–539

Thakkar MG, Ngangom M, Thakker PS, Juyal N (2012) Terrain response to 1819 Allah Bund earthquake in Western Great Rann of Kachchh, Gujarat, India. Curr Sci 103:208–212

Tyagi AK, Shukla AD, Bhushan R, Thakker PS, Thakkar MG, Juyal N (2012) The mid-Holocene sedimentation and landscape evolution in the western Great Rann of Kachchh, India. Geomorphology 151–152:89–98

Wynne AB (1872) Memoir on the geology of the Kutch to accompany a map compiled by A.B. Wynne and F. Fedden during the seasons 1867–68 and 1868–69. Mem Geol Surv India IX:394

The Sambhar Lake: The Largest Saline Lake in Northwestern India

Rajiv Sinha

Abstract

The Sambhar Lake is a playa (seasonal lake with a flat bottom) located at the eastern fringe of the Thar Desert in Rajasthan. For more than four decades, this lake has attracted enormous attention due to its hypersalinity. The hypotheses to explain the geological evolution of the Sambhar have ranged from marine connection, river blockage by dunes and tectonics. This chapter presents the current understanding of the evolution of the Sambhar based on modern tools such as remote sensing and GIS and geochemical data such as evaporite mineralogy, elemental chemistry and stable isotopes. Available data suggest that the Sambhar Lake basin evolved as a pull-apart (extensional) basin and was further extended by geomorphic processes such as deflation. The hypersalinity of the Sambhar brine is related to progressive evaporation of fresh water in a semi-arid climate and the necessary chemical inputs for a hypersaline brine are derived from the catchment rocks. However, the salinity of the Sambhar Lake water has changed over time in response to variations in hydrological flux induced by climate change. Evaporite mineralogy and chemical composition of Sambhar Lake sediments have been used for reconstructing palaeoclimatic fluctuations in Thar region for the last 30,000 years.

Keywords

Playa • Thar desert • Salina • Hypersalinity • Palaeoclimate

1 Introduction

The Sambhar Lake is the largest playa in the Thar Desert located about 80 km northwest of Jaipur (Rajasthan) in northwestern India (Fig. 1a). The Sambhar is one of the important sites covered under the Ramsar Convention in 1990 due to its biological and biotic importance namely flamingoes habitat. A playa is a flat-bottom depression found in interior desert basins and adjacent to coasts within arid and semi-arid regions. Playas are periodically covered by water that slowly infiltrates into the groundwater system or evaporates into the atmosphere, causing the deposition of salt, sand, and mud along the bottom and around the edges of the depression. A saline playa may be called a salt flat, salt marsh, salada, salar, salt pan, alkali flat, or salina. A salt-free playa may be termed a clay pan, hardpan, dry lake bed, or alkali flat. In Australia and South Africa, small playas are generally referred to as pans. The terms *takyr, sabkha*, and *kavir* are used in Central Asia, Saudi Arabia, and Iran, respectively. Playas have no vegetation and are among the flattest geomorphological features in the world. The Thar Desert in NW India hosts, apart from Sambhar, several playas such as Lunkaransar, Didwana, Bap Malar, Tal Chapar, and Parihara.

R. Sinha (✉)
Department of Earth Sciences, Indian Institute of Technology, Kanpur, 208016, India
e-mail: rsinha@iitk.ac.in

V. S. Kale (ed.), *Landscapes and Landforms of India*, World Geomorphological Landscapes,
DOI: 10.1007/978-94-017-8029-2_26, © Springer Science+Business Media Dordrecht 2014

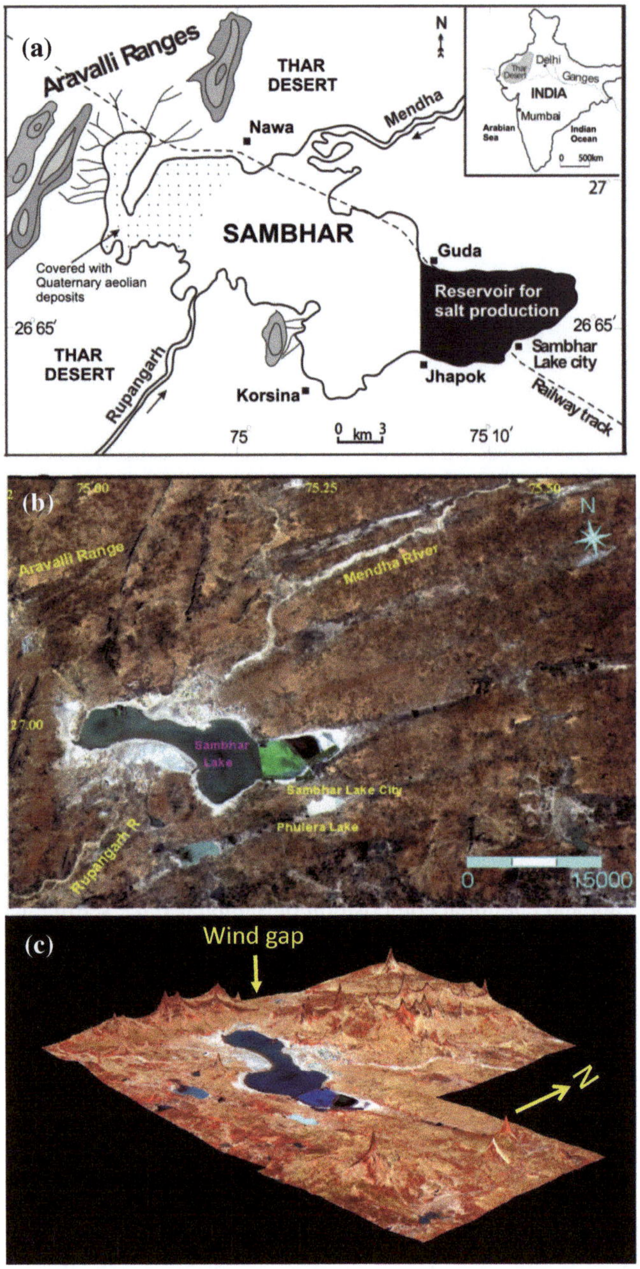

Fig. 1 **a** Geographic location of the Sambhar Lake in north-western India. **b** The Sambhar Lake as seen on a True Color Composite prepared from the IRS LISS II image of 1995. The natural part of the lake appears green on the image separated from the reservoir by a dam. The reservoir is used for storing water for salt production and the difference in color of the water arises from the difference in salinity. Two main rivers, Mendha from the north and Rupangarh from the south, feed the lake and appear as white line and dry sand/salt deposits around the lake appear as bright white patches. **c** Three-dimensional view of the Sambhar Playa and the surrounding terrain. Transverse wind gaps in the Aravalli Hill Ranges NW of Sambhar are conspicuous

The Sambhar Lake (Fig. 2a) is a classical example to understand the interaction of physical and geochemical processes in generating a distinctive landform and sedimentary fill. The abnormally high salinity (hypersalinity) of the Sambhar water has been a topic of debate for over close to a century now. Salinity of brine water from Sambhar has been reported as ~115–140 ppt of Cl, which is 3–4 times higher than that of sea water (30–50 ppt). The origin and evolution of the Sambhar Basin has been attributed to a number of tectono-geomorphic processes such as riverine connection, natural blocking of streams by dunes, and tectonic activity. The Sambhar Lake has been studied for high-resolution palaeoclimatic (past climate) reconstruction in the Thar region for the last 30 ka.

2 Geological and Geomorphological Setup

The Sambhar Lake is located southeast of the Aravalli Ranges, at an elevation of ~365 m a.s.l., comprising of folded rocks of early and middle Proterozoic age (>500 Ma) extending NE-SW for over 700 km. The range is occupied and surrounded in its immediate neighbourhood by the Pre-Cambrian rocks and forms a part of the Indian Shield. The major rocks in the Sambhar Lake catchment are mainly quartzites and mica schists with occasional amphibolite dykes. These rocks contain the primary minerals like feldspar, biotite and amphibole, which weather to clay and iron oxyhydroxides, and contribute the necessary constituents such as K^+, Na^+, Ca^{2+}, Mg^{2+} to river and lake water to eventually precipitate as salts in the playa during evaporation. As a result, the Sambhar is a major hub of salt production (Fig. 2c).

The Sambhar Lake is bordered by sand ridges formed by wind activity through the gaps in the Aravalli Range, which intervenes between the playa and the western desert (Fig. 1b, c). The playa is spread over a transitional climatic zone of semi-arid in the west and semi-humid in the east. The annual precipitation over this region ranges from 100 to 500 mm. The Sambhar Basin has centripetal drainage pattern as streams drain towards the lake. The lake is fed by two major ephemeral streams, namely the Mendha flowing NE-SW and the Rupangarh flowing SW-NE (Figs. 1b and 2b). The maximum length of the Sambhar Lake is ~22 km, whereas the width ranges from 3 to 11 km, with its main axis along the NE-SW direction. The total catchment area of the lake is ~7,560 km^2. It is a shallow lake with a maximum depth of only 3 m and an average depth of 0.61 m, The

Fig. 2 **a** Flat playa surface **b** V-shaped rejuvenated gullies in the Mendha catchment **c** Salt production in Sambhar Playa **d** A pothole and laminated mud deposit at the top of a ~20 m high cliff in the Mendha River catchment

Sambhar Lake contains water only during the monsoon months. Many other small ephemeral streams debouch from the Aravalli Hills and then disappear in aeolian sand cover.

3 Tectono-Geomorphic Evolution of Sambhar Lake

The Sambhar is a structural depression formed as a pull-apart (extensional) basin linked with strike-slip faults. Neogene-Quaternary indentation tectonics caused by westward movement of the Rajasthan craton has been cited as the primary factor for the evolution of the basin. Tectonic movements are manifested in the overall geomorphic expression of the area such as offsetting of the Mendha and Rupangarh River channels, linear stretches of the Mendha River, swarms of structurally controlled palaeochannels (former channels), ponding and formation of smaller salt lakes across disorganized channels, truncated hill fronts with fault scarps, broken ridges with rotational effects to form transverse wind gaps, etc. Most of the lineaments (linear structures indicating underlying geological structure) in the region are oriented in ENE, SE, and NNE, and the Sambhar is located at the intersection of several lineaments. After the formation of structural depression through tectonic processes (Fig. 3a), the Sambhar Basin was emphasized by large-scale deflation through the wind gap transverse to the Aravalli Ranges.

The next stage of evolution of the Sambhar Lake involved stream capture, blockage and ponding of streams (Roy 1999). On satellite images, the area surrounding the lake shows numerous old and abandoned rivers channels (palaeochannels) (Sinha et al. 2004) indicating strong fluvial activity in the geological past, in contrast to the present-day dry climate. The development of centripetal drainage due to a pre-existing depression allowed water and sediment to accumulate and a full-fledged lake developed (Fig. 3b). Drill cores raised from the Sambhar Playa show that the deeper (>10–15 m) sediments consist of carbonate nodules embedded in sand/silt and point to an alluvial nature of sedimentary fill before the lake started filling up (Sinha et al. 2006).

The present-day playa setting of the Sambhar must have evolved through gradual drying of the lake (Fig. 3c). It is generally agreed that the onset of aridity in the Thar Desert

(a)

(b)

(c)

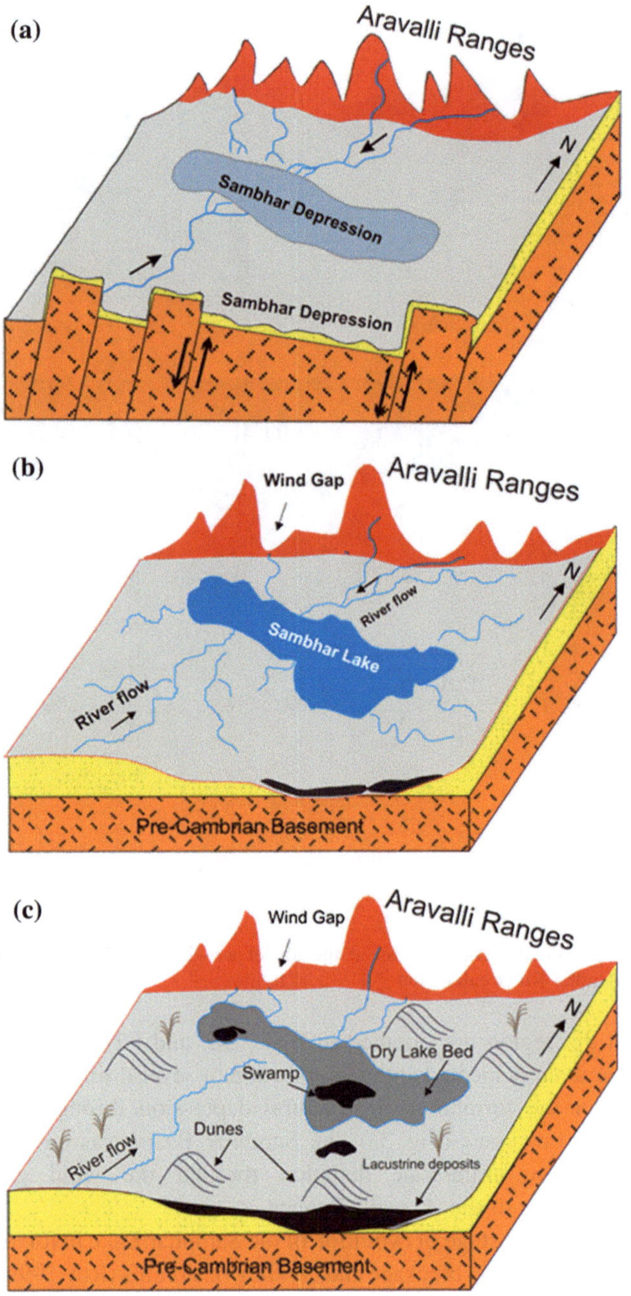

Fig. 3 Block diagram depicting tectono-geomorphologic evolution and palaeohydrological changes in the Sambhar Playa region; **a** Tectonic subsidence and formation of Sambhar depression (pre-25 ka) **b** Wind gaps, deflation and lake filling (post 25 ka) **c** Drying out phase (post 6 ka). Note that there is a huge time gap between phase 'a' and phase 'b'

started around ~5 ka triggered by weakening of monsoon (discussed later). This climatically driven event caused the lowering of the water level in most lakes and drying out of rivers in Thar. Although the Sambhar Lake now remains dry for most parts of the year, it has experienced extensive flooding as recently as in 1977–1978 and 1995. The playa presently receives only occasional flow through seasonal rainfall via ephemeral rivers and subsurface flow but does not generally attain a high stand.

4 Geochemical Evolution of the Sambhar and Palaeoclimatic History

A number of hypotheses have been proposed to explain the pronounced hypersalinity of the Sambhar viz. wind-borne source of salt from the Rann of Kachchh, an inland sea (Tethys) during the Paleogene-Neogene, and dissolution of halite bed in the lake area. However, recent work, based on isotopic analysis of lake water, and evaporite mineralogy has clearly refuted the marine origin. It has been suggested that the lake brine (solution of salts) is completely replenished by meteoric input (groundwater/precipitation) and surface runoff and hypersalinity evolves through progressive evaporation (Sinha and Raymahashay 2000, 2004).

Sediment mineralogy of the playas such as the Sambhar allows us to reconstruct lake phases (ephemeral, perennial) as well as depositional conditions (e.g. salinity, brine composition). These sediments essentially consist of rock fragments, derived from the catchment, and the evaporite fraction, formed by precipitated from aqueous solution. The evaporite fraction, including calcite, gypsum, halite, etc., is significantly influenced by lake water salinity and brine composition, and therefore, these minerals provide very important clues for reconstruction of past lake environments. Evaporites form in a predictive sequence with increasing evaporation (aridity) (Eugster and Hardie 1978). The first mineral to precipitate in most cases is calcite ($CaCO_3$). Subsequent precipitation of a mineral sequence of sulfates (e.g. gypsum, $CaSO_4 \cdot 2H_2O$), silicates (e.g. smectite) and chlorides (e.g. halite, NaCl) is controlled by relative concentrations of calcium, magnesium, bicarbonate, sulfate and chloride in the brine. It follows therefore that calcite would represent the onset of salinity and aridity, and halite would form in late stages of evaporation under high salinity conditions. Gypsum would represent an intermediate stage. The presence of minerals like thenardite, kieserite and polyhalite, carnallite and sylvite indicates hypersaline conditions.

Variety of data from a ~23 m deep core from the Sambhar have allowed a reconstruction of the past hydrological and lake conditions induced by climatic fluctuations in the Thar Desert margin for the last 30,000 years (Sinha and Raymahashay 2000, 2004; Sinha et al. 2006). Investigation of sediment samples has indicated three distinct lithological units (Fig. 4) characterized by evaporite mineralogy. The lowermost unit (I) is dominated by carbonates (mainly calcite and dolomite), the middle unit (II) is sulfate-rich (gypsum, thenardite, and polyhalite) and the uppermost unit (III) consists of chlorides as a major fraction (halite) (Sinha and Raymahashay 2004). This clearly indicates that

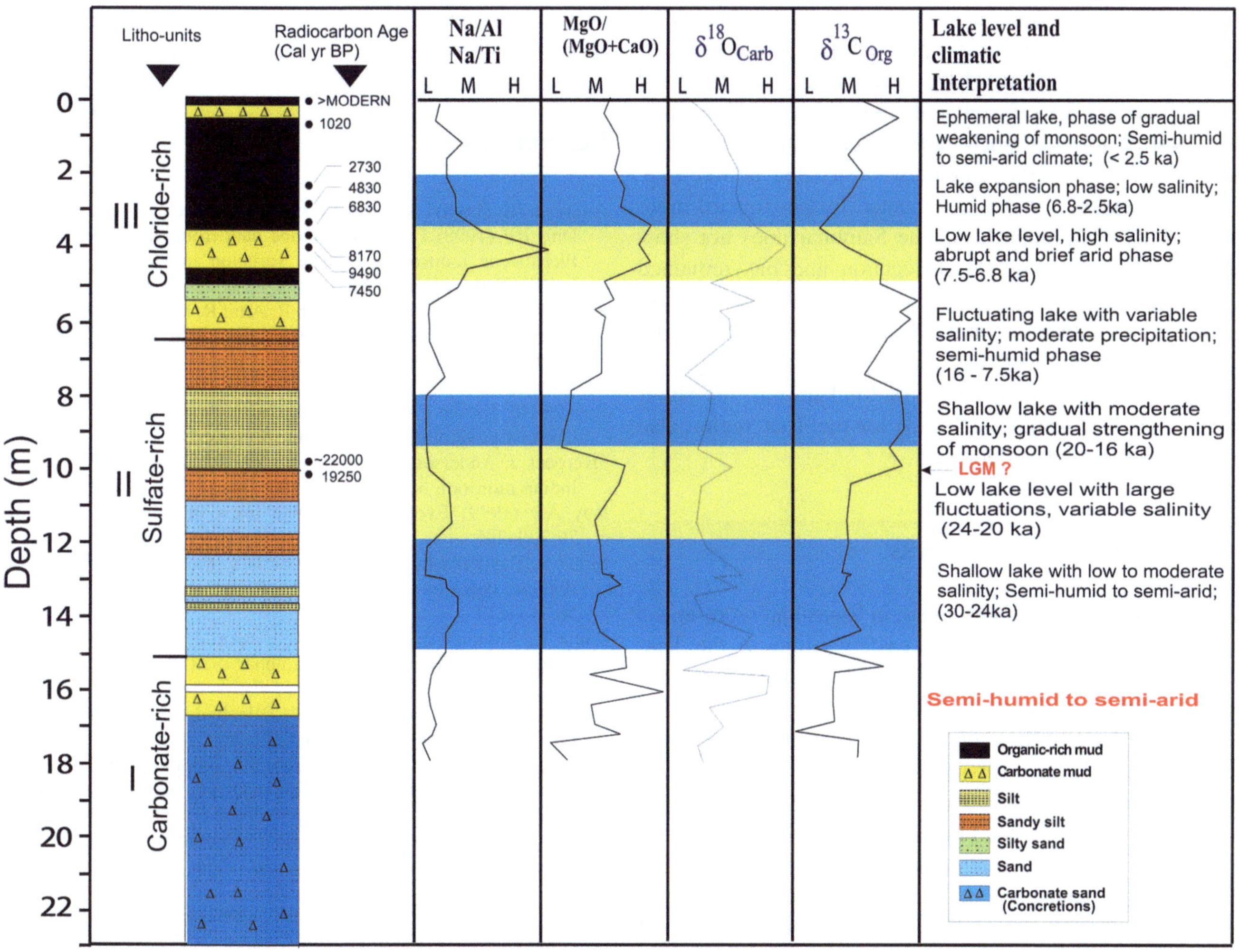

Fig. 4 Palaeoclimatic interpretations from the mineralogical, geochemical and stable isotope data from Sambhar playa core, Thar Desert (modified after Sinha et al. 2006). Three major stratigraphic units (I, II, and III) were recognized based on lithologcal and evaporate mineralogy. Several periods of humid and arid phases are recognized during the last ~30,000 years of history of the Sambhar playa. The LGM aridity is marked by lower lake level and another brief and abrupt arid phase is picked up between ~7.5 and 6.8 ka. No evidence of complete desiccation is recorded at Sambhar throughout its history. *L*, *M* and *H* indicate, low, medium and high values, respectively

salinity of the Sambhar Lake has changed through time in response to hydrological flux into the lake which is in turn governed by climatic conditions (discussed later).

Interpretation of sediment core data and radiometric dates by Sinha et al. (2006) has facilitated the palaeoclimatic reconstruction of the history of the Sambhar Playa Lake. The lower (below ~12 m depth) and older lacustrine sediments represent a shallow lake with low to moderate salinities and correspond to the time period of 30–24 ka. A relatively arid phase between ~24 and 20 ka is indicated by a noteworthy change in evaporate mineralogy and geochemistry of the sediments. This corresponds to the weakening of the monsoon during the Last Glacial Maximum (LGM). Other playas of the Thar Desert such as the Didwana also record hypersaline conditions during the same period (Wasson et al. 1984). Weakening of the monsoon and increased aridity during LGM in Thar Desert have also been denoted by marine records from the Arabian Sea (e.g. Overpeck et al. 1996) and the calcrete records from the Thar Desert (Andrews et al. 1998).

A gradual strengthening of SW Indian monsoon after 20 ka reaching a maxima around 10 ka is manifested in a shallow lake condition at Sambhar. The period ~16–7.5 ka in the Sambhar is marked by semi-humid conditions. The period of high lake level overlaps partly with the peak monsoon predictions during ~9.5–5.5 ka by most workers (Swain et al. 1983) and also with the high lake levels in Lunkaransar Playa Lake (~7.2–5.6 ka; Enzel et al. 1999) and Didwana Playa Lake (~6–4 ka; Wasson et al. 1984). Evidence for slight amelioration of climate during 6.8–2.5 ka, following a brief and abrupt arid phase (7.5 and 6.8 ka) is provided by the Sambhar core sediment data. The lake level at Sambhar was lowered again after 2.5 ka due to increase in aridity and gradual weakening of monsoon. It is important to note here that the onset of this aridity at Sambhar lags by more than 1,500–2,500 years in comparison to the Didwana ($\sim$4 ka; Wasson et al. 1984) and Lunkaransar ($\sim$5 ka; Enzel et al. 1999) playas where water level started to drop much earlier. Further, both Didwana and Lunkaransar were completely desiccated between ca. 3–4 ka. These two playas are located further toward more arid west. On the other hand, the Sambhar does not show any evidence of complete desiccation and has remained ephemeral until present. Such spatial differences are expected at Sambhar due to its location at the desert margin. Not only is the present-day mean precipitation higher at Sambhar compared to Didwana or Lunkaransar, the simulation models (Swain et al. 1983) for the Thar region also predict such spatial variation.

5 Concluding Remarks

The Sambhar Lake, a Ramsar site, in Rajasthan represents a unique landform located at the eastern fringe of the Thar Desert. This largest playa in Thar has been in existence for at least 30,000 years or more. The playa owes it origin first to tectonic activity that created a depression, followed by large-scale deflation by wind and accumulation of water through surface runoff and groundwater. Hypersalinity of the Sambhar is understood to have resulted from long-term weathering processes in the catchment rather than derived from marine sources or a halite bed underneath as believed earlier. The Sambhar has provided for a large proportion of salt production in the country and is well-known for its biological importance. The fluctuations in the lake levels during the last several millennia were linked to changes in the monsoon strength. Unlike other playas in more arid parts of the Thar Desert, the Sambhar does not show any desiccation phase throughout its geological history and still carries water seasonally.

References

Andrews JE, Singhvi AK, Kailath AJ, Kuhn R, Dennis PF, Tandon SK, Dhir RP (1998) Do stable isotope data from calcrete record late Pleistocene monsoonal climate variation in the Thar Desert of India? Quatern Res 50:240–251

Enzel Y, Ely LL, Mishra S, Ramesh R, Amit R, Lazar B, Rajaguru SN, Baker VR, Sandler A (1999) High resolution Holocene environmental changes in the Thar Desert, northwestern India. Science 284:125–128

Eugster HP, Hardie LA (1978) Saline lakes. In: Lerman A (ed) Lakes: Chemistry, Geology, Physics, Springer-Verlag, pp 237–293

Overpeck J, Anderson D, Trumbore S, Prell W (1996) The southwest Indian monsoon over the last 18000 years. Clim Dyn 12:213–225

Roy AB (1999) Evolution of saline lakes in Rajasthan. Curr Sci 76:290–295

Sinha R, Raymahashay BC (2000) Salinity model inferred from two shallow cores at Sambhar lake Rajasthan. J Geol Soc India 56:213–217

Sinha R, Raymahashay BC (2004) Evaporite mineralogy and geochemical evolution of the Sambhar Salt Lake, Thar Desert, Rajasthan, India. Sed Geol 166:59–71

Sinha R, Stüben D, Berner Z (2004) Paleohydrology of the Sambhar Lake Playa, Thar Desert, India using geomorphological and sedimentological evidences. J Geol Soc India 64:419–430

Sinha R, Smykatz-Kloss W, Stueben D, Harrison SP, Berner Z, Kramar U (2006) Late Quaternary palaeoclimatic reconstruction from the lacustrine sediments of the Sambhar playa core, Thar Desert margin, India. Palaeogeogr Palaeoclimatol Palaeoecol 233:252–270

Swain AM, Kutzbach JE, Hastenrath S (1983) Estimates of Holocene precipitation for Rajasthan, India based on pollen and lake level data. Quatern Res 19:1–17

Wasson RJ, Smith GI, Agarwal DP (1984) Late Quaternary sediments, minerals and inferred geochemical history of Didwana Lake. Palaeogeogr Palaeoclimatol Palaeoecol 46:345–372

Part IV

Geoheritage and Geotourism

Geomorphosites and Geoheritage Sites in India

Vishwas S. Kale

Abstract

India is a storehouse of fascinating and exquisite landforms and landscapes. On account of varied geology, structure, tectonic history, climatic variability and a long coastline there is immense diversity of landforms (geodiversity). In the Indian context, notwithstanding the fact that a detailed inventory of geomorphosites/geoheritage sites important from the viewpoint of geotourism is awaited, the subcontinent's outstanding geodiversity and extraordinary geoheritage is very well recognized all over the world. The Himalaya, the Indian Peninsula and the Deccan Traps have a special place in Earth's geological history. These outstanding landscapes, along with other prominent terrains and landforms have special conservation value that needs to be recognized and conserved as geoheritage to meet the present and future scientific, aesthetic, cultural, and socio-economic needs of over a billion people living in the subcontinent.

Keywords

India • Geomorphosites • Geoheritage sites • Geodiversity • World Heritage Sites • National Geological Monuments • Hill resorts • Hill forts • Waterfalls • Landscapes

1 Introduction

The Indian subcontinent is dotted with numerous ancient temples, medieval forts and hill resorts. Virtually all of these *tourist and/or pilgrimage* sites are places of unparalleled natural beauty and more often than not are associated with specific, sometimes outstanding, landforms or landform assemblages (landscapes). From historical times these landforms or landscapes have acquired great value for cultural/historical, defense, aesthetic, ecological, socio-economic and/or scientific reasons. Such sites or areas are designated as "geomorphosites" (Panizza 2001; Reynard and Panizza 2005). Further, as these natural features have been transmitted or inherited from the past, they are described as geoheritage (Brocx and Semeniuk 2007).

On account of varied geology, structure, tectonic history, climatic variability, altitudinal variations and a long coastline there is immense diversity of landforms or geodiversity. Geodiversity, as an indicator of the abiotic richness, has been defined by Gary (2004) as *the natural range (diversity) of geological (rocks, minerals, fossils), geomorphological (landform, processes) and soil features. It includes their assemblages, relationships, properties, interpretations and systems*. The Indian landmass is a storehouse of fascinating landforms and landform assemblages (landscapes). Examples of outstanding and prominent landscapes in the subcontinent include the tallest and youngest mountain ranges in the world (Himalaya), the hot and cold deserts (Thar and Ladakh, respectively), glacial landforms (e.g. Nubra Valley, Lahul-Spiti Valley), vast riverine plains (Ganga Plains), an outsized river (Brahmaputra), large volcanic province (Deccan Traps),

V. S. Kale (✉)
Department of Geography, University of Pune,
Pune, 411007, India
e-mail: vskale.unipune@gmail.com

V. S. Kale (ed.), *Landscapes and Landforms of India*, World Geomorphological Landscapes,
DOI: 10.1007/978-94-017-8029-2_27,

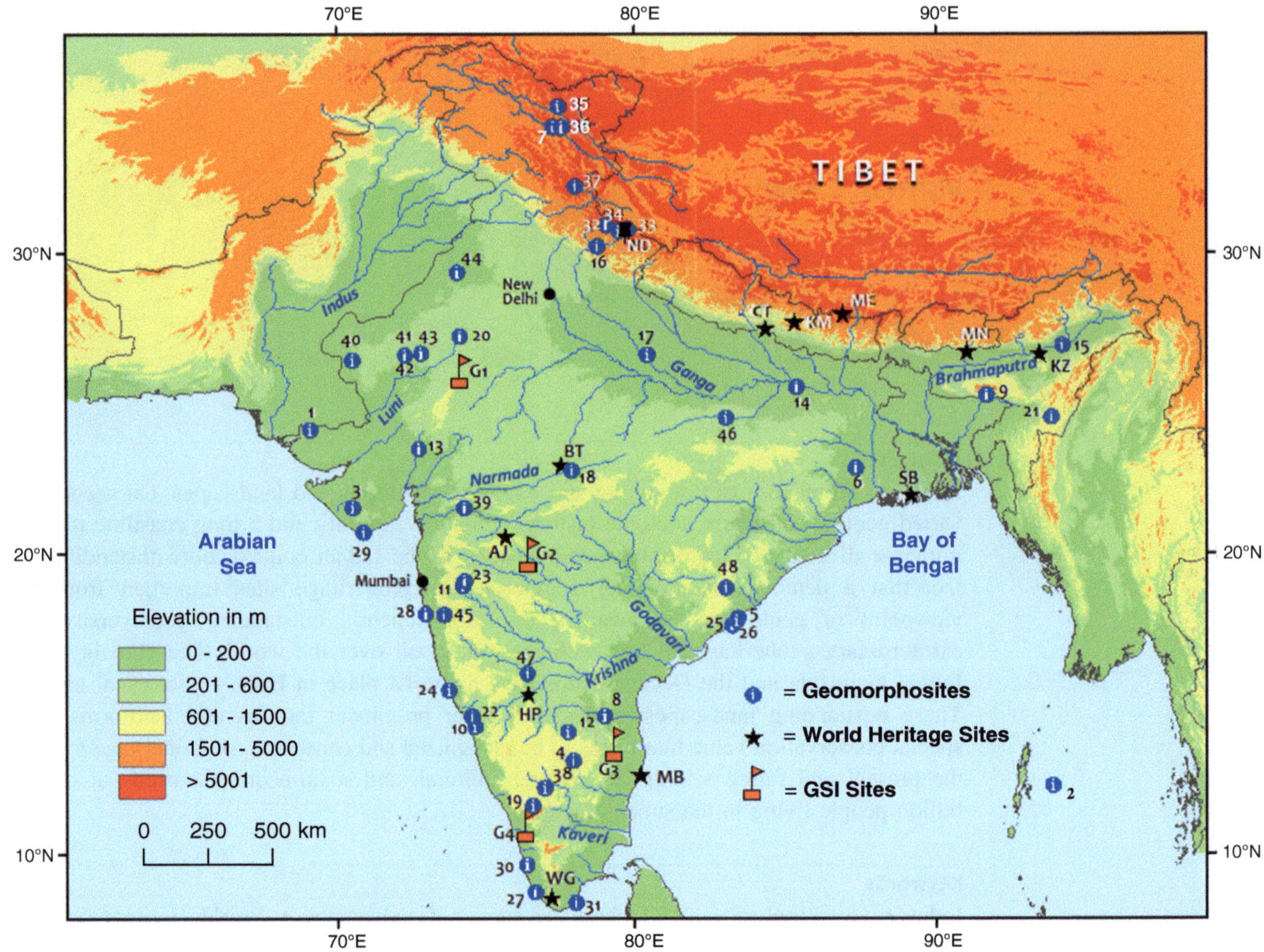

Fig. 1 Map showing the locations of potential geomorphosites in India. The geosites identified by GSI (as National Geological Monuments) and the World Heritage Sites in India and Nepal and described in the text are also shown. Numbers correspond with geomorphosite numbers given in Table 3. UNESCO World Heritage Sites in India and Nepal—*ND* Nanda Devi and Valley of Flowers National Parks, *CT* Chitwan National Park, *KM* Kathmandu Valley, *ME* Sagarmatha National Park (Mt. Everest), *MN* Manas Wildlife Sanctuary, *KZ* Kaziranga National Park, *SB* Sundarbans National Park, *WG* Western Ghat (*Sahyadri*), *BT* Bhimbetka, *AJ* Ajanta Caves, *MB* Mahabalipuram and HP Hampi. GSI's National Geological Monuments—*G1* Sendra, *G2* Lonar Lake, *G3* Silathoranam and *G4* Angadipuram

the Great Escarpment (Western Ghat), picturesque lakes and lagoons (e.g. Dal, Nainital, Loktak, Chilka, Pulicat, Vembanad), a large saline marshland (Rann of Kachchh), and widespread granite landforms (e.g. Hampi, in Karnataka). In addition, there are also coastal, duricrusted, and karstic landforms. This makes India one of the most fascinating places for studying all types of landforms and associated processes, except features of continental glaciation.

2 Current Listings of Geoheritage Sites in India

Listings of the noteworthy Indian heritage sites have been prepared over the last few decades by the UNESCO and the Geological Survey of India (GSI).

2.1 UNESCO World Heritage Sites

Currently (January 2014), there are 30 properties from India inscribed on the UNESCO World Heritage List. Of these, 24 are cultural properties and 6 are natural properties. Amongst the natural properties, at least four properties are also geoheritage sites (Fig. 1), because they are associated with distinctive landforms or landscapes. These are (DI = Date of Inscription):

- Nanda Devi and Valley of Flowers National Parks (DI = 1988)
- Manas Wildlife Sanctuary (DI = 1985)
- Sundarbans National Park (DI = 1987)
- Western Ghat (DI = 2012).

The Himalaya Mountains have some incredible landforms. Nanda Devi, the 2nd highest peak in India (8,590 m a.s.l. Kanchenjunga is the highest peak), along with Sagarmatha (or Mt. Everest) National Park in Nepal Himalaya, have been classified as areas of exceptional natural beauty and aesthetic importance. Chitwan (Nepal), having untouched vestiges of the 'Terai' region, a moist zone of marsh and forest at the foot of the Himalaya, is another geoheritage site from the subcontinent listed in the World Heritage List. The Manas Wildlife Sanctuary in northeast India has been categorized as an outstanding example of a significant geomorphic or physiographic feature. The Manas River is the largest Himalayan tributary of the Brahmaputra and delivers huge amount of sediment load to the river. Like the Kosi Basin, the mountainous area of Manas Basin is several times larger than the plains area. The river forms a fan in the narrow space between the Himalaya and the Brahmaputra River. Further, the Kaziranga National Park World Heritage Site (ID = 1985) in northeastern India is a vast grassy wetland on the floodplain of the Brahmaputra River at the foot of the Mikir Hills, habited by the one horned rhinoceros. The Sundarbans National Park (India and Bangladesh) are characterized by an intricate network of tidal channels, mud flats and small islands covered by salt tolerant mangrove forests (Chapter "The Sundarbans and Bengal Delta: the World's Largest Tidal Mangrove and Delta System", this monograph). The 1,500-km long Western Ghat (*Sahyadri*) is one of the most spectacular great escarpments in the world (Kale 2010) and is one of the world's major biodiversity hotspots. Although the Great Himalayan National Park (GHNP) in Kullu District of Himachal Pradesh was not inscribed on the World Heritage List in 2013, it is likely to be inscribed in the coming years.

Four other cultural World Heritage Sites are also geoheritage sites because they are associated with distinctive landforms or landscapes. These are (Fig. 1)

- Ajanta Caves (DI = 1983) dated from 2nd and 1st centuries CE—sinuous, narrow gorge of the Waghora/Waghur River cut in Deccan Traps (Ajanta Range) (Fig. 2)
- Monuments at Mahabalipuram (DI = 1984)—charnockite outcrops on sandy coast. A shore temple complex, including rock-cut chariots, sculptured scenes on open rocks and caves (7th and 8th centuries) (Fig. 3).
- Temples and palaces (14–16th centuries) at Hampi (DI = 1986)—granitic area surrounded by boulder-strewn hills (Fig. 4).
- Rock Shelters of Bhimbetka (DI = 2003)—area with interesting forms in sandstones. Huge overhangs and alcoves in massive sandstones in the foothills of the Vindhyan Ranges, displaying rock paintings that date from the Mesolithic to historical period.

2.2 National Geological Monuments: GSI Geosites

The Geological Survey of India (GSI) has enlisted 26 geosites and has declared them as National Geological Monuments (Anantharamu et al. 2001). Whereas majority of these sites are geologically significant sites, there are at least four sites that fall in the category of geomorphosites/geoheritage sites. These are (Fig. 1):

- Sculpted forms in granite at Sendra, Pali District of Rajasthan
- Silathoranam (meaning rocky arch) Natural arch in quartzite of Cuddapah Supergroup in Tirumala Hills, Chittoor District, Andhra Pradesh
- Laterite in Angadipuram, Malappuram District, Kerala
- Lonar Lake, a meteoric impact crater in Buldhana District, Maharashtra ("The Lonar Crater- The Best Preserved Impact Crater in the Basaltic Terrain", this monograph)

In January 2013, a travel brochure on National Geological Monuments of India: Region South was released by Geological Survey of India (GSI 2012). The brochure, basically meant for tourists, gives description and location map of about a dozen interesting geosites in southern India. Apart from Silathoranam and Angadipuram laterite geosites, the Varkala Cliffs (geopark in Kerala) and Karai Badlands (Perambalur, Tamil Nadu) are also included in the travel brochure.

Many of the ancient monuments and archaeological sites under the protection and care of the Archaeological Survey of India (ASI) are oftentimes associated with remarkable landforms or landscapes. These sites include historical forts and ancient temples on hills, rock-cut caves and ruins of monuments at scenic locations.

3 Geotourism in India

The GSI website contains a section on 'geotourism', which includes photographs of interesting geological features (geophotos) and information about the National Geological Monuments in India. The photographs of glacial striations and erratics, talus cones, and colourful landscape of Lahul and Spiti Valley are most fascinating.

Geotourism, that is, tourism related to geomorphological sites and landscapes, is not a new phenomenon in India. Many well-known tourist destinations are, in fact, geoheritage sites. A good number of hill stations (hill resorts), first identified and developed by the British during the colonial period, are nestled at high elevations and provide spectacular views of the surrounding landscapes. Every year thousands of tourists throng to these hill stations (Table 1) to escape the scorching heat and dust during hot Indian summer for leisure and/or adventure.

Fig. 2 The Ajanta Caves overlooking the narrow, sinuous gorge of the Waghur (Waghora) River. The World Heritage Site, with nearly 30 caves, is located about 110 km north of Aurangabad. The caves are excavated in Deccan Traps basalts, exposed along the steep wall (~80 m high) of the horseshoe-shaped bend of the Waghur River gorge. The caves were excavated from ~2nd century BCE to 6th century CE. The caves are famous for their sculptures as well as paintings/frescoes. (Photo by Francesca Lugeri)

Further, within the Indian subcontinent in general and the Indian Peninsula in particular, numerous historical forts were built on top of precipitous hills and escarpments for strategic reasons. Leading examples include Kangra Fort in Himachal Pradesh, Jhansi Fort in Uttar Pradesh, Hill Fort of Mandu in Madhya Pradesh, Raigad Fort in Maharashtra, Golkonda Fort in Andhra Pradesh, Bellary Fort in Karnataka, Rajagiri and Krishnagiri Forts, Gingee (Fig. 4, in chapter "Granite Landforms of the Indian Cratons" this monograph) in Tamil Nadu, etc. In 2013, six such hill forts of Rajasthan have been included in the UNESCO's World Heritage List.

Likewise, innumerable temples or shrines have been built from ancient times atop hills and mountains or in caves throughout the subcontinent. Examples include Amarnath Cave shrine in Higher Himalaya, Vaishnodevi Shrine and Mansa Devi temple on Siwalik Hills, Dilwara temple at Mount Abu, the Great Stupa at Sanchi, the great stone statue of Gomateshwara Bahubali in Sravanabelagola in Karnataka, and Old Mahabaleshwar temple (source of Krishna River) and Talkaveri temple (source of Kaveri River) in the Western Ghat. Many more examples can be found in almost every region. Thousands of pilgrims and devotees have been visiting these picturesque geosites for centuries.

Among the most visible and outstanding elements of the landscape, that have attracted locals as well as outsiders alike from times immemorial, are the waterfalls. In the Indian subcontinent there is no dearth of cataracts and cascades. Although there are virtually countless waterfalls in the subcontinent, which acquire their most captivating form during the monsoon season, there are a few that have attained recognition as national landmarks due to unparalleled scenic beauty. Table 2 gives the list of some of the most popular and tourist-thronged waterfalls in India. All these waterfalls are geomorphosites because of their great aesthetic, cultural, economic (touristic) and scientific value.

Fig. 3 Historical monuments at Mahabalipuram (also Mamallapuram). The World Heritage Site is located about 65 km south of Chennai. **a** The seventh century shore temple overlooking the sea, **b** monolithic temples known as Five Rathas (chariots) carved from single chunks of exposed rocks, and **c** Athiranachanda Cave in the Tiger's Cave complex area, near Mahabalipuram. The caves and temples are cut in outcrops of Archean charnockites (~2,500 Ma) on the sandy shore. There are reports that the December 2004 tsunami uncovered several ancient structures buried under the beach sand. (Photos by Vishwas S. Kale)

4 Status of Geomorphosite Research

Geomorphosite research in India is still in its infancy. Notwithstanding the fact that a systematic and detailed inventory of geomorphosites and geoheritage sites important from the viewpoint of geotourism is awaited, a few efforts have already been made in this direction in the last few years. The first noteworthy attempt to record the geomorphosites/geoheritage sites in south India was made by the Karnataka Geologists' Association (KGA), Bengaluru. The KGA has published an impressive monograph on "Geomorphology of Karnataka" (Ranganathan and Jayaram 2006). The volume contains over hundred colour photographs of exquisite landforms ranging from caves and stalagmites to inselbergs, bornhardts, nubbins, colourful badlands, waterfalls, beaches, caves and high-level erosional surfaces in the state of Karnataka. The pictures of three natural bridges, namely the Bheemana Kindi Arch in granite near Halagur, the Sidlapadi Arch in sandstone in Badami Taluka, and an arch in laterite near Talaguppa-Sirsi Cross are most interesting.

An inventory of glaciers in Indian Himalaya was published by Raina and Srivastava (2008). This "Glacier Atlas of India" includes maps, photographs and description of the geomorphological features from the entire length and breadth of Indian Himalaya. A collection of photographs of the landforms of the Himachal Himalaya was published by Bhargava et al. (2010). The book contains a good number of photographs with some description of lithology and structure controlled landforms, glacial and fluvial landforms, glaciers, lakes and other features in this part of Himalaya. The fourth noteworthy attempt was made by Parthasarathy et al. (2012) to prepare a photographic inventory of local landforms in Vaigai-Vaippar Basins in Tamil Nadu.

In the last week of May 2013, a 2-day national workshop on 'Geoheritage—need for an Indian activism' was jointly organized by the Department of Geology, Andhra University, the GSI, and the Indian National Trust for Art and Cultural Heritage (INTACH) at Visakhapattnam, to identify and preserve unique geological sites. An abstract volume was brought out on this occasion that gives information about numerous geological sites and some geomorphosites in southern, northern, northeastern, central, western and eastern India (Reddy 2013).

Even though a few attempts have been made in the last few years to record the geomorphosites/geoheritage sites in some parts of the subcontinent (Singh and Anand 2013), there is still

Fig. 4 The ornate stone chariot at Hampi. The historical monuments at Hampi are situated on the southern bank of the Tungabhadra River, a tributary of the Krishna River. Once the seat of the mighty Vijayanagar Empire, the Vijayanagar city was built between ~1336 and 1570 CE. The undulating rocky terrain around the World Heritage Site is characterized by numerous block- and boulder-strewn granite nubbins and koppies. Two such boulder-strewn hills are seen in the background. (Photo by H. R. Mungekar)

Table 1 Popular hill resorts in India that are also geomorphosites

Geomorphic Province	Hill stations/resorts	Geographic setting
Peninsular India (PI)	Mount Abu	Aravalli Hills
	Saputara, Khandala, Lonavala, Mahabaleshwar, Matheran, Panchgani, Panhala, Coorg	Western Ghat
	Coonoor, Ooty, Kodaikanal, Munnar, Peermade	Nilgiri–Palni Hills
	Mainpat, Chirmiri, Netarhat, Hazaribagh	Chhotanagpur Plateau
	Pachmarhi	Satpura Ranges
	Ananthagiri, Araku Valley	Eastern Ghat
	Shillong	Meghalaya Plateau
Himalaya Mountains (HM)	Almora, Dalhousie, Mussoorie, Nainital, Bhimtal, Chamba, Gangotri, Khajjiar, Kullu-Manali, Mussoorie, Ranikhet, Shimla, Kashmir Valley	Lesser Himalaya (west)
	Badrinath	Higher Himalaya (west)
	Darjeeling, Gangtok, Kalimpong, Tawang	Lesser Himalaya (east)
		Higher Himalaya (east)

Table 2 Famous waterfalls as geomorphosites in India

Name of the waterfall	Latitude and longitude	River	State	Special geomorphic characteristics
Chitrakoot Falls (~30 m)	19.207138°N 81.700081°E	Indravati	Chhattisgarh	Horse-shoe shaped falls
Dhuandhar Falls (~10 m)	23.125259°N 79.813235°E	Narmada	Madhya Pradesh	At the head of Marble Canyon
Dudhsagar Falls (~310 m)	15.315726°N 74.314501°E	Dudhsagar	Goa	Multi-tiered waterfall, over the Western Ghat Escarpment, near Sonaulim
Gokak Falls (~53 m)	16.192185°N 74.777526°E	Ghataprabha	Karnataka	Gorge downstream
Jog Falls (~253 m)	14.246865°N 74.672869°E	Sharavathi	Karnataka	Gorge downstream (Fig. 8)
Nohkalikai Falls (~198 m)	25.275409°N 91.686292°E	Pynjngithuli	Meghalaya	Near Cherrapunji. Over the high escarpment of Meghalaya Plateau (Fig. 7)
Shivanasamudra Falls (~91 m)	12.294196°N 77.169167°E	Kaveri	Karnataka	At the head of Kaveri Canyon. Hogenakal Falls located at the downstream end of the Kaveri Canyon

lack of serious research dealing with the definition, mapping, assessment and conservation of geomorphosites and geoheritage sites (Reynard and Panizza 2005) in the subcontinent.

5 Potential Geomorphosites and Geoheritage Sites in India

The term 'geoheritage' is applied to sites or areas of geologic features, landforms, and culturally significant natural features and landscapes. Several international bodies are involved in this activity, such as the UNESCO's Division of Ecological and Earth Sciences, the International Union for Conservation of Nature (IUCN), the International Union of Geological Sciences (IUGS), and the International Association of Geomorphologists (IAG).

The 18 landforms and landscapes described in some detail in the Part III (Chapters "The Siachen Glacier: The Second Longest Glacier Outside the Polar Regions", "Ladakh: The High-Altitude Indian Cold Desert", "The Vale of Kashmir: Landform Evolution and Processes", "Duns: Intermontane Basins in the Himalayan Frontal Zone", "The Chambal Badlands", "The Kosi Megafan: The Best-Known Himalayan Megafan", "The Sikkim-Darjeeling Himalaya: Landforms, Evolutionary History and Present-Day Processes", "The Brahmaputra River in Assam: The Outsized Braided Himalayan River", "The Meghalaya Plateau: Landscapes in the Abode of the Clouds", "The Sundarbans and Bengal Delta: the World's Largest Tidal Mangrove and Delta System", "The Spectacular Belum and Borra Caves of Eastern India", "Granite Landforms of the Indian Cratons", "The Andaman Archipelago", "Teri Red Sands, Tamil Nadu", "The Laterite-Capped Panchgani Tableland, Deccan Traps", "The Lonar Crater: The Best Preserved Impact Crater in the Basaltic Terrain", "The Great Rann of Kachchh: The Largest Saline Marshland in India", and "The Sambhar Lake: The Largest Saline Lake in Northwestern India") of this monograph are so prominent and outstanding in the Indian context that they deserve inclusion in the national geoheritage list, as national landmarks. These sites/areas have international level of significance and appeal. Apart from these, there are countless sites or areas of scientific, cultural and socio-economic significance that could be designated as geomorphosites.

The IUCN, the advisory body on natural heritage to the World Heritage Committee, has recommended thirteen geothematic areas for identifying geological World Heritage. The relevant themes (geomorphic) could be used as guidelines to identify geomorphosites/geoheritage sites in India. Accordingly, a modest endeavor has been made to list some of the potential geoheritage sites within the Indian region (Table 3, Fig. 1) that are geological and geomorphological elements of nature, worthy of being conserved. The list is certainly not an exhaustive one but just a representative one. There are undoubtedly many more fascinating and exquisite places of natural wonders that deserve the geoheritage status and merit preservation for future generations.

Table 3 Potential geomorphosites/geoheritage sites in India. See Fig. 1 for location

Geo No.	IUCN theme	Geothemes	Geomorphic features/landforms	Latitude (N)	Longitude (E)	Locality	Name of the State/Union Territory	Reported/ suggested by
1	1	Fault scarp	Allah-Bund fault scarp (1.6 m) developed in Rann of Kachchh sediments following 1819 earthquake	24.125280°	69.117468°	80–100 km northwest of Khavda, near Bhuj	Gujarat	Navin Juyal
2	2	Volcano	Only active volcano in India	12.279181°	93.860480°	Barren Island ~150 km NE of Port Blair	Andaman and Nicobar Islands	Jyotiranjan S. Ray
3	2	Volcano	Volcanic Plug (1,117 m high). The dome-shaped circular hill complex has radial drainage	21.524762°	70.531848°	Mount Girnar, near Junagadh, Kathiawad (Fig. 5)	Gujarat	L.S. Chamyal Nilesh P. Bhatt
4	6	Fluvial	Badland topography, gullies within deeply-lateritized and saprolitic rocks	13.113343° 13.248128°	77.966467° 77.929512°	near H-cross, Kolar District and near Nandagudi	Karnataka	Ranganathan and Jayaram (2006)
5	6	Fluvial	Badland topography, gullies within coastal red sand dunes	17.873587°	83.428292°	Bheemunipatnam, Erra Matti, Dibbalu at Visakhapatnam	Andhra Pradesh	R. Vaidyanadhan
6	6	Fluvial	Badland topography, gullies within alluvium. Site rich in ichnofossil and petrified woods	22.860892°	87.346958°	Garhbeta, West Medinipur district (Fig. 6)	West Bengal	Sunando Bandyopadhyay
7	6	Fluvial	River terraces, and terraced alluvial fan	34.191627°	77.312319°	Basgo-Nimmoo, west of Leh, at the confluence of Zanskar and Indus Rivers	Jammu and Kashmir	Navin Juyal
8	6	Fluvial	Alluvial fan; near-symmetric fan in Peninsula (7 km long and 10 km wide)	14.605922°	79.015831°	10 km SW of Badvel (Kale 2009)	Andhra Pradesh	Vishwas S. Kale
9	6	Fluvial	Waterfall	25.275409°	91.686292°	Nohkalikai (Fig. 7)	Meghalaya	Vishwas S. Kale
10	6	Fluvial	Waterfall, gorge and river capture	14.246865°	74.672869°	Gersoppa/Jog, Sharavathi River (Fig. 8)	Karnataka	Vishwas S. Kale
11	6	Fluvial	Multiple potholes and spectacular erosional forms in basalt	18.931999°	74.262799°	Kukdi River, Nighoj (Fig. 9)	Maharashtra	Vishwas S. Kale
12	6	Fluvial	Knick zone on a tributary of the Pennar, with a ruined temple on a bedrock island in the falls	14.064124°	77.795802°	Peddapalli, about 15 km southwest of Puttaparthi	Andhra Pradesh	Yanni Gunnell
13	6	Fluvial	High alluvial bank and bank gullies	23.493054°	72.796393°	Mahudi, Sabarmati River	Gujarat	Navin Juyal
14	6	Fluvial	Huge channel bar	25.544785°	85.384651°	Patna, Ganga River	Bihar	Robert Wasson
15	6	Fluvial	Largest channel Island	26.950551°	94.169788°	Majuli Island, Brahmaputra River	Assam	J. N. Sarma
16	6	Fluvial	Six river terraces	30.220810°	78.779045°	Srinagar, Alaknanda Valley (Fig. 10)	Uttarakhand	Navin Juyal
17	6	Fluvial	Large abandoned meander	26.627592°	80.429047°	Ganga River, Bithur near Kanpur (Kale 2009)	Uttar Pradesh	Vishwas S. Kale
18	6	Fluvial	Large abandoned meander	22.766240°	77.925436°	Narmada River near Hoshangabad (Kale 2009)	Madhya Pradesh	Vishwas S. Kale
19	6	Fluvial	Deep linear gorge ~20 km long that follows a shear zone. Incised meander (tributary)	11.610623°	76.621666°	Moyar, Mudumalai National Park	Tamil Nadu	Vishwas S. Kale
20	6	Fluvial	Gravel ridge in Thar Desert	27.240891°	74.188351°	~50 km undulating ridge of coarse gravel between Ladnun and Jayal	Rajasthan	S. N. Rajaguru

(continued)

Table 3 (continued)

Geo No.	IUCN theme	Geothemes	Geomorphic features/landforms	Latitude (N)	Longitude (E)	Locality	Name of the State/Union Territory	Reported/ suggested by
21	6	Lacustrine	Lake in an intermontane valley. A Ramsar site	24.547222°	93.808119°	Loktak Lake, Imphal	Manipur	Vishwas S. Kale
22	7	Karst	Tower karst	14.590079°	74.567281°	Yana, about 50 km northeast of Honavar	Karnataka	Yanni Gunnell Ranganathan and Jayaram (2006)
23	7	Karst	Calc tufa of early Holocene in Deccan Traps Region	19.092953°	74.336504°	Wadgaon Darya near Kanhur	Maharashtra	Vishwas S. Kale
24	8	Coastal	Wave-cut platform in laterite	15.451438°	73.802850°	Dona Paula, Panaji	Goa	Vishwas S. Kale
25	8	Coastal	Wave-cut platform and sea cliffs in khondalites	17.613417°	83.238100°	Gangavaram, SW of Visakhapatnam	Andhra Pradesh	R. Vaidyanadhan
26	8	Coastal	Perfectly crescent-shaped beach	17.784016°	83.385453°	Rushikonda near Visakhapatnam	Andhra Pradesh	R. Vaidyanadhan
27	8	Coastal	Cliff in Mio-Pliocene sediments overlooking a beach	8.736025°	76.703137°	Varkala Beach, about 40 km NW of Thiruvananthapuram	Kerala	Vishwas S. Kale
28	8	Coastal	Wave-cut platform in basalt and honeycomb weathering on cliff face	17.990811°	73.019257°	Harihareshwar near Shrivardhan	Maharashtra	Vishwas S. Kale
29	8	Coastal	Modern and raised cliffs and tidal notches in miliolite/aeolinites	20.7018883°	70.892908	Diu Island, Saurashtra (Fig. 11)	Diu	Navin Juyal Nilesh P. Bhatt
30	8	Coastal	Vembanad Lagoon/ backwaters. A Ramsar site	9.640861°	76.425512°	Kumarakom near Kottayam	Kerala	A. C. Narayana
31	8	Coastal	Cliffs of calcretized aeolianite, sandy spits, high coastal dune systems. Teri sands in the hinterland	8.367653°	78.063891°	Manapad, about 3 km south of Kulasekharapatnam	Tamil Nadu	Yanni Gunnell
32	10	Glacial	Terminal moraines	30.744822°	79.492910°	Badrinath, Chamoli District	Uttarakhand	Robert Wasson
33	10	Glacial	Moraines and remnants of a proglacial lake	30.800989°	79.804042°	Goting Basin in Himalaya (Fig. 12)	Uttarakhand	Navin Juyal
34	10	Glacial	Moraines of the last glacial cycle	30.936627°	79.070567°	Gangotri Glacier and Bhagirathi River	Uttarakhand	Lewis A. Owen
35	10	Glacial	Moraines, glacial valley, hanging valleys, alluvial fans	34.887845°	77.489082°	Nubra Valley	Jammu and Kashmir	Mahendra Bhutiyani
36	10	Glacial	Crescent-shaped terminal moraines	34.187847°	77.596396°	Khardung Glacier, Ganglas near Leh (Fig. 13)	Jammu and Kashmir	Pradeep Srivastava, Lewis A. Owen
37	12	Glacial	Desert mountain valley. Glacio-fluvial features, terraces, talus cones, etc.	32.223954°	78.056591°	Lahul and Spiti Valley	Himachal Pradesh	Navin Juyal
38	12	Aeolian	Riverine sand dunes	12.191308°	77.025464°	Talakad on left bank of Kaveri River	Karnataka	S. V. Srikantia
39	12	Aeolian	Riverine sand dunes	21.518019°	74.321444°	Prakasha on left bank of Tapi River	Gujarat	Vishwas S. Kale
40	12	Aeolian	Linear dunefield	26.449902°	70.5338302°	Bersiyala; ~63 km SW of Jaisalmer, on Myajlar road	Rajasthan	Amal Kar
41	12	Aeolian	Transverse dunefield	28.580962°	72.401738°	~11 km W of Pugal; 95 km NW of Bikaner	Rajasthan	Amal Kar
42	12	Aeolian	Compound parabolic dunes	26.675254°	72.885467°	~7 km S of Osian; 70 km NW of Jodhpur	Rajasthan	Amal Kar

(continued)

Table 3 (continued)

Geo No.	IUCN theme	Geothemes	Geomorphic features/landforms	Latitude (N)	Longitude (E)	Locality	Name of the State/Union Territory	Reported/ suggested by
43	12	Aeolian	Compound parabolic dunes under stabilization program	26.600116°	72.319591°	Shaitrawa Crossing; 20 km S of Dechu; ~105 km NW of Jodhpur on Jaisalmer road	Rajasthan	Amal Kar
44	12	Aeolian	Linked star dunefield	29.352961°	74.101303°	2 km E of Baropal; 20 km E of Suratgarh on Rawatsar road	Rajasthan	Amal Kar
45	14	Landform	Escarpment	17.977730°	73.637669°	Western Ghat Scarp, Arthur's Seat, Mahabaleshwar (Fig. 14)	Maharashtra	Vishwas S. Kale
46	14	Landform	Escarpment	24.518235°	83.023344°	Kaimur Scarp, from Chopan village in Sonbhadra District	Uttar Pradesh	K.N. Prudhvi Raju
47	14	Landforms	Boulder inselbergs in granite	16.010921°	76.440499°	Mudgal town, Lingasugur Taluka, Raichur District	Karnataka	Ranganathan and Jayaram (2006)
48	14	Landforms	Duricrusted, Laterite tableland/plateau	18.861428°	83.024107°	Panchpatmali	Odisha	R. Vaidyanadhan

GeoNo. Geomorphosite number. IUCN Themes—*1* Tectonic and structural features, *2* Volcanoes/volcanic system, *6* Fluvial, lacustrine and deltaic systems, *7* Caves and karst system, *8* Coastal systems, *10* Glaciers and ice caps, *12* Arid and semi-arid systems, *14* Other landforms not directly listed under IUCNs themes

Fig. 5 Mount Girnar (1,117 m a.s.l.), a volcanic plug. View of the imposing hill from the base showing Ambaji Peak. The dome-shaped circular Girnar Hill Complex forms a prominent geomorphic feature of the Deccan Traps Region in Saurashtra that rises abruptly from the Kathiawad Plains. There are several ancient Jain and Hindu temples on top of Girnar. Junagadh town is located on the west of the complex. The hill complex (~ 175 km^2) has four distinct geomorphic units (see inset Google Map) **a** the circular outer ranges of hills, **b** the central high hill, **c** radial ridges joining the central high hill to the outer hills, and **d** low-lying plains in the intervening area (Mathur et al. 1926). The Girnar Massif is made up of differentiated magmatic rocks like granophyre, basalt, dolerite, gabbro, pyroxenite, nepheline syenite and tonalite. The complex is believed to be pene-contemporaneous with Deccan Traps ($\sim$56–63 Ma). Besides Girnar, there are four other volcanic plugs in Saurashtra/Kathiawad, namely Barda, Alech, Chogat and Chamardi. (Text and photo by L. S. Chamyal and Nilesh P. Bhatt)

Fig. 6 Garhbeta Badlands, West Medinipur District, West Bengal. Reaching up to 24 m in height locally and covering an area of 0.8 km^2, the area is characterized by intense gullying. Gullying of the region was initiated by southward shift of a meander of the Shilai River into a Pleistocene lateritic terrain. The site is also an important storehouse of ichnofossils and petrified woods. (Text and photo by Sunando Bandyopadhyay)

Fig. 7 The Nohkalikai Falls is developed on a cliff that retreated some 11 km along the course of the Pynjngithuli River from the southern edge of the Meghalaya Plateau. Initiating at 1,203 m a.s.l., the fall is a sheer drop of ~198 m on the Eocene sandstones of local Cherra Formation. Although the size of Nohkalikai's catchment is just ~14 km^2, its situation at 3.5 km east of Cherrapunji, one of the wettest places of the world with 11,780 mm of average annual rainfall, explains the high amount of discharge received by it. (Text and photo by Sunando Bandyopadhyay)

Fig. 8 The Gersoppa or Jog Falls (~253 m) on Sharavathi River in Karnataka. This west-flowing Sharavathi River flows over the Mysore Plateau before leaping over the edge of the Western Ghat and entering the ~25 km long Gersoppa Gorge. The gorge provides evidence of the recession of the knickpoint associated with a river capture. The area is underlain by banded gneisses. The falls acquire their most captivating form when water is released from the Linganamakki Dam located upstream. (Text by Vishwas S. Kale and photo by H. R. Mungekar)

Fig. 9 Bizarre wallforms in basalt seen on the gorge walls of the Kukdi River, near Nighoj, Maharashtra. The slot-canyon-like gorge (~250–300 m long and >15–25 m deep) exhibits intricately undulating walls, with remnants of multiple potholes (coalesced) of various dimensions. The remnants of potholes seen in this picture are several meters in diameter. Abrasional faces, flute marks, grooves, natural bridges and multiple isolated and coalesced potholes are the prominent features seen in the area. (Text and photo by Vishwas S. Kale)

Fig. 10 Multiple river terraces (*T1–T6*) on the bank of Alaknanda River, Srinagar, Uttarakhand. The Alaknanda River drains through three major morpho-tectonic zones. These are the (i) Trans-Himalaya located north of the South Tibetan Detachment System (STD), the Higher Himalaya located between the STD and the Main Central Thrust (MCT) and the Lesser Himalaya located south of the MCT. The valley has preserved a spectacular sequence of fill and strath terraces particularly south of the MCT. The presence of both fill and strath terraces suggests that the fluvial landforms owe their genesis to a combination of climate (monsoon) and tectonics. (Text and photo by Navin Juyal)

Fig. 11 Raised intertidal terrace and wave-cut notch, Diu Island, Saurashtra coast, Gujarat. The Saurashtra coast in western India consists of both relict and active geomorphic components. The cuspate coastline has bays, coves, creeks, ridges and lagoons. The ridges run parallel to the coast and invariably have curvilinear forms which probably represent relict barriers, bars or spits. At places, longitudinal lithified dunes (aeolinites) appear on top of the ridges. The ridges have a steep cliff face towards the sea, are backed by a set of parallel rows of aeolinite, and slope down towards the mainland Saurashtra plunging into relict/active lagoonal depressions. The cliffy coasts at places have preserved raised intertidal terraces and wave-cut notches which can be used to reconstruct the past sea-land level changes. (Text and photo by Navin Juyal)

Fig. 12 Moraines and remnants of a proglacial lake (varve and rythemite), in Goting Basin in Himalaya, Uttarakhand. The basin lies in the transitional zone between a sub-humid south and semi-arid north of the Uttarakhand Himalaya. The southern extent of the basin is delimited by NW–SE trending South Tibetan Detachment System (STDS). The basin was carved by valley glaciers which have deposited the terminal moraines against the rising footwall of the STDS in the south. Following this, a proglacial lake came into the existence in which deposition of varve and rythemites occurred. Finally the pro-glacial lacustrine environment was terminated with the deposition of outwash gravel terraces. (Text and photo by Navin Juyal)

Fig. 13 This landscape in Ladakh preserves some of the finest examples of alluvial fans and moraines of great antiquity. The Indus has drained down this valley since the Eocene (for >30 Ma) depositing the Indus molasse and has helped shape this mighty landscape. The Zanskar Range in the distance recorded a long history of continental–continental collision and has exceptional exposures showing numerous thrust duplexes. The concentric moraine in the middle foreground is one of the best preserved old moraines in the Himalaya and is dated to be about 150 ka. Recent exposure dating on summit tors in the Ladakh Range from where this photo was taken date to several million years and show that erosion of the summits is negligible. What is impressive is that this is an ancient landscape within the world's most active mountain belt. (Text and photo by Lewis A. Owen)

Fig. 14 The Western Ghat (*Sahyadri*) Escarpment in horizontally bedded Deccan Traps basalt flows (~65 Ma) at Mahabaleshwar. This photo taken near Auther's Seat (~1,347 m a.s.l.), shows the western edge of the laterite-capped Mahabaleshwar Plateau (~1,220 and 1,438 m a.s.l.) (plateau not seen in the photo). The scarp at places has a sheer drop of many hundred meters (up to 900 m) and is embayed. The sinuous nature of the scarp, the deeply entrenched valley heads, the elongated offshoots and the occurrence of the beheaded valley of the Krishna River (behind this Ghat offshoot) suggest that the escarpment have receded east from an earlier position westward. All this geomorphic evidence indicates that the present scarp is erosional. (Text by Vishwas S. Kale and photo by H. R. Mungekar)

6 Concluding Remarks

The Indian subcontinent has extraordinary geoheritage. Whereas the Himalaya Mountains and the Western Ghat Escarpment are iconic landscape features, there are countless other fascinating and exquisite landforms and landscapes that have immense cultural, socio-economic and scientific value. All the geomorphosites or geoheritage sites described in Part III of this monograph as well as those listed in Tables 1, 2 and 3 of this chapter are important not only from the standpoint of public interest and economic benefits, but also from the viewpoint of geo-science studies. Such geomorphosites are critical to advancing knowledge about landscape evolution, geomorphic processes, geohazards, impact of climate change and the history of Earth. The need for promotion and conservation of such geomorphosites need not be overemphasized. To achieve this, first, the goal should be to identify, document and assess geomorphosites through a rigorous peer-review process. Various earth science associations and organizations in India, such as the Geological Society of India, the Indian Institute of Geomorphologists (IGI), etc. and the Ministry of Earth Sciences (MoES) as well as the Geological Survey of India have a critical role to play in this overdue activity. Second, one of the most challenging tasks in this endeavor is to convince the local communities, concerned government authorities (local, state and national) and geo-asset users (geo-tourists and service providers) about the geo-environmental value of these geomorphosites or geoheritage sites. The active participation of all these stakeholders is critical for the management and conservation of these geoheritage sites. Finally, environment-friendly and sound geo-conservation work and legislation that protect geoheritage sites are exceedingly vital from the standpoint of the future of geotourism in India.

References

Anantharamu TR, Bellur D, Bhasker AA (2001) National Geological Monuments. Geological Survey of India, Special Publication No. 61, vol vi, 98 p

Bhargava ON, Kumbkarni S, Ahluwalia AD (2010) Geomorphology and landscapes—illustrations from Himalayas. Technology Publications, Dehradun 211 p

Brocx M, Semeniuk V (2007) Geoheritage and geoconservation—history, definition, scope and scale. J R Soc West Aust 90:53–87

Gray M (2004) Geodiversity: valuing and conserving abiotic nature. Wiley, Chichester 434 p

GSI (2012) Travel brochure on National Geological Monuments of India: Region South. Geological Survey of India, Hyderabad, 12 p

Kale VS (2009) DEM-based studies of some outstanding fluvial landforms in India. In: Sharma HS, Kale VS (eds) Geomorphology of India. Prayag Pustak Bhavan, Allahabad, pp 59–68

Kale VS (2010) The Western Ghat: the great escarpment of India. In: Migon P (ed) Geomorphological landscapes of the world. Springer, New York, pp 257–264

Mathur KK, Dubey VS, Sharma NL (1926) Magmatic differentiation in Mount Girnar. J Geol. 34:289–307

Panizza M (2001) Geomorphosites: concepts, methods and examples of geomorphological survey. Chin Sci Bull 46(suppl):4–6

Parthasarathy GR, Rengan V, Arivalagan V (2012) Understanding geomorphology through inventory of landforms. Report submitted to University Grants Commission under UGC-UPE Project, 139 p

Raina VK, Srivastava D (2008) Glacier Atlas of India. Geological Society of India, Bangalore 315 p

Ranganathan N, Jayaram S (2006) Geomorphology of Karnataka. Karnataka Geologists' Association, Department of Mines and Geology, Bangalore 350 p

Reddy DR (ed) (2013) Proceedings of national workshop and brainstorming session on geoheritage: need for an Indian activism. Department of Geology, Andhra University, Visakhapatnam, 30–31 May 2013, 94 p

Reynard E, Panizza M (2005) Geomorphosites: definition, assessment and mapping. An Introduction. Géomorphol Relief Process Environn 3:177–180

Singh RB, Anand S (2013) Geodiversity, geographical heritage and geoparks in India. Int Jour Geoheritage 1:10–26

Short Biodata of Authors

Vishwas Shripad Kale is a Professor and Head in the Department of Geography, University of Pune, India. His research interests include palaeohydrology, fluvial and flood geomorphology, Quaternary geomorphology, and landscape evolution. He has published over two dozen research papers in these topics. He has jointly authored a textbook on "Introduction to Geomorphology" with Avijit Gupta and edited four books on geomorphology and related fields.

Mahendra R. Bhutiyani is presently working as Director, Defence Terrain Research Laboratory, New Delhi, India. Earlier he was working as a Senior Scientist, Associate Director and Head of the Hazard Assessment and Forecasting Division in Snow and Avalanche Study Establishment (SASE), a Defence R&D Organization Laboratory, at Chandigarh, India. His research mainly deals with climate change, mass balance, sediment transport and natural hazards in NW Himalaya. He has studied many Himalayan glaciers, including the Siachen Glacier. He has over 40 research publications, some in high impact factor journals, on the subject of glaciers, river hydrology and the environment in Himalaya Mountains.

Makarand S. Bodas is with the Geological Survey of India and has been associated with the studies on the Deccan Province for a long time. He has notable contributions in the fields of stratigraphy, geochemistry and study of slope failures in the Deccan Volcanic Province.

Partha Pratim Chakraborty is an Associate Professor at the Department of Geology, University of Delhi, India. His research interest includes sedimentology, sequence stratigraphy and basin modeling. He has worked extensively on Proterozoic basins of India.

Farooq A. Dar is a Research Fellow at the Centre for Earth and Space Sciences, University of Hyderabad, India. He is working on processes and potential of groundwater in karstified terrains.

Rajinder Kumar Ganjoo is a Professor of Quaternary Geology, Head of the Department of Geology and Director of the Institute of Himalayan Glaciology, University of Jammu, India. He has been chiefly interested in Quaternary palaeoclimate of High Himalaya, climate change and glaciers, and landscape evolution in cold deserts. He has 70 papers in peer-reviewed national and international research journals. He has contributed to the official document of the Indian Ministry of Environment & Forests entitled "Snow and Glaciers of the Himalaya: Inventory and Monitoring: Discussion Paper" refuting the claims made on the Himalayan glaciers in IPCC-AR4.

Steven L. Goodbred, Jr is an Associate Professor in the Department of Earth and Environmental Sciences at Vanderbilt University in Nashville, TN, USA. His research has centered on processes and development of the Ganga-Brahmaputra River Delta and Bengal Basin from modern to Quaternary timescales.

Yanni Gunnell is a Professor of Physical Geography and Environmental Science at the University of Lyon, France. He studies the archives of environmental change and the timescales of continental denudation and material transfer across the landscape. His comparative approach ranges from cratonic environments to mountain belts and covers different climatic settings. He has worked in parts of India, Africa, Brazil, Peru, North America, Madagascar and Europe.

R. Jayangondaperumal is a scientist at the Wadia Institute of Himalayan Geology, Dehradun, India. His research interests are active tectonics, palaeoseismology, tectonic geomorphology, and coastal geomorphology. For more than a decade he has worked extensively in northwestern and northeastern Himalaya on aspects of active tectonics and paleoseismology. He has published over 25 research papers in high impact factor research journals.

V. S. Kale (ed.), *Landscapes and Landforms of India*, World Geomorphological Landscapes,
DOI: 10.1007/978-94-017-8029-2, © Springer Science+Business Media Dordrecht 2014

Veena U. Joshi is a Professor in the Department of Geography, University of Pune, India. Her research concerns gully erosion and badland development in the Deccan Traps region.

Navin Juyal is a Senior Scientist in the Geoscience Division of Physical Research Laboratory, Ahmedabad, India. His research interest involves Quaternary geology, palaeoclimate and recent environmental changes. He has published more than 60 research papers in peer reviewed journals.

Amal Kar was working as Principal Scientist and Divisional Head with the Central Arid Zone Research Institute (CAZRI), Jodhpur, India till May, 2012. He is involved in desert research for almost four decades, especially in the fields of geomorphology, desertification, natural resources assessment, and application of geomatics. He was a Visiting Professor at the University of Tokyo in 2001. He also contributed to the UNCCD/UNDP as a resource person/trainer on desertification at different times.

S. M. Karisiddaiah is an Emeritus Scientist at the National Institute of Oceanography, Goa, India. His research interests are in the field of methane and gas hydrates in marine sediments from the eastern Arabian Sea, their distribution, origin and their possible impact on climate change, seismic and sequence stratigraphy of the central western continental margin of India and Late Quaternary evolution.

R. Krishnan is the Executive Director of the Centre for Climate Change Research (CCCR) at the Indian Institute of Tropical Meteorology (IITM), Pune. His interests include: Monsoon dynamics and variability, phenomenon of monsoon-breaks and droughts, large-scale organization of monsoon convection, global climate change and impacts on the Asian monsoon and regional climate extremes. He is a member of the Asia Australia Monsoon Panel (AAMP) of the World Climate Research Programme (WCRP) and also a member of the Science Advisory team of the WCRP Coordinated Regional Downscaling Experiment (CORDEX) programme. He has been a recipient of the Frontier Outstanding Research Award by FRSGC, Japan, Prof. Anandu Vernekar Award of the Indian Meteorological Society (IMS) and the IITM Silver Jubilee Award.

Ranadhir Mukhopadhyay is the Chief Scientist at the National Institute of Oceanography, Goa, India. His scientific interests are in Gondwanaland dismemberment and shelf-slope architecture of the Indian continent, dynamics of formation and growth of oceanic manganese nodules, vulnerability index and ecosystem management, non-uniform growth style of seamounts, and petrology and tectonics of the plate boundaries. He is the recipient of Dr. J Coggin Brown Gold Medal, MS Krishnan Gold Medal, CV Raman Fellowship, Asiatic Society Medal and was the Director of Mauritius Oceanography Institute, Mauritius.

A.C. Narayana Professor at the Centre for Earth and Space Sciences, University of Hyderabad, India, has research interest in the field of geomorphology, palaeoclimate and water resources management.

Lewis A. Owen is a Professor and Head of the Department of Geology at the University of Cincinnati, Cincinnati, Ohio, USA. He received his Ph.D. in Geomorphology from the University of Leicester, U.K., in 1988. Before joining the University of Cincinnati in 2004, he held positions at the Hong Kong Baptist University, Royal Holloway—University of London, and the University of California—Riverside. Professor Owen's research focuses on the Quaternary geology and geomorphology of tectonically active mountain belts and their forelands, particularly in the Himalayan-Tibetan orogen. In 2011, Professor Owen was awarded the Busk Medal from the Royal Geographic Society for his field research in Quaternary history and geomorphology in tectonically active areas. He has published >120 research papers and edited over a dozen volumes of papers and articles.

Pawel Prokop is a physical geographer at the Department of Geoenvironmental Research of the Institute of Geography and Spatial Organization of the Polish Academy of Sciences. His research interests concerns land degradation and natural hazards (extreme rainfalls, soil erosion, floods). He is author and co-author of several peer-reviewed papers and chapters in books related to environmental problems of northeast India with special emphasis on the Meghalaya Plateau and piedmont of the Eastern Himalaya.

Jyotiranjan S. Ray is an Associate Professor at the Physical Research Laboratory, Ahmedabad, India. He is an isotope geochemist and his research interests include understanding of subduction zone processes, chemical evolution of Earth's mantle through time, and provenance of sedimentary deposits.

R. Ramesh is a Senior Professor at the Geosciences Division of the Physical Research Laboratory, Ahmedabad, India. He is actively involved in speleothem based paleomonsoon reconstructions. He is a fellow of all the three science academies of India and also the Third World Academy. He is recipient of SS Bhatnagar Award, 1998 and TWAS Prize, 2006. As a lead author of IPCC AR4 & 5, he received a special citation for Nobel Peace Prize 2007 awarded to IPCC.

Kimberly G. Rogers is a Research Associate at the Institute for Arctic and Alpine Research at the University of Colorado in Boulder, CO, USA. Her research interests include fluvial-to-shelf sediment transport, and predicting the response of river delta morphology to climate and anthropogenic change.

Subir Sarkar is a Professor and former Head of Department of Geography and Applied Geography at the North Bengal University, India. He is a geomorphologist and physical geographer. Most of his research is devoted to the study of landslides and fluvial processes in the Darjeeling Himalaya.

Jogendra Nath Sarma is a Professor in the Department of Applied Geology, Dibrugarh University, Assam, India. His research interest includes fluvial geomorphology, sedimentology, Quarternary geology and neotectonics. He has been working on the rivers of Assam for the last three decades. He has over fifty research papers in peer-reviewed journals such as Geomorphology, Geografiska Annaler, International Geology Review, Developments in Sedimentology, Communicacoes Geologicus, etc.

Bibhas Sen is with the Geological Survey of India and has been associated with the studies on the Deccan Province for a long time. He has important contributions in identification of lava transportation systems and geochemistry of Deccan dykes in Satpura Region.

Vimal Singh is an Assistant Professor at the Department of Geology, University of Delhi, India. His research interest includes landscape evolution and fluvial geomorphology.

Rajiv Sinha is a Professor and Head of the Department of Earth Sciences, at the Indian Institute of Technology, Kanpur, India and has worked on various aspects of Quaternary Geology. His main research interests include river dynamics and floods, fluvial remote sensing, and river response to climate change. Some of his major contributions include source area classification of Gangetic river systems, hydro-geomorphic approach to understanding floods in the Himalayan rivers and paleohydrologic and paleoclimatic reconstructions in the Ganga Plains using sedimentary archives and drill cores. He is currently involved in several projects on changing water cycle, river health, and environmental flow assessment of the Indian rivers.

Ashok Kumar Singhvi works at the Physical Research Laboratory, Ahmadabad and holds a J. C. Bose National Fellowship of the Department of Science and Technology, India. His scientific interests are in Quaternary Geochronology, Paleoclimatology and Luminescence dating. He has pioneered applications of luminescence method for the dating of desert sands, carbonates, gypsum, fulgurites and has made significant methodological contributions. He has extensively contributed to the study of global deserts and dust sequences and, has dated most of the Indian surface. He is a fellow of the Indian National Science Academy, the Third World Academy of Sciences, the Geological Society of London and was a member, Executive Committee of IGBP- PAGES during 2000–2005.

Leszek Starkel is the former Head of Department of Geomorphology and Hydrology of Mountains and Uplands in Krakow, branch of the Institute of Geography and Spatial Organisation, Polish Academy of Sciences. He is the founder of INQUA Commission on Global Continental Palaeohydrology. Most of his studies have been devoted to the evolution of fluvial systems, present day extreme events and palaeogeography of the Holocene in the Carpathians and in the whole of the temperate zone. From 1968 to 2009 he carried out research in the Darjeeling Himalaya and Meghalaya Plateau as well organised collaborative research between Polish and Indian geomorphologists.

Sampat K. Tandon currently holds the D. N. Wadia Chair Professor at the Department of Earth Sciences, Indian Institute of Technology, Kanpur, India. Earlier he has been Professor in Geology at the Department of Geology, University of Delhi for several decades and has wide ranging interest in earth surface processes including fluvial and arid zone environment.

Rajagopala Vaidyanadhan is a retired Professor and Head of the Department of Geography of the Andhra University, Visakhapatnam, India. He served as a Professor of Geology at the University of Dar-es-Salam, Tanzania for some time. Trained as a photogeologist and geomorphologist, he has extensively worked on the geomorphology and denudation chronology of Peninsular India. Much of his research has been published in the form of articles (over 70) in reputed journals and in the form of a monograph. From 1992 to 1995, he was the Editor of the Journal of the Geological Society of India, Bangalore. He has jointly authored the two set volume "Geology of India" first published by the Geological Society of India, Bangalore in 2008.

M. G. Yadava is In-charge of the Radiocarbon Dating Laboratory at the Physical Research Laboratory, Ahmedabad, India. He is actively involved in speleothem based paleomonsoon reconstructions.

MIX
Papier aus verantwortungsvollen Quellen
Paper from responsible sources
FSC® C105338

If you have any concerns about our products,
you can contact us on
ProductSafety@springernature.com

In case Publisher is established outside the EU,
the EU authorized representative is:
Springer Nature Customer Service Center GmbH
Europaplatz 3, 69115 Heidelberg, Germany

Printed by Libri Plureos GmbH
in Hamburg, Germany